312
Current Topics in Microbiology and Immunology

C. Boshoff and R. A. Weiss (Eds.)

Kaposi Sarcoma Herpesvirus: New Perspectives

With 29 Figures and 14 Tables

Springer

Chris Boshoff, MRCP, Ph.D.

University College London
Wolfson Institute for Biomedical Research
Gower Street
London WC1E 6BT
UK

e-mail: c.boshoff@ucl.ac.uk

Robin A. Weiss, Ph.D.

Wohl Virion Centre
Division of Infection and Immunity
University College London
46 Cleveland Street
London W1T 4JF
UK

e-mail: r.weiss@ucl.ac.uk

Cover Illustration:
Juan Chang and Patrick Moore receiving the Koch Prize for their discovery of Kaposi's sarcoma-associated herpesvirus in 1994. Also shown is a KSHV virion from chapter by O'Connor and Kedes, p. 43. and a photograph of Moritz Kaposi. Background shows an histological image of a Kaposi's sarcoma tumor.

Library of Congress Catalog Number 72-152360

ISSN 0070-217X
ISBN-10 3-540-34343-1 Springer Berlin Heidelberg New York
ISBN-13 978-3-540-34343-1 Springer Berlin Heidelberg New York

Springer is a part of Springer Science+Business Media
springeronline.com

Editor: Simon Rallison, Heidelberg
Desk editor: Anne Clauss, Heidelberg
Production editor: Nadja Kroke, Leipzig
Cover design: WMX Design, Heidelberg
Typesetting: LE-TEX Jelonek, Schmidt & Vöckler GbR, Leipzig
Printed on acid-free paper SPIN 11560036 27/3150/YL – 5 4 3 2 1 0

Preface

Moritz Kaposi (born Moritz Kohn 1837, Kaposvár, Hungary; died 1902, Vienna) succeeded his father-in-law, Ferdinand von Hibra (1816–1880) to become one of the foremost dermatologists in the German-speaking world. His remarkable clinical acumen is evident from his descriptions of systemic lupus erythematosus (1869), idiopathic multiple pigmented sarcoma of the skin (1872 Kaposi sarcoma), and xeroderma pigmentosum (1882). His name change from Kohn to Kaposi (1871) reflects his professional ambitions to be distinguished from the other *Kohns* working in the Vienna Medical Faculty at the time. Although we now associate Kaposi sarcoma occurring in eastern European and Mediterranean populations (*classic Kaposi*) with an indolent disease, the cases first described by Kaposi had an aggressive clinical course, including some with fatal systemic disease. Underlying illness (*krankheit*) might have contributed to a more immunosuppressive state (Kaposi 1872).

From the 1940s onwards, various workers observed similar vascular tumours in sub-Saharan Africa, including an aggressive lymphadenopathy in children (endemic Kaposi) (Ackerman 1962; Oettle 1962). It remains to be established whether this lymphadenopathic disease in African children represents primary infection with Kaposi sarcoma-associated herpesvirus (KSHV). In the 1970s, the introduction of organ transplantation with its related iatrogenic immunosuppression resurrected this tumour as an important cause of morbidity and even mortality (Penn 1979). The Italian dermatologist Giraldo suggested in 1972 that herpesviruses could be involved in the pathogenesis of Kaposi sarcoma, after they showed by electron microscopy herpes-type viral particles in cells cultured from these tumours (Giraldo 1972). They subsequently identified these virions as cytomegalovirus.

The reports from New York City and Los Angeles in 1981 of Kaposi sarcoma and *Pneumocystis carinii* pneumonia by the Centers for Disease Control in Atlanta heralded the AIDS pandemic (Friedman-Kien 1981). In 1981, about 48% of gay men with AIDS presented with or developed Kaposi sarcoma during the course of their illness. The incidence of AIDS-related Kaposi sarcoma in 1990 in this population was less than 15%. It is still unclear whether this is mainly due to a fall in the seroprevalence of Kaposi sarcoma-associated herpesvirus (KSHV or HHV8) in this population, or whether these were related to other disease-associated risk factors. The particular symmetrical distribution of AIDS–related Kaposi sarcoma on the skin, high rate of mucosal membrane, lymph node and visceral involvement, provoked various hypotheses regarding the pathogenesis of this disease: tissue-specific temperature or oxygen tensions might favor KS tumour cell proliferation, or attraction of infected cells by specific chemokine-rich environments could favor these sites of disease. The demonstration that solid epithelial cancer micrometastases could be attracted to specific organs by chemokines corresponding to the specific chemokine receptors expressed on such cancer cells, favors a hypothesis where circulating Kaposi tumour cells, or KSHV-infected cells, home-in on specific organs. Underlying HIV infection might influence the distribution of such chemokines and certainly promotes the aggressiveness of the disease.

Robert Gallo and his team were among the first to culture cells from AIDS-related Kaposi sarcoma, and they introduced the concept of an autocrine and paracrine driven tumour, where cytokines stimulate the growth of this malignancy (Salahuddin 1988; Ensoli 1989). A contemporary of Kaposi, Virchow proposed in 1863 that cancerous tumours occur at sites of chronic inflammation (Balkwill 2001) and in 1986 Dvorak speculated that tumours are wounds that do not heal (Dvorak 1986). The model of a tumour landscape, where inflammatory cells produce cytokines and other growth factors which contribute to the initiation and maintenance of tumour growth has since become a major focus of cancer research. Interestingly, Kaposi sarcoma lesions also display the Kübner phenomenon (tumour develops at sites of an old injury).

Epidemiological studies by Harold Jaffe and Valerie Beral in the early years of the AIDS epidemic suggested that an infectious agent other than HIV was the culprit causing Kaposi sarcoma (Beral 1990). However, their suggestion that such an infection is mainly transmitted by the faecal-oral route was probably incorrect.

Patrick Moore (epidemiologist to the City of New York), and his wife Yuan Chang (pathologist at Columbia University), used representational difference analysis to identify sequences of a new herpesvirus in 1994 (Chang 1994). The technique of representational difference analysis was first described by Lisitsyn and Wigler (Lisitsyn 1993). They employed this polymerase chain reaction subtraction technique to successfully identify a sequence that clearly belonged to a herpesvirus genome, but was distinct from previously characterized herpesviruses. KSHV is the second human oncogenic virus to be identified by molecular techniques, the first being human cervical papillomavirus in 1983 by Harald zur Hausen and colleagues.

Time Line: Important events in the history of Kaposi sarcoma and its associated virus

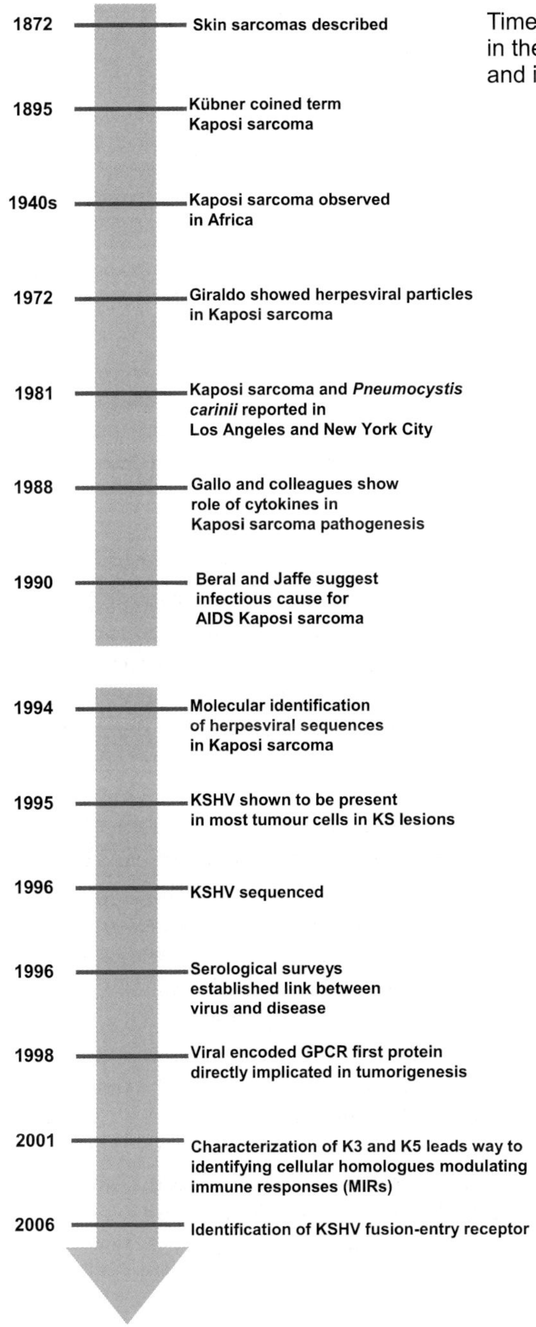
1872
Skin sarcomas described
1895
Kübner coined term Kaposi sarcoma
1940s
Kaposi sarcoma observed in Africa
1972
Giraldo showed herpesviral particles in Kaposi sarcoma
1981
Kaposi sarcoma and *Pneumocystis carinii* reported in Los Angeles and New York City
1988
Gallo and colleagues show role of cytokines in Kaposi sarcoma pathogenesis
1990
Beral and Jaffe suggest infectious cause for AIDS Kaposi sarcoma
1994
Molecular identification of herpesviral sequences in Kaposi sarcoma
1995
KSHV shown to be present in most tumour cells in KS lesions
1996
KSHV sequenced
1996
Serological surveys established link between virus and disease
1998
Viral encoded GPCR first protein directly implicated in tumorigenesis
2001
Characterization of K3 and K5 leads way to identifying cellular homologues modulating immune responses (MIRs)
2006
Identification of KSHV fusion-entry receptor

Since the identification of two sequences from KSHV, we have learned much about this pathogen as reflected in this book. As it turned out, KSHV itself encodes for a number of cytokines, and induces cellular cytokine secretion, contributing to tumour growth. Moreover, KSHV vFLIP targets the IKK-NFκB axis to encourage the inflammatory microenvironment observed by Gallo and his colleagues. KSHV continues to elucidate both mechanisms of viral oncogenesis, and cellular and immune pathways involved in non-viral driven neoplasia.

We bring together here various experts in their respective fields, addressing some key aspects of KSHV biology, and we are grateful for their contributions.

University College London *Chris Boshoff and Robin A. Weiss*

Table 1 Twenty of the most cited articles on Kaposi sarcoma herpesvirus

	Article	Estimated number of citations (2006)
1.	Chang Y et al. (1994) Identification of herpesvirus-like DNA sequences in AIDS-associated Kaposi's sarcoma. Science 266:1865–1869	1950
2.	Cesarman E et al. (1995) Kaposi's sarcoma-associated herpesvirus-like DNA sequences in AIDS-related body-cavity-based lymphomas. NEJM 332:1186–1191	880
3.	Russo JJ et al. (1996) Nucleotide sequence of the Kaposi sarcoma-associated herpesvirus (HHV8). PNAS 93:14862–14867	735
4.	Whitby D et al. (1995) Detection of Kaposi sarcoma associated herpesvirus in peripheral blood of HIV-infected individuals and progression to Kaposi's sarcoma. Lancet 346:799–802	570
5.	Moore PS et al. (1995) Detection of herpesvirus-like DNA sequences in Kaposi's sarcoma in patients with and without HIV infection. NEJM 332:1181–1185	555
6.	Gao SJ et al. (1996) KSHV antibodies among Americans, Italians and Ugandans with and without Kaposi's sarcoma. Nat Med 2:925–928	527
7.	Moore PS et al. (1996) Molecular mimicry of human cytokine and cytokine response pathway genes by KSHV. Science 274:1739–1744	505
8.	Renne R et al. (1996) Lytic growth of Kaposi's sarcoma-associated herpesvirus (human herpesvirus 8) in culture. Nat Med 2:342–346	499

Table 1 (continued)

	Article	Estimated number of citations (2006)
9.	Kedes DH et al. (1996) The seroepidemiology of human herpesvirus 8 (Kaposi's sarcoma-associated herpesvirus): Distribution of infection in KS risk groups and evidence for sexual transmission. Nat Med 2:918–924	464
10.	Boshoff C et al. (1995) Kaposi's sarcoma-associated herpesvirus infects endothelial and spindle cells. Nat Med 1:1274–1278	415
11.	Simpson GR et al. (1996) Prevalence of Kaposi's sarcoma associated herpesvirus infection measured by antibodies to recombinant capsid protein and latent immunofluorescence antigen. Lancet 348:1133–1138	403
12.	Bais C et al. (1998) G-protein-coupled receptor of Kaposi's sarcoma-associated herpesvirus is a viral oncogene and angiogenesis activator. Nature 391:86–89	390
13.	Lennette ET et al. (1996) Antibodies to human herpesvirus type 8 in the general population and in Kaposi's sarcoma patients. Lancet 348:858–861	365
14.	Gao SJ et al. (1996) Seroconversion to antibodies against Kaposi's sarcoma-associated herpesvirus-related latent nuclear antigens before the development of Kaposi's sarcoma. NEJM 335:233–241	351
15.	Moore PS et al. (1996) Primary characterization of a herpesvirus agent associated with Kaposi's sarcoma. J Virol 70:549–558	351
16.	Rettig MB et al. (1997) Kaposi's sarcoma-associated herpesvirus infection of bone marrow dendritic cells from multiple myeloma patients. Science 276:1851–1854	346
17.	Nador RG et al. (1996) Primary effusion lymphoma: A distinct clinicopathologic entity associated with the Kaposi's sarcoma-associated herpesvirus. Blood 88:645–656	346
18.	Staskus KA et al. (1997) Kaposi's sarcoma-associated herpesvirus gene expression in endothelial (Spindle) tumor cells. J Virol 71:715–719	292
19.	Boshoff C et al. (1997) Angiogenic and HIV-inhibitory functions of KSHV-encoded chemokines. Science 278:290–294	279
20.	Kledal TN et al. (1997) A broad-spectrum chemokine antagonist encoded by Kaposi's sarcoma-associated herpesvirus. Science 277:1656–1659	249

References

1. Ackerman LV, Murray JF (1962) Symposium on Kaposi's sarcoma, Acta Union Internationalis Contra Cancrum. Karger, Basel
2. Bais C, Santomasso B, Coso O, Arvanitakis L, Raaka EG, Gutkind JS, Asch AS, Cesarman E, Gershengorn MC, Mesri EA (1998) G-protein-coupled receptor of Kaposi's sarcoma-associated herpesvirus is a viral oncogene and angiogenesis activator. Nature 391:86–89
3. Balkwill F, Mantovani A (2001) Inflammation and cancer: back to Virchow? Lancet 357:539–545
4. Beral V, Peterman TA, Berkelman RL, Jaffe HW (1990) Kaposi's sarcoma among persons with AIDS: a sexually transmitted infection? Lancet 335:123–128
5. Boshoff C, Endo Y, Collins PD, Takeuchi Y, Reeves JD, Schweickart VL, Siani MA, Sasaki T, Williams TJ, Gray PW, Moore PS, Chang Y, Weiss RA (1997) Angiogenic and HIV-inhibitory functions of KSHV-encoded chemokines. Science 278:290–294
6. Boshoff C, Schulz TF, Kennedy MM, Graham AK, Fisher C, Thomas A, McGee JO, Weiss RA, O'Leary JJ (1995) Kaposi's sarcoma-associated herpesvirus infects endothelial and spindle cells. Nat Med 12:1274–1278
7. Cesarman E, Chang Y, Moore PS, Said JW, Knowles DM (1995) Kaposi's sarcoma-associated herpesvirus-like DNA sequences in AIDS-related body-cavity-based lymphomas. N Engl J Med 332:1186–1191
8. Chang Y, Cesarman E, Pessin MS, Lee F, Culpepper J, Knowles DM, Moore PS (1994) Identification of herpesvirus-like DNA sequences in AIDS-associated Kaposi's sarcoma. Science 266:1865–1869
9. Davies JNP, Loethe F (1962) Kaposi's sarcoma in African children. Acta Union Internationale Contra Cancrun 18:394–399
10. Dvorak HF (1986). Tumours: wounds that do not heal. Similarities between tumor stroma generation and wound healing. N Engl J Med 315:1650–1659
11. Ensoli B, Nakamura S, Salahuddin SZ, Biberfeld P, Larsson L, Beaver B, Wong-Staal F, Gallo RC (1989) AIDS-Kaposi's sarcoma-derived cells express cytokines and autocrine and paracrine growth effects. Science 243:223–226
12. Friedman-Kien AE, Laubenstein L, Marmor M et al. (1981) Kaposi's sarcoma and pneumocystis pneumonia among homosexual men – New York and California. Morbidity Mortality Rep 30:305–308
13. Gao SJ, Kingsley L, Hoover DR, Spira TJ, Rinaldo CR, Saah A, Phair J, Detels R, Parry P, Chang Y, Moore PS (1996) Seroconversion to antibodies against Kaposi's sarcoma-associated herpesvirus-related latent nuclear antigens before the development of Kaposi's sarcoma. New Engl J Med 335:233–241
14. Gao SJ, Kingsley L, Li M, Zheng W, Parravicini C, Ziegler J, Newton R, Rinaldo CR, Saah A, Phair J, Detels R, Chang Y, Moore PS (1996) KSHV antibodies among Americans, Italians and Ugandans with and without Kaposi's sarcoma. Nat Med 8:925–928
15. Giraldo G, Beth E, Haguenau F (1972) Herpes-type virus particles in tissue culture of Kaposi's sarcoma from different geographic regions. J Natl Cancer Inst 49:1509–1513

16. Kaposi M (1872) Idiopatiches multiples pigment sarcom der Haut. Archic fur Dermatologie Syphilis 4:265–272
17. Kedes DH, Operskalski E, Busch M, Kohn R, Flood J, Ganem D (1996) The seroepidemiology of human herpesvirus 8 (Kaposi's sarcoma-associated herpesvirus): distribution of infection in KS risk groups and evidence for sexual transmission. Nat Med 2:918–924
18. Kledal TN, Rosenkilde MM, Coulin F, Simmons G, Johnsen AH, Alouani S, Power CA, Luttichau HR, Gerstoft J, Clapham PR, Clark-Lewis I, Wells TN, Schwartz TW (1997) A broad-spectrum chemokine antagonist encoded by Kaposi's sarcoma-associated herpesvirus. Science 277:1656–1659
19. Lennette ET, Blackbourn DJ, Levy JA (1996) Antibodies to human herpesvirus type 8 in the general population and in Kaposi's sarcoma patients. Lancet 348:858–861
20. Lisitsyn N, Lisitsyn N, Wigler M (1993) Cloning the differences between two complex genomes. Science 259:946–951
21. Moore PS, Boshoff C, Weiss RA, Chang Y (1996) Molecular mimicry of human cytokine and cytokine response pathway genes by KSHV. Science 274:1739–1744
22. Moore PS, Chang Y (1995) Detection of herpesvirus-like DNA sequences in Kaposi's sarcoma in patients with and without HIV infection. N Engl J Med 18:1181–1185
23. Moore PS, Gao SJ, Dominguez G, Cesarman E, Lungu O, Knowles DM, Garber R, Pellett PE, McGeoch DJ, Chang Y (1996) Primary characterization of a herpesvirus agent associated with Kaposi's sarcomae. J Virol 70:549–558
24. Nador RG, Cesarman E, Chadburn A, Dawson DB, Ansari MQ, Sald J, Knowles DM (1996) Primary effusion lymphoma: a distinct clinicopathologic entity associated with the Kaposi's sarcoma-associated herpesvirus. Blood 88:645–656
25. Oettle AG (1962) Geographical and racial differences in the frequency of Kaposi's sarcoma as evidence of environmental or genetic causes. Acta Union Internationale Contra Cancrum 18:330–363
26. Penn I (1979) Kaposi's sarcoma in organ transplant recipients. Transplantation 27:8–11
27. Renne R, Zhong W, Herndier B, McGrath M, Abbey N, Kedes D, Ganem D (1996) Lytic growth of Kaposi's sarcoma-associated herpesvirus (human herpesvirus 8) in culture. Nat Med 2:342–346
28. Rettig MB, Ma HJ, Vescio RA, Pold M, Schiller G, Belson D, Savage A, Nishikubo C, Wu C, Fraser J, Said JW, Berenson JR (1997) Kaposi's sarcoma-associated herpesvirus infection of bone marrow dendritic cells from multiple myeloma patients. Science 276:1851–1854
29. Russo JJ, Bohenzky RA, Chien MC, Chen J, Yan M, Maddalena D, Parry JP, Peruzzi D, Edelman IS, Chang Y, Moore PS (1996) Nucleotide sequence of the Kaposi sarcoma-associated herpesvirus (HHV8). Proc Natl Acad Sci USA 93:14862–14867
30. Salahuddin SZ, Nakamura S, Biberfeld P et al. (1988) Angiogenic properties of Kaposi's sarcoma-derived cells after long-term culture in vitro. Science 242:430–433

31. Simpson GR, Schulz TF, Whitby D, Cook PM, Boshoff C, Rainbow L, Howard MR, Gao SJ, Bohenzky RA, Simmonds P, Lee C, de Ruiter A, Hatzakis A, Tedder RS, Weller IV, Weiss RA, Moore PS (1996) Prevalence of Kaposi's sarcoma associated herpesvirus infection measured by antibodies to recombinant capsid protein and latent immunofluorescence antigen. Lancet 348:1133–1138
32. Staskus KA, Zhong W, Gebhard K, Herndier B, Wang H, Renne R, Beneke J, Pudney J, Anderson DJ, Ganem D, Haase AT (1997) Kaposi's sarcoma-associated herpesvirus gene expression in endothelial (spindle) tumor cells. J Virol 71:715–719
33. Whitby D, Howard MR, Tenant-Flowers M, Brink NS, Copas A, Boshoff C, Hatzioannou T, Suggett FE, Aldam DM, Denton AS et al. (1995) Detection of Kaposi sarcoma associated herpesvirus in peripheral blood of HIV-infected individuals and progression to Kaposi's sarcoma. Lancet 346:799–802

List of Contents

List of Contributors

(Addresses stated at the beginning of respective chapters)

CTMI (2006) 312:1–42

Modern Evolutionary History of the Human KSHV Genome

G. S. Hayward (✉) · J.-C. Zong

Viral Oncology Program, The Sidney Kimmel Comprehensive Cancer Center, The Johns Hopkins University School of Medicine, 1650 Orleans Street, Baltimore, MD 21231, USA
HAYWAGA@jhmi.edu

Abstract The genomes of several human herpesviruses, including Kaposi sarcoma (KS) herpesvirus (KSHV), display surprisingly high levels of both genetic diversity and clustered subtyping at certain loci. We have been interested in understanding this phenomenon with the hope that it might be a useful diagnostic tool for viral epidemiology, and that it might provide some insights about how these large viral genomes evolve over a relatively short timescale. To do so, we have carried out extensive PCR DNA sequence analysis across the genomes of 200 distinct KSHV samples collected from KS patients around the world. Here we review and summarize current understanding of the origins of KSHV variability, the spread of KSHV and its human hosts out of Africa, the existence of chimeric genomes, and the concept that different segments of the genome have had different evolutionary histories.

1 Introduction

Kaposi sarcoma herpesvirus (KSHV) (or HHV8) is a gamma-2 or rhadinovirus subfamily herpesvirus (Chang et al. 1994; Russo et al. 1996) that is believed to be the etiological agent of Kaposi sarcoma (KS), primary effusion lymphoma (PEL), and some forms of multicentric Castleman disease in humans. Importantly, both classic KS and long-term inapparent latent infection by KSHV are relatively rare in extant human populations within the USA, Europe, and Asia (approx 1% seropositivity) but are significantly more common in the Mediterranean area (5%–15%) and most common in sub-Saharan Africa (50% or more). Presumably because of reduced immunosurveillance, the incidence of KS increases 500-fold in iatrogenic organ transplant patients and up to 20,000 fold in homosexual AIDS patients in the USA and Europe, and KS is now the most common human tumor found in southern Africa (Chokunonga et al. 1999; Dedicoat and Newton 2003).

Our study of genotypic variation among KSHV genomes derived from different human population groups on different continents has proceeded in three stages. First, we generated a simplistic "ethnic migration" model based only on sequencing of the hypervariable ORF-K1 or VIP genes at the LHS end of the genome (Kajumbula et al. 2006; Zong et al. 1999; Zong et al. 2002). Second, we reported the discovery and analysis of additional complexity and chimerism at the extreme RHS of the genome, associated with the concept of multiple alleles of the ORF-K15 or TMP gene (Poole et al. 1999; Zong et al. 2002) (Zong, J.C., Su, I.J., Morris, L., Alagiozolou, L., Sitas, F., Kajumbula, H., Katange-Mbide, E., Boto, W and Hayward, G.S., manuscript in preparation). Third, we have also now carried out an exhaustive analysis of multiple loci across the central constant segment of the genome (Zong, J-C and Hayward, G.S., unpublished data). Each stage and segment gave rather different results but a reasonably complete picture of the overall pattern of variability within and across the whole length of modern human KSHV genomes has now emerged. These data permit deductions about the history and origin of different parts of the KSHV genome and provide for the first time some important insights into how a large DNA virus genome evolves within the relatively short evolutionary time frame of the divergence of both premodern humanoid species and of modern *Homo sapiens*. In particular, we will introduce the concept of both "old" and "new" segments of the genome and of the role of rare interspecies chimerism, as well as how founder and bottleneck effects associated with modern human migration patterns have impacted the genome.

A critical aspect of this story for the KSHV genome is the realization that the situation observed here is dramatically different from that found in most other more typical herpesviruses (e.g., HCMV). In particular, multiple infections and intragenomic recombination and chimerism are both very rare events in KSHV, presumably because of preferential familial transmission and low horizontal transmission rates, which have in general maintained high levels of linkage between multiple loci across the genome. In contrast, similar multisite analysis of variability and cluster patterns across the HCMV genome has revealed much higher levels of genetic scrambling, with a nearly total loss of subtype linkage between nonadjacent variable loci, apparently resulting from rampant intratypic recombination and extensive penetrance of subtypes across ethnic and geographic boundaries (Zong, J-C et al., unpublished data).

2
KSHV Genetic Variability

2.1
Early Studies of VIP Hypervariability Patterns

Two very different types of variability occur within the VIP (K1) and TMP (K15) proteins encoded at the extreme LHS and RHS, respectively, of the KSHV genome. VIP is a 289-amino acid lytic cycle membrane-associated and ITAM-containing IgG family glycoprotein that displays up to 30% amino acid differences between isolates and clusters into five major subtypes (A, B, C, D, and E) plus numerous minor variants that correlate well with ethnic and geographic ancestry in different human host populations (Biggar et al. 2000; Cook et al. 1999; Fouchard et al. 2000; Hayward 1999; Kadyrova et al. 2003; Lacoste et al. 2000a, 2000b; Lagunoff and Ganem 1997; Meng et al. 2001, 1999; Nicholas et al. 1998; Treurnicht et al. 2002; Whitby et al. 2004; Zhang et al. 2001; Zong et al. 1999, 2002). VIP is unique to KSHV and functions as a Tyr-kinase signaling protein that has antiapoptotic properties, can either mimic or block B-cell receptor activation, and appears likely to play an important role in initial establishment or control of latency (Damania and Jung 2001; Lee et al. 2003, 2005; Wang et al. 2004).

Variability within the VIP gene displays very high nonsynonymous rates, with up to 85% of nucleotide changes creating amino acid changes, implying that some as yet unknown but powerful biological selection process has been involved. Furthermore, this variability can be divided into three distinct levels. First, the oldest level occurs predominantly between subtypes and is distributed relatively evenly across the whole gene. Second, a more recent level is observed even within subtypes but is largely limited to two 40-amino acid blocks referred to as VR1 and VR2. The third level is found even within specific variants but is restricted to a 20-amino acid Cys bridge bounded hypervariable domain (VR*) that resembles the VR3 loop in the HIV ENV protein and appears to correlate with strong T-cell epitopes (Stebbing et al. 2003; Zong et al. 1999, 2002). However, unlike the generation of rapidly evolving pseudospecies of HIV or HCV genomes within a single patient, we have never detected any convincing evidence for multiple variants of KSHV VIP either between multiple KS lesions and PMBC samples or within multiple PEL cell lines derived from a single patient. We have concluded that our evidence for highly ethnically restricted clusters of closely related VIP proteins indicates that (unlike most other herpesviruses) KSHV transmission is predominantly familial, with very low rates of horizontal or multiple infections.

We previously interpreted the patterns of distribution of specific VIP clusters as indicating that the principal VIP subtypes arose during the migration of modern humans out of east Africa first into sub-Saharan Africa (B branch), then into south Asia, Australia, and the Pacific Rim (D/E branch), and finally into the Middle East, Europe, and north Asia (A/C branch), with very little subsequent remixing (Hayward 1999; Zong et al. 1999, 2002). The divergence of the B branch from the progenitor of all the other branches is judged to have occurred approximately 100,000 years ago, with the split of D/E from A/C occurring close to 70,000 years ago. The current estimated times of divergence of D from E and of A from C are 50,000 and 35,000 years ago, respectively, based on both the overall evidence about human migrations and the lengths of the branches in the VIP phylogenetic trees. Within the A and C branches, individual variant clades (A1 to A10 and C1 to C7) probably arose between 10,000 and 12,000 years ago as human populations expanded out from a limited number of Eurasian Ice Age refuges.

2.2
TMP Allelic Variability and Chimeric Genes

TMP(K15), also known as LAMP, is a 500-amino acid latent-state membrane protein with twelve transmembrane domains encoded by an eight exon spliced mRNA and related to the LMP2 latency protein of EBV. TMP is also a Tyr-kinase signaling protein, although it lacks the ITAM motifs present in both KSHV VIP and EBV LMP2, and by analogy with LMP2 probably contributes to maintenance of the latent state (Brinkmann et al. 2003; Sharp et al. 2002). TMP genes from different KSHV isolates have been described to fall into two alternative allelic subtypes, referred to as P (prototype) and M (minor), that have diverged by 70% at the amino acid level but otherwise show little variability (Choi et al. 2000; Glenn et al. 1999; Hayward 1999; Kakoola et al. 2001; Nicholas et al. 1998; Poole et al. 1999). It is important to appreciate that only a very small segment at the extreme RHS of those genomes that carry the TMP-M allele show this high level of divergence: In fact, it is limited to just the TMP gene itself. However, the patterns of divergence in the region adjacent to the TMP gene (ORF75E and UPS75) are very interesting and complex and provide critical insights and information about the origin and history of the TMP-M allele. Furthermore, about 20% of all human KSHV genomes examined have the TMP-M allele, but they can be found associated with all three of the major branches of VIP genes (A/C, B, or D/E). In essence, there are two overall classes of KSHV genomes, the predominant P class that are considered to have the modern P pattern throughout their length and a second less prevalent class that are chimeric hybrids of P genomes with just a small highly diverged M allele segment at the RHS.

We originally proposed that the TMP-M alleles are of nonhuman origin and are carried only in relatively rare genomes that consist of a predominantly modern human KSHV genome joined to an exotic Old World primate-derived TMP gene at the RHS (Poole et al. 1999). These chimeras were evidently generated by a two-step recombination process, which we suggested involved an initial chimera that was created by a single original cross-species recombination event from an exotic highly diverged primate KSHV-like virus, followed by a second event from a much less diverged premodern humanoid lineage virus, such as might be expected to have been present in Neanderthals, for example. However, there is an alternative possibility that the parent M-type virus from which they originated was instead an anciently diverged true second species of human KSHV virus, which has since either become extinct or has not yet been detected as an intact genome within modern humans (Hayward 1999; Zong et al. 2002).

2.3
More Recent Comprehensive Analyses Including the Central Constant Segment

Subsequent to those original reports, we have continued to extend our detailed analysis of KSHV genome variability in three ways. First, we have sequenced a large number of additional VIP genes (now totaling 180), especially from two major new collections in sub-Saharan Africa, including PBMC from 35 KS patients in Uganda and KS biopsies from 20 patients in South Africa. Second, we have sequenced a total of 60 complete TMP genes from a selected subset of the same samples evaluated previously for VIP, as well as from many of the new African samples. Third, we have also carried out PCR sequencing at up to 10 additional internal genomic loci in many of these same genomes (averaging between 60 and 150 samples each).

Interpretation of the results of these new data from all three segments of the genome now permits refinement and expansion of our previous models, leading especially to the concepts of both chimeric herpesvirus evolution and differences in the recent history of the "new" LHS and "old" RHS ends of modern human KSHV genomes. Based on the overall picture that has emerged, we can now classify known KSHV genomes into a model involving "12 principal genotypes" (see Fig. 1), in which we view each genotype class as being composed of variable-length segments with different estimated ages of divergence that correspond to the different degrees of variability or sequence conservation found at the 12 PCR loci studied. Unlike the hypervariable VIP and TMP genes, the level of variation at the 10 internal constant region loci is much smaller (on the order of 1% to 5% nucleotide differences only), and in this case the polymorphisms rarely affect the primary protein structure,

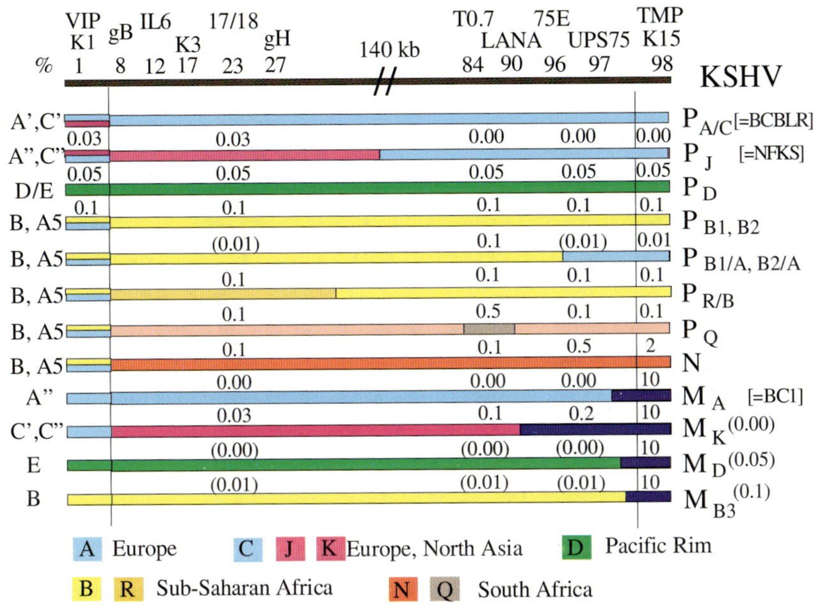

Fig. 1 The 12 major KSHV genotypes model. The diagram illustrates the deduced segmental and chimeric patterns of the 12 principal KSHV genotype structures. The linear genomic positions (%) of the PCR loci used are depicted at the top above the genome lines. For simplicity, the central segments (including the RAP/K8.1 locus at position 37) have been omitted (//). The allelic and subtype descriptions are given to the *right*, and the nine segment subtype color codes are denoted *below* the diagram. In addition, the exotic RHS M allele segments are designated by the *dark purple bars*. The estimated ages of divergence from the prototype Eurasian P genome on the *top line* are given for each major segment of each genotype in millions of years ago (MYD values)

but nevertheless they still show parallels to the subtype clustering patterns observed in VIP and TMP. Where possible, we have tried to obtain data for the complete VIP gene, not just fragments of it, and to analyze nearly all of the internal loci for each genome to obtain a complete and representative sampling picture across each genome.

3
Phylogenetics of the Hypervariable VIP or K1 Genes

A stylized generic radial phylogenetic tree showing the major branching and cluster patterns and divergence ages of VIP proteins from major human

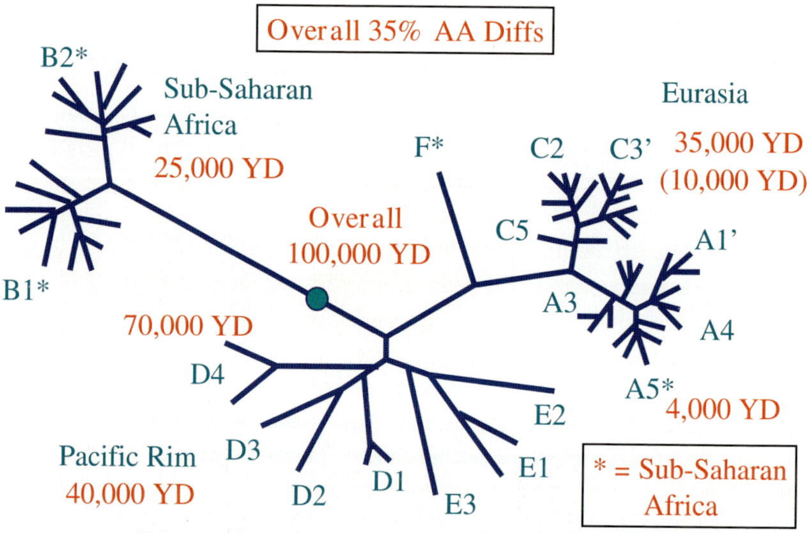

Fig. 2 Overall radial tree branching pattern for the KSHV VIP(K1) hypervariable ITAM membrane signaling protein. The diagram summarizes the major features of clustering and relative distance and branching patterns for samples from Eurasia, the Pacific Rim, and sub-Saharan Africa. The *asterisks* denote the four VIP subtypes found only in sub-Saharan Africa (B1, B2, A5, and F)

Fig. 3 Detailed linear phylogenetic tree for all 180 VIP proteins analyzed in our studies. This tree is based on the complete 289-amino acid VIP/K1 ORF. The entire tree is judged to contain only "modern" VIP subtypes, representing no more than 100,000 years of divergence. Overall, the VIP protein displays 30% amino acid variability, with 85% of the nucleotide polymorphisms being nonsynonymous. This gives an estimated rate of change of approximately one amino acid per 1,000 years. *Color patterns* highlight the samples from Uganda (*yellow*), South Africa (*blue*), Chinese Taiwan and Korea (*pink*), and American samples with African connections (*green*). The two Taiwan samples in the D and E branches were from aboriginal Hwalian samples. HKS22 from Uganda represents the sole example of an F subtype VIP gene found. In general, the A and C subtypes are found only in European and Asian samples, whereas B subtypes are found only in sub-Saharan African populations, and D and E subtypes are found only in Austronesian/Austroasian branch populations including Polynesians and Amerindians

population groups is presented in Fig. 2. The actual linear phylogenetic VIP protein tree for all 180 KSHV samples that we have sequenced plus a few representative samples from other studies is shown in Fig. 3, with key ethnic subgroup collections indicated in the color key.

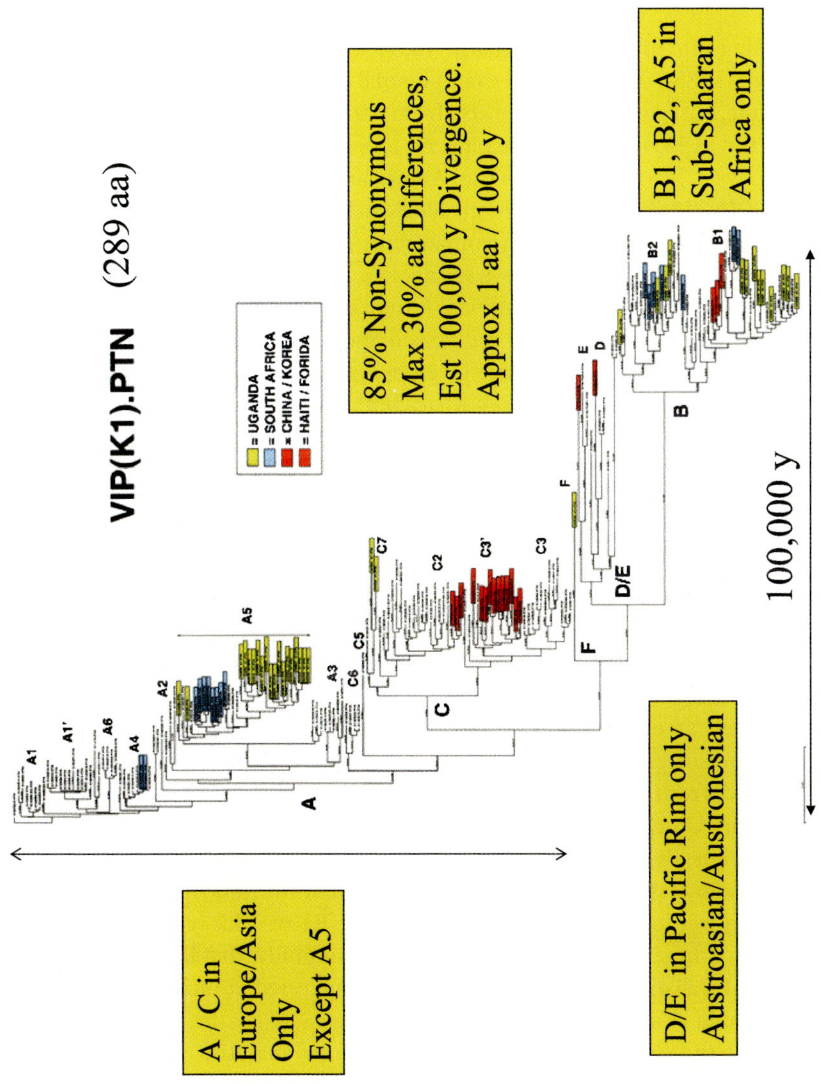
VIP(K1).PTN (289 aa)
= UGANDA
= SOUTH AFRICA
= CHINA / KOREA
= HAITI / FORIDA
A / C in Europe/Asia Only Except A5
85% Non-Synonymous Max 30% aa Differences, Est 100,000 y Divergence. Approx 1 aa / 1000 y
B1, B2, A5 in Sub-Saharan Africa only
D/E in Pacific Rim only Austroasian/Austronesian
100,000 y
A1
A1'
A6
A4
A2
A5
A3
C6
C5
C7
C2
C3'
C3
F
E
D
B2
B1
A
C
F
D/E
B

3.1
Age and Origin of B-Subtype VIP Genes in Sub-Saharan Africa

In our original analysis of 60 complete VIP genes from around the world (Zong et al.), eight of nine samples from sub-Saharan African were B subtypes, including our original B prototype (431K), and one was the prototype A5 variant (OKS3). Only five other examples of B-subtype VIP genes were found at that time, but four of these were also from patients of African origin, including ZKS6 and JKS20 from African emigrants to New Zealand and the United States, respectively, JKS15 from an African American patient in Baltimore, and OKS7 from a Haitian patient in Miami, Florida. OKS8, which closely resembled OKS7, came from a Mexican woman with AIDS in Miami, and it remains the only exception that we have found of a B-subtype VIP gene without an obvious direct connection to Africa. A total of 30 additional complete B-subtype VIP genes have now been sequenced and the overall results are shown in the partial VIP phylogenetic tree given in Fig. 4. All but one of the B-subtype genomes came from sub-Saharan Africa (either Uganda or South Africa), with the exception being the VG1 PEL sample from Haiti. Data for an incomplete internal segment of the VIP genes from the same 23 HKS samples described here were also presented recently in an analysis of just the VIP genes from patients with different tribal affiliations from Uganda (Kajumbula et al. 2006).

Overall, the 39 B-subtype VIP proteins that we have analyzed, although all differing from the A/C Eurasian subtype by 30%, show only moderate intratypic variations (totaling 8% at the amino acid level), and they cluster into two very distinctive major branches that we refer to as B1 or B2 variants. 431K is designated as our prototype for the B1 VIP gene, and SAPB3 is designated as our prototype B2 VIP gene. There were 13 B1s and no B2s among our original set of 14 examples of African VIP genes, which included five of five from Gambia, three of four from Tanzania, two from Uganda, and three from USA or Haiti (Zong et al., 1999). However, we have now found that four of eight Bs from South Africa and five of 15 Bs from Uganda proved to be B2 variants rather than B1 variants. Several of the new African VIP genes display novel in-frame deletions within the VR1 or VR2 regions. From the analysis of the partial Uganda VIP genes, the possibility that HKS9 and HKS27 represent a distinct third B3 subgroup was suggested in our earlier analysis (Kajumbula et al. 2006), and this remains plausible for the complete genes, including the South African samples in our larger analysis.

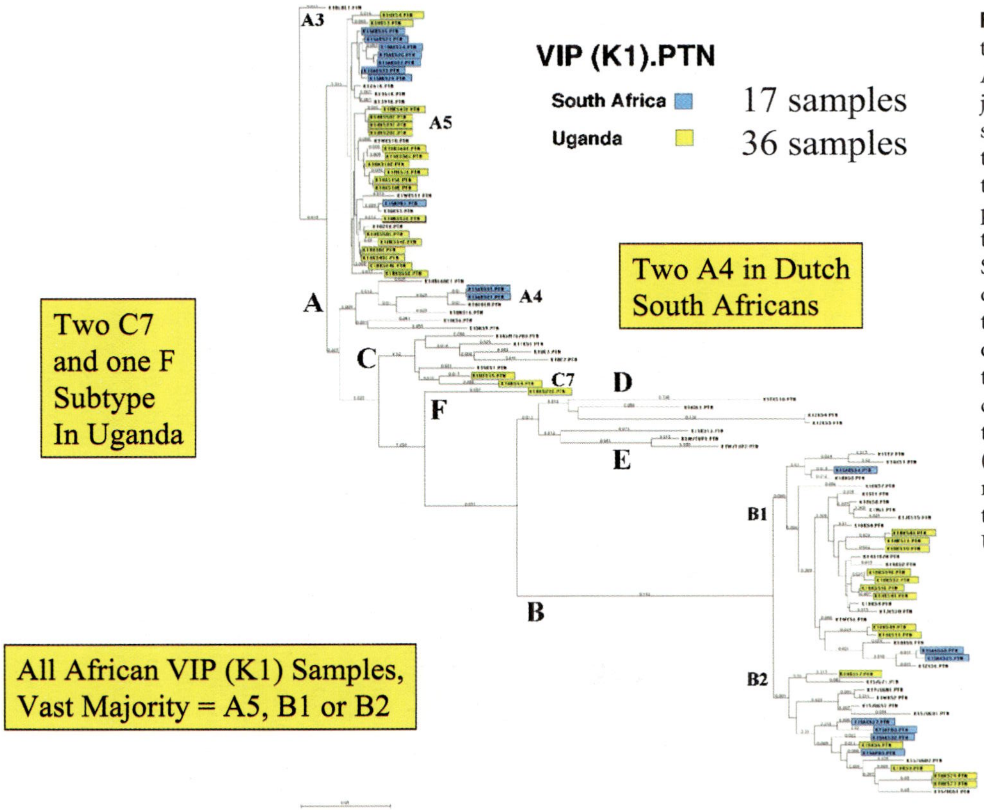

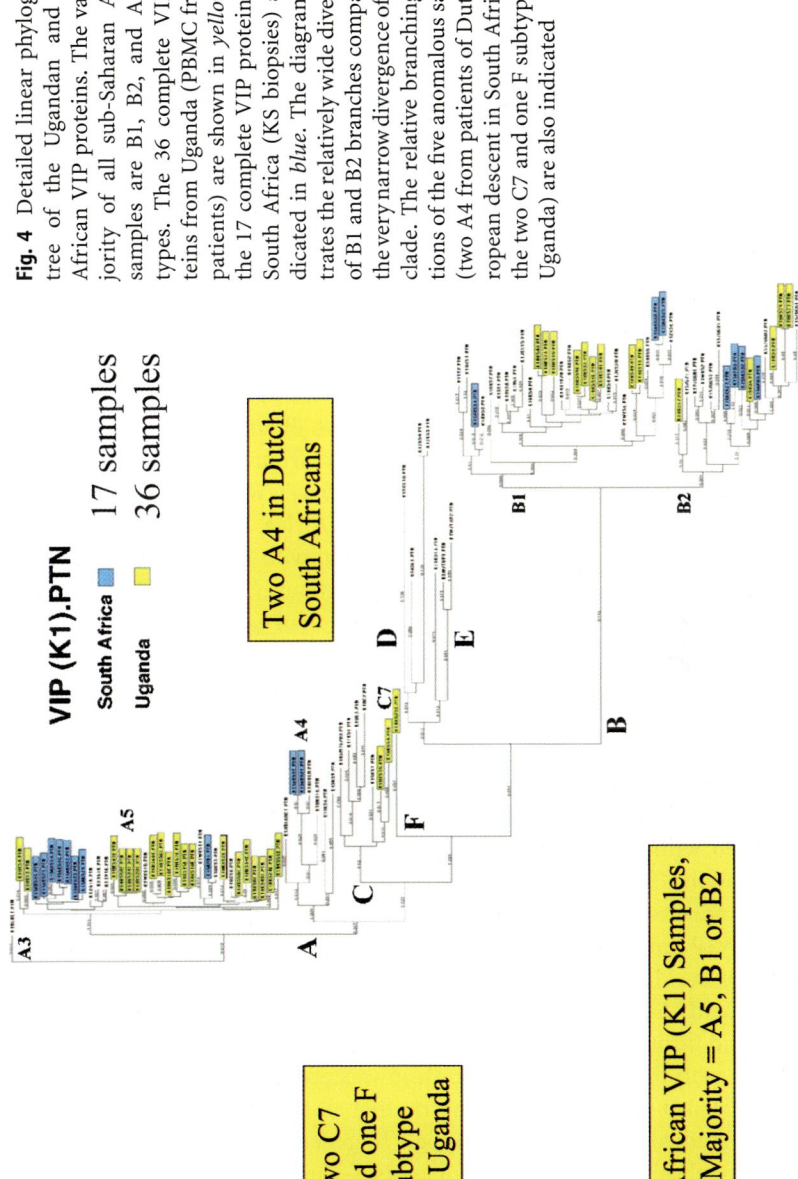

Fig. 4 Detailed linear phylogenetic tree of the Ugandan and South African VIP proteins. The vast majority of all sub-Saharan African samples are B1, B2, and A5 subtypes. The 36 complete VIP proteins from Uganda (PBMC from KS patients) are shown in *yellow*, and the 17 complete VIP proteins from South Africa (KS biopsies) are indicated in *blue*. The diagram illustrates the relatively wide divergence of B1 and B2 branches compared to the very narrow divergence of the A5 clade. The relative branching positions of the five anomalous samples (two A4 from patients of Dutch European descent in South Africa and the two C7 and one F subtype from Uganda) are also indicated

3.2 The African A5 Cluster of VIP Genes

In addition to the characteristic African B-subtype VIP genes, we have also now detected 34 more of the A5 subtype, plus one novel F- and two novel C7-subtype VIP genes from Uganda and South Africa (Fig. 4). In addition, two A4 examples from South Africa represent the only classic European variants that we have found in Africa, but these are the exceptions that prove the rule, because both came from the only known Caucasian patients among the group, who were both South African men of Dutch descent. In contrast, the rare prototype examples of VIP-F (HKS22) and the C7 subtypes (HKS35, HKS54) from Uganda all appear to be of genuine endemic African origin. Surprisingly, A5 variants have been detected relatively uniformly throughout Africa, including within a majority of our samples from South Africa as well as many from Uganda.

The 35 A5 VIP genes that we have analyzed all fall into a very tight and narrowly diverged cluster that is quite distinct from the closest European-north African variant clusters A3 and A7. However, we have never found an A5 subtype VIP gene outside of sub-Saharan Africa. Overall, the clustering of A5-subtype VIP proteins is extremely high, with branch lengths on the phylogenetic tree that are very short, being less than 1/5th that of both the B1s and B2s, suggesting a common origin no more than 4,000 years ago, with the B1 versus B2 (Fig. 4) divergence probably occurring about 25,000 years ago.

One other early study reported only A5-subtype VIP genes in PBMC among 15 children with symptomatic fevers in Zambia (Kasolo et al. 1997), but in retrospect it seems extremely unlikely to us that the particular primers used in that study would have been able to detect B-subtype VIP genes. Therefore, the implication that A5s may predominate in currently circulating viruses associated with lymphadenopathy and febrile illnesses in children, rather than in adult KS (although plausible), needs to be re-evaluated. The A5 VIP proteins are clearly much more similar to Eurasian VIP genes than to the B-subtype VIP genes, and they must have had a single-source introduction from Europe or the Middle East presumably via North Africa (backwards flow). The A5s evidently have been (and presumably continue to be) spreading rapidly across and throughout sub-Saharan Africa over a time frame that appears to coincide with the Bantu expansion from western Africa over the past 4,000 years. It is also important to emphasize that all KSHV genome samples that we have studied that have A5-subtype VIP genes are recombinant chimeras in which only a very small LHS segment is A-like, and the remainder of their genomes consist of typical distinctive sub-Saharan B, Q, R, or N subtypes (Zong et al. 2002) (Zong, J-C. et al, manuscript in preparation).

3.3 Source and Origin of the Novel Ugandan VIP-F and VIP-C7 Genes

Among the total of 69 complete VIP genes that we have studied with clear sub-Saharan African origins, only three, namely HKS35 (C7), HKS54 (C7), and HKS22 (F), which are all from Uganda, do not fit into the B1, B2, or A5 patterns described above. HKS35 and HKS54 cluster together between the two major groups of subtype-C VIP proteins, namely, C″(C3) and C′(C1/C2), and are referred to as the first examples known of C7 variants. These are closest to the C4 and C5 variants from Saudi Arabia (with an estimated divergence from other Cs approximately 15,000 to 20,000 years ago). In contrast, HKS22 lies between the A/C and D/E branches on the VIP protein phylogenetic trees (Figs. 2–4) and is the only sample that we have ever detected that falls outside of the three major branches. It is therefore tentatively referred to as an F-subtype VIP gene, suggesting divergence from the other main branches approximately 50,000 years ago. However, the C terminus of HKS22 is very similar to the C7 pattern, including the diagnostic five-amino acid deletion common to all Cs, and this sample clearly represents a complex F/C intragenic recombinant. In addition to HKS22, there are several other samples known (all from central Africa) that also do not fit readily into any of the currently defined A to E classification patterns. These include SAN2 (Whitby et al. 2004) and UGD23 (Kakoola et al. 2001), plus 43/Ber and 8/Dem (Lacoste et al. 2000b). Although they do differ very significantly from one another, we suggest that these, together with HKS22, are all members of a new fourth main branch that also splits into two subgroups designated F and G.

Finally, there is the question of the source or origin of these rare C7, F, and G VIP variants. In fact, all three genomes containing C7- or F-subtype VIP genes are extremely unusual at all loci tested, being regarded as novel intermediate F or G subtypes at several internal constant region loci (see Fig. 8; Zong, J-C et al., unpublished data) and either having unusual divergent B4- and B5-subtype ORF75 and TMP genes (HKS54 and HKS22) or, in the case of HKS35, having N-subtype ORF75 and TMP genes. We presume that the VIP C7 genes were relatively recently reintroduced into sub-Saharan Africa from the Middle East or Europe, whereas the novel F (and G) subtypes probably represent one or more genuine independent branches that may either never have left Africa or more likely returned into Africa from an early Middle Eastern branch that diverged after the south Asia/Pacific Rim branch but before the Eurasian A/C branch.

3.4
A Novel E-Subtype VIP Gene from a Taiwan Aborigine

Several rare KSHV VIP protein sequences have been described from indigenous populations around the Pacific Rim. Although diverged from one another by up to 24%, these nevertheless cluster into either of two distinctive D and E branches within a single D/E superbranch that is intermediate between the Eurasian A/C and African B branches (Figs. 1–3). Examples of D-subtype VIP genes include one (D1) from a Taiwan Hwalian aborigine (Zong et al. 1999), two (D2) from Pacific Island Polynesians seen in New Zealand (Zong et al. 1999), one (D3) from Australia that is presumed to be of aboriginal origin (Meng et al. 1999), and three (D4) of presumed Ainu origin from Hokaido in Japan (Meng et al. 2001). In addition, two VIP genes from Tupi Amerindians in Brazil were designated as the first examples of a new VIP subtype E1 (Biggar et al. 2000). More recently, Whitby et al. (Whitby et al. 2004) have also described five more partial VIP sequences (E2) from two Ecuadorean Amerindian tribes that branch close to the Tupi samples.

Interestingly, a second Hwalian sample (TKS13) that we have analyzed also clustered with the nearly identical D1 (TKS10)- and D2 (ZKS3)-subtype genomes in both the ORF26 and T0.7 constant region loci (Zong, J-C. et al., manuscript in preparation) but proved to have a VIP protein that is more similar to the E subtypes than to TKS10 or any other Ds. Therefore, we propose to designate TKS13 as E3 in contrast to the Brazil (E1) and Ecuador (E2) Amerindian subtypes. Furthermore, on the RHS, TKS13 proved to have a TMP-M allele, whereas TKS10 and the two Polynesian samples (ZKS3, ZKS4) have TMP-P alleles (see Fig. 5 and below). Unfortunately, sequence data for the constant region loci and TMP are not available for comparison from any of the other known Pacific Rim KS genomes with D or E VIP genes. Overall, we interpret that these 14 genomes with D/E pattern VIP genes were all derived from human populations that have common ancestors who initially migrated out of Africa approximately 70,000 years ago as a single distinctive south Asian branch, giving rise to all subsequent native Australian, Austroasian, Austronesian, and Polynesian populations, as well as Ainu and Amerindian branches, that is, collectively representing a Pacific Rim branch of modern human KSHV. Furthermore, all seven individual variant lineages within the D and E subtypes exhibit much longer branches than any of the variant clusters within the A, C, or B branches (Figs. 1–3), and therefore we suggest that all of these diverged from one another at least 40,000 years ago.

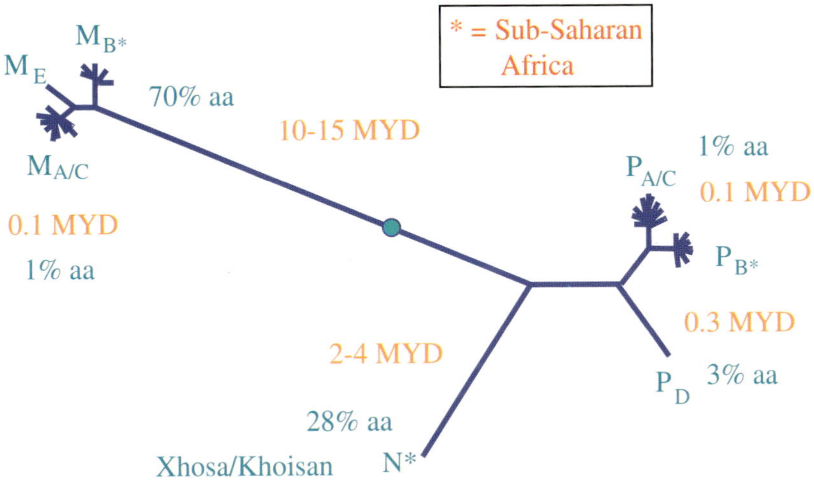

Fig. 5 Overall radial tree branching pattern for the KSHV TMP(K15) membrane signaling protein. The diagram summarizes the clustering, relative distance, and branching patterns for samples from Eurasia, the Pacific Rim, and sub-Saharan Africa. Estimated divergence times in millions of years (MYD) are given in *red*, and the overall % amino acid difference levels are shown in *blue*

4
Variable Sequence Loci within the Constant Region

To contrast with the hypervariable LHS and RHS termini of the KSHV genome, we refer to the remainder of the genome as the constant region. However, there are patches of low-level nucleotide variability all the way across the genome, and we have selected known sites of variability within the constant region as likely to be the most informative. The specific internal loci chosen for PCR sequencing analysis range in size from 640 bp up to 1,900 bp and include similar or extended versions of four previously utilized loci (ORF26E, T0.7, ORF75E, and UPS75) as well as five new loci encompassing segments of the vIL6, vMIR2/K3, ORF18/19, RAP/K8.1, and LANA1 genes. Some data have also been obtained for the UL22(gB) locus, but other potential loci such as the ORF4, vIRF1, vCYC, and vGPCR genes showed much less variability still. All of these nine chosen PCR loci together represent a sampling of 7,823 nucleotides (or 5.6% of the 137,000-bp unique segment genomes). Across this sampled part of the constant region we have detected a total of 409 chimeric nucleotide positions (5.2%), with the differences between a prototype Eurasian P(A/C) genome and the prototype P(B1) genome totaling 86 nucleotides (1.1%).

4.1 Clustering, Subtyping, and Linkage

We then interpreted these results in terms of clustering and subtyping at each locus, as well as in terms of maximal levels of linkage between loci and minimal inferred chimerism. Based on the concept of prototypic nonchimeric genomes assigned for each major subtype [e.g., BCBL-R as a average typical P(A/C) genotype, 431K as an average typical P(B)-subtype genome, and ASM72 as the most typical prototype P/M chimera], we have been able to define matching linked subtype patterns at all loci with considerable confidence. Major clustered deviations from these prototype subtype patterns at each locus were then assigned additional subtype status and matched between loci, using the assumption of maximal linkage. Deviations from the standard average or commonest linkage patterns are then considered evidence for additional chimeric recombination events.

A major feature of the results at all variable gene loci for KSHV (as well as for HCMV) is that the sequences show a very high degree of clustering, whereby almost all samples fall into one or another of a relatively small number of subtypes at each locus that may differ from each other quite extensively but themselves show very low levels of variability. There are occasional examples of intragenic chimerism with very obvious recombination junctions, but it is quite clear that there is not a continuum of variability or an infinite number of subtypes. However, the level of variability between subtypes can vary significantly between the different loci and the number of distinctive subtype branches found can also vary (especially in HCMV).

4.2 Subtype Distribution Patterns Within the Constant Region

One of the most puzzling aspects of the original VIP gene analysis (Zong et al. 1999), was that although it very neatly matched the expected patterns of strain divergence corresponding to founder effects from the migrations out of Africa, it did not display the anticipated patterns of greater diversity within sub-Saharan Africa than elsewhere that would have been expected based on the overall mitochondrial and Y chromosome phylogenetic trees of modern humans. However, this was not the case for the constant region loci, where much greater and older divergence is indeed found within the samples from sub-Saharan Africa. For example, we now recognize and define nine distinct subtype clusters within the T0.7 locus. With a few key exceptions, three of these are almost exclusively found only within Eurasian KSHV genomes (A/C, J, and K/M), whereas the D/Es are found only in aboriginal Pacific Rim samples and the other five (B, F/G, R, Q, and N) are found exclusively within sub-Saharan

African samples. All other constant region loci are also much more similar to this cluster distribution pattern than they are to the VIP patterns.

A summary of the observed levels of nucleotide divergence between subtypes at each of nine different constant region loci is presented in Table 1. Note that this analysis omits the hypervariable VIP and TMP loci and therefore greatly underestimates the overall level of variability between subtypes. There are three initial points to be made. First, within the constant region sites (especially at the RHS) there is essentially no distinction at all between genomes that have the A- and C-subtype VIP genes, and therefore we refer to them all as A/C. Second, there are nevertheless two other distinct subtypes within Eurasian constant region loci, which we refer to as J and K. Both are intermediately diverged compared to A/C and B, but there is a pattern whereby the "older" J and K constant region subtypes are preferentially linked to the VIP superpatterns referred to as A′ and C′, whereas the "newer" A/C subtype constant regions are preferentially linked to the A″ and C″ VIP superpattern genes (Zong et al. 2002). Furthermore, essentially only the K subtypes show direct linkage to the RHS TMP-M alleles. Third, it is evident that although the F- and G-subtype constant region genes are generally less diverged from A/C than the B patterns, the novel R, Q, and N subtypes are most often even further diverged from the Eurasian subtype than the typical African B pattern genomes.

4.3
The 12 Major Genotype Patterns in KSHV

Combining the results of our analyses for all three regions (i.e., the LHS VIP gene, the central segment constant region loci, and the RHS TPM alleles), leads to an overall picture that can be summarized in a highly simplified format by dividing the genotypes into 12 very distinctive and representative patterns as shown in Fig. 1. The subtype designations of different domain lengths of each genome are illustrated by the colored bars, with dark purple bars denoting the exotic TMP-M alleles. We envisage each of the different genotypes here as being composed of multiple segments with different ages of divergence. Numerical estimated ages (in millions of years) for each segmental domain are given in terms of the timing of its original divergence from the predominant European A prototype pattern, which is presented on the top line. Overall, there is a dramatic trend from left to right across the genome of finding relatively young segments only at the extreme LHS, toward retention of medium-age to older segments as well in the center, and then with some much older segments included as well toward the extreme RHS of the genome. The major features that contribute to this model of the 12 prototype genomes are described in the sections below.

Table 1 Comparison of the number of constant region nucleotide polymorphisms detected among KSHV genotypes

						Eurasia			Pacific Rim	Sub-Saharan Africa									
Locus name	Map position	Length (bp)	Sample No.	Polymorphisms Total	Polymorphisms Eurasia	A/C vs. CM J-A/C	A/C cdvs. K-M	J-A/C vs. K-M	A/C vs. D	A/C vs. F	A/C vs. G	A/C vs. B	A/C vs. R	A/C vs. Q	A/C vs. N	B vs. N	B vs. K-M	K-M vs. N	B vs. G
vIL6	13.3%	1826	69	75	19	11	13	2	5	19	21	22	20	17	20	19	26	23	21
K3	14.4%	920	71	48	12	3	3	2	12	NA	NA	10	9	12	12	5	6	9	NA
18/19	23.5%	930	51	28	8	8	8	0	11	NA	NA	13	13	8	13	10	11	7	NA
26E	34%	950	73	36	15	5	9	8	4	3	NA	11	8	9	11	12	15	13	NA
K8/8.1	58%	1900	48	62	31	NA	29	NA	13	NA	32	35	23	23	28	15	13	8	12
T0.7	84%	597	149	65	19	3	10	8	11	7	14	11	12	24	18	11	11	12	10
LANA**	90%	980	95	61	15	2	9	9	6	1	6	5	6	3	15	12	11	20	8
75E	96%	850	116	34	16	0	12	12	6	7	9	7	9	12	13	10	10	2	6
UPS75*	96.5%	770	116	66	18	0	16	16	17	7	6	7	NA	8	44	37	26	52	3
Total		9723	161	475	153	32	109	57	85	44	88	121	90	116	175	131	129	146	60

*Truncated before the hypervariable TMP(K15) C terminus; **nonrepetitive N terminus; NA, not known yet

4.4 Geographic Distribution of Constant Segment Subtypes

It is also dramatically evident that, just as for VIP, there are major differences in the distribution of constant region subtypes between the Eurasian samples (Fig. 6) and those from the Pacific Rim (Fig. 7, upper group) and sub-Saharan Africa (Figs. 7 and 8). In fact, in most cases there are also major differences

KSHV – EURASIAN GENOTYPES

		K1	gB	vIL6	K3	18/19	26E	T0.7	LNA	75E	UPS	K15	
BCBL-R		A1	A/C	A/C	A/C	A/C	A/C	A/C	A/C	A/C	A/C	P	
BCP1	*	A1	A/C	A/C	A/C	A/C	A/C	A/C		A/C	A/C	P	
BCBL-B		A4	A/C	A/C	A/C	A/C	A/C	A/C		A/C	A/C	P	
2x SAKS21		A4	A/C	A/C	A/C	A/C	A/C	A/C	A/C	A/C	A/C	P	HOLLAND
EKS1	*	o C2					A/C	A2		A/C	A/C	P	EGYPT
BKS12	*	o C2					A/C	A2	A2	M	M	MX	
BC1		A2	A/C	A/C	A/C	A/C	A/C	A2	A2	M	M	MX	
BKS16		A6						A6	A2	M	M	MX	
BC2		C3	A/C	A/C	A/C		J	A/C	A/C	A/C	A/C	P	
JSC1		C3	A/C	A/C	A/C		J	A/C		A/C		P	
BKS15		C3					J	A/C		A/C		P	
2x RGKS3/4	*	C3			K		A/C	A/C		A/C	A/C	P	SICILY
2x BKS3/4	*	C3					J	A/C				P	
ASM-K3		A1		J	A/C		A/C	A/C	A/C	A/C	A/C	P	
OKS13		C3					A/C	A3	M	--	M	MX	
SKS3	*	o C2					K	A3	M	M	M	MX	SAUDI
SKS9	*	o C5					K			M		MX	SAUDI
BC3		C3	A/C	J	K	J/K	J	A/C		A/C		P	
ARK1	*	o A7	J/K	J	J		A/C	A/C	A/C		A/C	P	ITALY
BCBL1	*	o A3	J/K	J	J	A/C	A3	A3	A3	A3	A/C	P	
OKS18	*	o C2						C2			A/C	P	
KSF		C3	J/K	J	K	J/K	A/C	A/C	A/C				GERMANY
BKS20	*	o A7					K	A/C			A/C	P	
11x KKS5	*	C3'			J	J/K	K	M	M	M	M	MX	CHINA/K
3x KKS6	*	o C2	J/K		K	J/K	K	M	M	M	M	MX	CHINA/K
BKS21	*	o C2				J/K		M		M	M	MX	
BKS13		o C1					K	M	M	M	M	MX	
ASM70		o C1	J/K	K	K	J/K	K	M	M	M	M	MX	

* = Classical, non HIV associated   o = Inner, older A' + C'

Fig. 6 Whole genomic structural summary for 43 Eurasian KSHV samples, including 23 P/M chimeras. Color coding distinguishes the M-associated segments (*red*) from P(A/C) patterns on the RHS, as well as the intermediate-aged P(J) and P(K) subtype patterns (*pink*) from the P(A/C) patterns on the LHS. *MX* indicates the exotic M TMP allele. The chart diagram is based on direct PCR data for up to 10 loci across each genome. Multiple related samples that have the same or very similar patterns are shown together on a single line

KSHV – PACIFIC RIM and SOUTH AFRICAN GENOTYPES

		K1	vIL6	K3	26E	T0.7	LNA	75E	UPS75	K15	
BCBL-R	o	A1	A/C	A/C	A/C	A/C	A/C	A/C	A/C	P[A/C]	USA
ASM70	o	C1	K	K	K	M	M	M	M	M[A/C]	USA
TKS10	o	D1		D	D/E	D/E	D	D	D	P[D]	TAIWAN/H
TKS13		E2			D/E	D/E			E/M	M[E]	TAIWAN/H
2x ZKS3/4		D2			D/E	D/E		D	D		NZ
431K	o	B1	B1	B1	B1	B1	B1	B1	B1	P[B]	ZAIRE
3x RKS3		B1	B2	B2	B2	B2	B1	B	B1	P[B]	ZAMBIA
RKS1		B1	Q	Q	B2	B2	B1	B	B1	P[B]	ZAMBIA
O21K		A5	Q	Q/B3	B2	B2	B1	B	B1	P[B]	ZAIRE
RKS2		B1	B2	B2	B2	B1	M'	A[B]	A[B]	P[AB]	ZAMBIA
391K		A5	B1	B1	M'	B1	B2	R/A	A[B]	P[B]	ZAIRE
RKS5		B1			M'	B1	M'	A[B]	A[B]	P[AB]	ZAMBIA
3x VG1		B1	B3	B3	B2	B3'	B2	A[B]	B3/M	M[B]	HAITI/FL
2x SAKS21/31		A4	A/C	A/C	A/C	A/C	A/C	A/C	A/C	P[A/C]	RSA/HOLLAND
SAKS32		B2	R	R	B2	B3	B1	B	B2	P[B]	RSA
SAKS24		A5	Q	Q	Q	B1	A[S]	A[S]	A[S]	P[AS]	RSA
3x SAKS29/33/35	o	A5	Q	Q	Q	QX	A[Q]	M[Q]	B[Q]	P[B]	RSA
2x SAKS25/30		B1'	Q	Q	Q/B	B1'	R	N	N	N	RSA
2x SAKS22/28		A5	N	N	N	B3	B1	B	B2	P[B]	RSA
SAKS23	o	A5	N	N	N	N	N	N	N	N	RSA
SAKS26		B2	N	N	N	B1	R	R/M'	M'	M[A]	RSA
SAKS27		B2	N	N	N	B3'	N	M[Q]	B[Q]	P[B]	RSA
ZKS6		B1'			Q	B1'		N			RSA/NZ

o = PROTOTYPES

Fig. 7 Whole genomic structural summary comparing four Pacific Rim KSHV samples (*upper group*) with the 15 South African samples (*lower group*) plus 14 other non-Ugandan samples (*central group*). Color coding distinguishes novel D, E, K, M, M′, and Q segments from prototype A/C or B patterns in the *top* and *center* groups, respectively, and distinguishes the A[S], A[Q], Q, R, M, and N segments from prototype A/C or B patterns in the *lower group*. *RSA*, Republic of South Africa. The three prototype PA/C(BCBL-R), PB(431K), and P/M(ASM70) genotypes are also included

between the subtype patterns found in our South African samples (Fig. 7, lower group) and those found in Uganda (Fig. 8) or other places in central Africa (Fig. 7, center group). In particular, both LHS and RHS N-subtype segments were found almost exclusively in South Africa, and LHS R-subtype segments were found primarily in Uganda, whereas LHS Q patterns were common throughout Africa.

One of the most notable features of sub-Saharan KSHV genomes compared to Eurasian and Pacific Rim genomes is an evidently much greater degree of chimerism within the central constant segment region, with many genomes that fall into just one of the "African" classes, in fact differing extensively

KSHV – UGANDAN GENOTYPES

	K1	vIL6	K3	26	T0.7	LNA	75E	UPS75	K15	
HKS22	F	R/B	B4/Q	K/R	A	A[Q]	F	B7	P[B]	GISU
HKS40	B1	F	K	J	B1	M'	A[B]	A[B]	P[AB]	G
HKS21	B1	F	K	J	B1'	M'	A[B]	A[B]	P[AB]	G
HKS54	C7	F'	K'	A/C	B4	A[B]	G	B5	P[B]	NYANKOLE
HKS35	C7	Q	Q	R	M	R	N	N	N	KAKWA (N)
HKS17	B2		Q		A2	R	H		P[AB]	LUO (N)
HKS41	B1	R	R	B2	B2	A[Q]	F	B3/M	M[B]	SOGA
HKS49	B1	B1	B1	B1	C5	A[Q]	F	B3/M	M[B]	NKOLE
HKS11	B1	B1	B1	B1	C4	A[Q]	F	B3/M	M[B]	NYORA
HKS56	A5	Q	Q	R	B1	R*	A/M	B6/M	M[A]	KIGA
HKS19	B1	R	R	B2	S2	B1	B	B1	P[B]	TORO
HKS27	B2	R	R	B2	B2	B1	B	B1	P[B]	G
HKS9	B2	R	R	B2	B1'	R	B	B1	P[B]	G
2x HKS20,47	A5	Q	R	B2	B2	B1	B	B1	P[B]	G,G
HKS23	A5	Q/R	R	B2	B2'	B1	B	B1	P[B]	G
HKS43	A5	Q	B1	R		B1	B			G
HKS50	A5	Q	R	B2	B2	B1	B	B1	P[B]	SOGA
HKS61	A5	Q	Q	B2	B2	R	B	B1	P[B]	G
2x HKS13,58	A5	Q	Q	B2	B2	B1	B	B1	P[B]	G, G
3x HKS24,34,52	A5	Q	Q	B2	B2	B1	B	B1	P[B]	TORO, LANGO(N),SOGA
2x HKS7,18	A5	Q	Q	R	B1	R	R/A	A[B]	P[AB]	G, NKOLE
2x HKS10,15	A5	Q	Q	R	B1	R'	R/A	A[B]	P[AB]	KIGA
HKS29	B1	Q	Q	R	B1	A[Q]	R/A	A[B]	P[AB]	G
HKS32	B1	Q	Q	R	B1	R	R/A	A[B]	P[AB]	G
HKS33	B1*	Q	Q	R	B1	R	R/A	A[B]	P[AB]	ALUR (N)
2x HKS36,60	A5	Q	Q	B2	B1	R	R/A	A[B]	P[AB]	G, G
HKS59	B1	Q	Q	B2	B1	R'	R/A	A[B]	P[AB]	G

G=GANDA

Fig. 8 Whole genomic structural summary for 34 Ugandan KSHV samples. Color coding distinguishes the segments designated as subtype F, G, H, J, K, Q, R, M′, and N from the commonest prototype sub-Saharan African B subtype pattern. Eurasian-like segments designated A, A[B], A[Q], and C are shown in *yellow*. M-allele and M-associated RHS segments are given in *red*. Other more complex chimeric segments are also indicated (see HKS22 especially as the most extreme example). Tribal affiliations are also listed on the RHS (Kajumbula et al. 2006). These genotypes are clustered into three distinct categories in the *upper, central, and lower groupings*, with the latter two being commonest within the Ganda tribe (*G*)

in the size, patterns and complexity of internal chimeric segments that mix Q, N, F, G, R, or even M′ loci into otherwise B genome backgrounds, for example. This is particularly common within a subset of Ugandan genomes derived from non-Ganda tribal sources (Fig 8, upper group). In contrast, the genomes from the Ganda tribe, the predominant Bantu group in Uganda, were of two major types, either those with mostly the common B-subtype patterns throughout except for R or Q at the LHS (Fig. 8, center group) or those with a more complex Q, B, R/A[B] chimeric pattern (Fig. 8, lower group).

5 The TMP-P and TMP-M Alleles and Adjacent Chimeric Regions

Variability in the KSHV TMP(K15) membrane protein mapping at the extreme RHS of the genome represents an entirely different situation than that for VIP. There are huge amino acid variations found between the P and M alleles, but these are not directly associated with the ethnic and geographic diasporas of modern humans. Rather, each allele itself displays the migration out of Africa patterns at a low variability level, and the differences between the P and M alleles (plus now a third newly recognized N allele as well) represent instead a vastly older evolutionary divergence in a gene that is not itself hypervariable like VIP but evidently evolves at the same slower rate as a typical constant region gene. Our reasons for reaching this conclusion are outlined below, and the current picture of TMP branching and divergence levels is summarized in Fig. 6 and Table 2.

5.1 Relative Lack of Variation Within UPS75 and the TMP Coding Regions of the Eurasian Subtypes of the TMP-P and TMP-M Alleles

Although the 15% differences between A and C VIP subtypes probably arose when modern humans first migrated as two major branches into Europe and

Table 2 Nucleotide differences and estimated divergence rates in TMP alleles

A. Other TMP-M Variants compared to the protype TMP-M strain (HBL6/BC1)

Name	Total Nucl diffs (2,100)	Exon Nucl diffs (1,500)	Intron Nucl diffs (605)	Total Amino acid diffs (500)	Est. years of divergence
VG1(B1)	27 (1.3%)	16 (1.0%)	10 (1.6%)	10 (2.0%)	100,000
HKS49(B2)	29 (1.4%)	18 (1.1%)	11 (1.8%)	12 (2.4%)	110,000
TKS13(E)	17 (0.85%)	8 (0.55%)	9 (1.5%)	6 (1.2%)	60,000

B. Other TMP-P variants and TMP-N compared to the prototype TMP-P strain (BCBL-R)

Name	Total Nucl diffs (2,080)	Exon Nucl diffs (1,470)	Intron Nucl diffs (600)	Total Amino acid diffs (489)	Est. years of divergence
431 K(B)	19 (0.95%)	13 (0.9%)	6 (1.0%)$^{+}$	7 (1.4%)	100,000
TKS10(D)	47 (2.3%)	33 (3.3%)	14 (2.3%)$^{+}$	25 (5.1%)	250,000
SAKS23(N)	425 (21.3%)	287 (19%)	138 (23%)*	140 (28%)	2,000,000

* Plus severalnon-contiguous blocks; + 10-bp insert

northwest Asia about 35,000 years ago, the 6% to 9% amino acid variations between and among individual variants of the subtype A and subtype C VIP branch has clearly arisen much more recently, probably within the past 10,000 to 12,000 years since expansion out of Ice Age refuges. In contrast, the rate of change within the TMP gene over this same time period has evidently been very slow and certainly is no greater than that in other typical constant region genes (Zong et al. 2002) (Zong, J.C. and Hayward, G.S., unpublished data). For example, there are very few differences at all (a total of only 8 nucleotide changes over 2500 bp) among nine representative TMP-P genes that we have sequenced from both European and Middle Eastern sources that are associated with a variety of A- or C-subtype VIP genes (Fig. 6). This group also includes the European GK18 sample described by Glenn et al. (Glenn et al. 1999). Similarly, across an 850-bp segment of the adjacent ORF75 gene (UPS75 locus) from a total of 17 genomes from Eurasian sources with TMP-P alleles, there were again no more than three total nucleotide changes detected, irrespective of what variant of either A- or C- subtype VIP genes were present (all those shown as P in Fig. 6). Similarly, among African UPS75 genes there was a surprisingly large subgroup of 22 examples with an A/C-like pattern, referred to as the A[B] subgroup (Figs. 7 and 8). Most of these display just a single distinctive additional nucleotide change from the Eurasian A/C pattern here, although at least nine have a common chimeric junction linking them to the African R subtype at the adjacent ORF75E locus.

In contrast, over 30% of all Eurasian KSHV samples examined, including 15 of 16 from Taiwan Chinese and South Korean KS patients, have the hugely different M-allele version of TMP (labeled M[X] and shown in red in Fig. 6), which is 70% diverged from the P-allele at the amino acid level (Poole et al. 1999; Zong et al. 2002). However, once again, sequencing of the complete coding regions plus adjacent 5′- and 3′-flanking areas from five TMP-M genes from both European and Asian sources showed no nucleotide differences at all over nearly 2,500 bp relative to the prototype BC1/HBL6 TMP-M gene. Similarly, within the nearby conserved UPS75 K/M locus (between positions 840 and 1300), 13 Eurasian genomes with associated TMP-M alleles displayed 16 common nucleotide differences (3%) from the TMP-P-associated UPS75 (A/C) prototype but again contained a total of only one variable position among them. Two other subtype K/M UPS75 genes derived from South Africa (SAKS26, FTKS2) were also identical to one another here but showed three differences from the Eurasian versions (Fig. 8 and see below).

5.2 Distinctive African Versions of TMP-P

Despite the nearly total lack of variation within and between A and C subtypes of both the Eurasian TMP-P and TMP-M alleles, there are consistent patterns of nucleotide differences within both the TMP-P and TMP-M allele versions that originate from sub-Saharan Africa and from south Asia/Pacific Rim sources. Our prototype African versions of TMP-P (431K) and TMP-M (VG1) were sequenced across the complete TMP coding and flanking region (2,500 bp) and proved to display a total of 18 and 29 nucleotide changes, respectively (representing 0.8% and 1.2%), compared to the prototype Eurasian TMP-P (C282) and TMP-M (BC1/HBL6) allele genes. Overall, with the exception of the few novel N-allele forms (see below), all 40 sub-Saharan African TMP-P allele genes examined proved to be either typical B subtypes similar to 431K (22 examples) or typical A/C-like Eurasian subtypes similar to BCBL-R (18 examples). However, there was considerably more sporadic variation among the African B-subtype samples (e.g., 48 variable positions over 2,500 bp) compared to both the African A/C subtypes of TMP-P with only eleven, or to only six across all eight tested Eurasian A/C subtype versions of TMP-P. Notably, all African A/C-like TMP-P genes (and the Saudi Arabian sample SKS1) have one common nucleotide change (431T) compared to all of the Eurasian versions, and therefore we shall refer to the African A/C-like group as the P[A/B1] subtype (Fig. 7).

Among the 23 examples sequenced of the African B subgroup of TMP-P allele genes (P[B] in Figs. 7 and 8), there were nine changes common to all of them. In addition, the predominant B1 subgroup (13 examples) all have six more common changes, whereas the B2 to B6 subgroups, which include eight of the nine examples from South Africa, lack those changes and are more variable. The six previously sequenced African TMP alleles (Kakoola et al. 2001) include three of our B1-subtype (Ugd2, Ugd12, Ugd19) and two of the A[B2] or A[B3] subset TMP-P alleles (Ugd4 and Ugd23) plus one B-subtype TMP-M allele (Ugd10) from Uganda.

5.3 Distinctive African Versions of TMP-M

Similarly, the seven sub-Saharan African TMP-M alleles that we have sequenced all proved to cluster into one of two groups, either those with typical B-subtype patterns similar to VG1 and Ug10 (5 examples) or those with typical Eurasian A/C-like patterns (2 examples). Therefore, although a subset of Eurasian TMP-P and TMP-M genes (like the A5 VIP genes) have seemingly returned from Eurasia to Africa, the majority of both are distinctive African

B subtypes that have not been found outside of either sub-Saharan Africa or related populations in America (e.g., Haiti and French Guyana). Interestingly, within African genomes there is little evidence for any linkage between VIP subtype genes (A5, B1, or B2) with any particular subtype of the TMP gene (whether $P_{A/C}$, P_B, $M_{A/C}$, or M_B), and examples of all combinations have been found (Figs. 7 and 8). Again, there was considerably more variation among the six known sequenced African TMP-M_B genes (six variable positions in all over 2,500 bp) compared to their invariant Eurasian TMP-$M_{A/C}$ counterparts, and SAKS26 and HKS56 both showed only one nucleotide difference from the five sequenced Eurasian TMP-$M_{A/C}$ genes. However, in addition, SAKS26 displays a 30-bp (10 amino acid) deletion within exon 8 (codon positions 389 to 398), and therefore represents the first and only example that we know about of a naturally deleted form of the TMP protein.

5.4 Distinctive Pacific Rim Versions of Both the TMP-P and TMP-M Alleles

We next considered whether, like the African specific versions of TMP-P and TMP-M, the KSHV genomes from the Pacific Rim branch might also show D/E subtype-specific TMP patterns. Although there were only four examples available in sufficient quantities to work with, namely, the two Hwalian samples from Taiwan of TKS10 and of TKS13 plus the two Polynesian ZKS3 and ZKS4 samples, these did indeed prove to have distinctive TMP-P(D) and TMP-M(E) patterns that are very different from those of any of the Eurasian A/C versions or the African B versions (Figs. 5 and 7). For the TMP-P(D) genes, TKS10, ZKS3, and ZKS4 were nearly identical across the entire 2,080-bp coding segment, but they displayed 43, 47, and 49 nucleotide differences (2.1%) from BCBL-R(A/C) and 33, 37, and 39 nucleotide differences (1.6%) from 431K(B), respectively. All three also included a 27-bp (9 amino acid) insertion after position 258 within exon 5. Similar to the B versions of TMP-P, TKS10, ZKS3, and ZKS4 also include a 10-bp insertion at the beginning of intron 1 compared to all A/C versions of TMP-P, but only TKS10, ZKS3, and ZKS4 also contain a 15-bp insertion in the noncoding region 93 bp upstream from the initiator codon. Out of 84 total variable nucleotide positions, the 431K(B1) and SAKS33(B2) variant TMP-P genes have nine differences from BCBL1(A/C) in common with TKS10(E) and ZKS4(D), but TKS10 and ZKS4 have 44 and 48 additional unique positions, respectively.

The TMP-M gene coding region from the Pacific Rim sample TKS13 shows 16 nucleotide differences from the prototype Eurasian version ASM70 (0.8%) and 22 from the prototype Africa version VG1 (1.1%). Out of a total of 33 variable nucleotide positions, HBL6(A/C) and TKS13(E) have 17 in common,

VG1(B) and TKS13(E) have 10 in common, and TKS13 has five unique changes. At the amino acid level, the 489-amino acid TMP-M proteins from HBL6(A/C) and TKS13(E) have six differences and VG1(B) and TKS13(E) have seven differences, whereas HBL6(A/C) and VG1(B) have 10 differences overall. VG1 and TKS13 have four amino acid differences from HBL6 in common, and TKS13 has two unique changes.

Therefore, the TKS13(E) subtype TMP-M gene parallels the VIP pattern by being intermediate between the HBL6(A/C) and VG(B) versions and more closely related to HBL6(A/C) than to VG1(B). In contrast, the TKS10 D-subtype TMP-P allele is instead much further diverged from BCBL1(A/C) than are the B versions, and it is 431K(B1) and SAKS33(B2) that are intermediate between TKS10(D) and BCBL1(A/C). For the TMP-P samples there are 25 amino acid differences between TKS10(D) and BCBL1(A/C) and 26 between TKS10(D) and 431K(B), compared to just five amino acid differences between BCBL1(A/C) and 431K(B1) or SAKS33(B2). Furthermore, among the TMP-P(D) genes, ZKS3 and ZKS4 differ from TKS10 at six and eight amino acid positions and between themselves at two positions. Therefore, the evolutionary distance of the TKS10/ZKS3/ZKS4(D) forms of the TMP-P protein appear to be between three and four times further diverged from both BCBL(A/C) and 431K(B1) or SAKS33(B2), compared to the distance that the TKS13(E) form of TMP-M has diverged from its BC1/HBL6(A/C) and VG1(B) counterparts.

5.5
Evidence for Three Distinct Chimeric Boundaries in African KSHV Genomes with TMP-M Alleles

We previously described a 500-bp transition region in HBL6 DNA encompassing the 3′-end of the TMP-M(K15) gene, a putative K14.1 gene, and the adjacent M-associated sequences at the 5′-end of the ORF75 gene (Poole et al. 1999). Within this region the level of sequence homology between the HBL6 M-prototype and the BCBL-R P-prototype changes gradually from 98% down to 80% and then abruptly becomes undetectable except for the coding sequences for several conserved SH3 and TRAF motifs in TMP. This clearly represents a complex ancient recombination junction between the modern genome constant region and a TMP-M allele of exotic origin. Our previously described diagnostic triple primer PCR test across the boundary here permits simple PCR product size-based discrimination between the standard P genomes and the less common P/M chimeric genomes (Poole et al. 1999).

Because of the patchwork homology over this region, the RHS of the chimeric domain is also readily detected within the sequence differences chart for UPS75, one of our previously described constant region PCR loci (Zong et al. 1997). The distinctive exotic M pattern begins at position 1357 in UPS75 and extends rightward. Importantly, all 13 European and Asian M allele genomes analyzed in the UPS75 locus have essentially the same adjacent, M-associated UPS75 subtype K/M sequences on the LHS, which are very distinctive from all A/C-, B-, and D-subtype UPS75 sequences.

In contrast, among the nine African genomes with TMP-M alleles, there are three distinctly different chimeric patterns. Six samples, including all three American or Haitian genomes (VG1, OKS7, and OKS8) (Fig. 7), and three Ugandan genomes (HKS11, HKS41, and HKS49) (Fig. 8), proved to have an identical recombination junction boundary mapping between UPS75 positions 1220 and 1280 and producing a chimeric B3/M pattern. The Ugd10 sequence reported by Kakoola et al. (Kakoola et al., 2001) also has this same B3/M junction. In contrast, HKS56 displays a second different chimeric pattern referred to as B6/M, with a junction between positions 1290 and 1350, whereas in the two B/R/M′ chimeras found in SAKS26 (Fig. 7) and FTKS2 (not shown), the UPS75 sequence pattern (referred to as M′) is very similar throughout to the standard Eurasian K/M pattern, although it is distinctive at four positions. These latter also have chimeric B to R/M′ boundaries further to the left inside the adjacent UPS75E locus.

Consistent with these patterns, all seven B3/M genomes have the distinctive M[B] African B type of TMP-M gene, whereas both HKS56 and SAKS26 have novel M[A] versions of the Eurasian A/C-subtype TMP-M alleles. Therefore, we suggest that the B3/M genomes all have a common origin from a different source of TMP-M than the Eurasian versions, whereas SAKS26 (plus FTKS2) and HKS56 each arose by separate and more recent secondary recombination events with Eurasian-derived TMP-M containing genomes. Whether the African TMP-Ms once had a common origin with their Eurasian counterparts before the divergence of their migrating human hosts 100,000 years ago, or instead represent more modern independent recombination events from two different Neanderthal K/M-like rhadinovirus genomes, for example, remains an interesting speculation.

Finally, both TKS10 and TKS13 also display complex patterns in UPS75 that suggest possible additional chimeric junctions. In fact, all three extensively analyzed D or E genomes (TKS10, TKS13, and ZKS3) contain similar but distinctive K/M-like segments between UPS75 positions 950 and 1150 or 950 and 1130, with TKS13 also having an additional novel upstream region between positions 720 and 950.

6 The Rare KSHV N Alleles in Southern Africa

6.1 Additional N-Subtype Genomes at the UPS75 Locus

Alagiozoglou et al. (Alagiozoglou et al., 2000) first described another distinctive KSHV subtype within the 850-bp UPS75 locus from among a set of KS lesion DNAs from South African patients seen in Johannesburg. This set of six samples proved to be almost twice as far diverged from the BCBL-R A/C prototype as were even the M allele-associated K/M sequences in UPS75. We initially confirmed and extended these results with two of the same N-subtype samples described by Alagiozoglou et al. (SALN7, SALN8) (Zong et al. 2002) and have subsequently also found five more examples, three from South Africa (SAKS23, SAKS25, SAKS30) and two from Uganda (HKS35, WKS14). The N-subtype sequence for the UPS75 region between positions 540 to 1300 has 42 nucleotide differences (5.5%) from the UPS75 P(A/C) prototype pattern of BCBL-R, although they vary at just one position among themselves. This compares to 16 differences in this region (2.1%) between BCBL-R and the Eurasian M-subtypes such as ASM70, which again show only one variant position among all 15 European, Asian, and African examples analyzed (Poole et al. 1999). By comparison, there are usually a total of only seven to nine nucleotide differences here from BCBL-R in each of the several different variants of African B-subtype UPS75 genes found. Finally, at this locus, the D1 and D2 subtypes found in ZKS3, ZKS4, and TKS10 were identical, but they all differ at 17 positions from BCBL-R, whereas TKS13(E2) is closely related to the Ds but is also more complex (22 differences from BCBL-R). The branching and divergence relationships among various different UPS75 (as well as LANA and ORF-75E) subtypes are illustrated in Fig. 9.

6.2 Novel N-Allele KSHV TMP Genes from Southern Africa

Obviously, the N-subtype UPS75 data suggested that the TMP genes from these KSHV samples might be very different from either the B-subtype TMP-P or B-subtype TMP-M alleles. Indeed, this proved to be correct for all four different KS lesions studied (SAKS23, SAKS25, SAKS30, and HKS35) that had N-subtype UPS75 genes and were available for further analysis. Two of these samples (SAKS25 and SAKS30) were derived a year apart from the same patient, but all four proved to be identical over the entire length of their TMP-N genes. Initially, we were unable to detect any PCR products directly with the standard triple primer reaction from this particular subset of

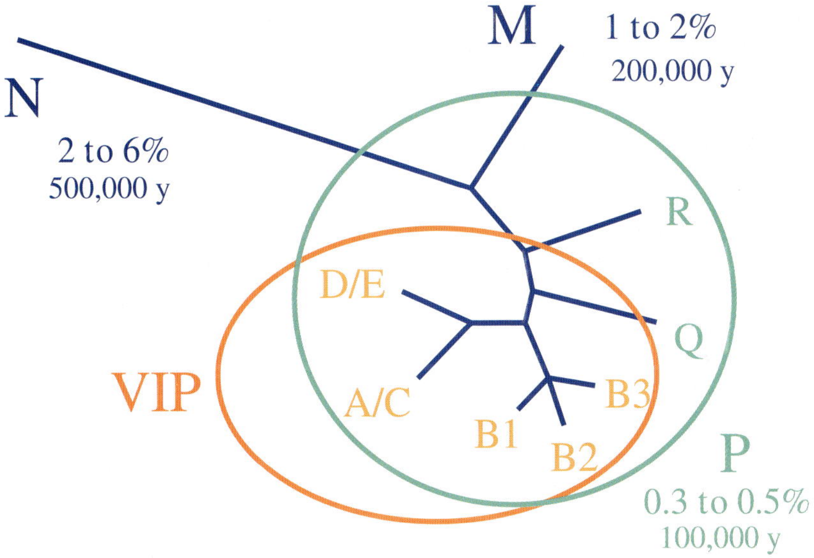

Fig. 9 Summary diagram illustrating the interpreted relative phylogenic ages for the subtype divergence patterns found within the RHS constant region segments in the LANA, 75E, and UPS75 loci. Note that in addition to the A/C, D/E, B1/B2/B3, Q, and R branches, which are all designated as the P-allele subset (encompassed by the *yellow circle*); there are also older more diverged M and N allele-associated segments with 1% to 2% (M) and 2% to 6% (N) nucleotide level variation from the PA/C pattern. The *red oval* encompasses the "modern human" subset of the P-allele VIP gene patterns (A/C, D/E, and B), which are the only subtypes now found at the extreme LHS of KSHV genomes

African samples that were known to have N-subtype UPS75 genes. They were still negative even after hybridization to Southern blots of the PCR DNA with either specific ^{32}P-labeled UPS75-P or ^{32}P-labeled UPS75-M DNA probes (data not shown). However, PCR products were ultimately recovered and sequenced from the SAKS23 prototype-N sample by using two rounds of redundant RHS 5′-primers whose design was based on weak residual DNA homology near the initiator codons of both TMP-P and TMP-M. Subsequently, a series of eight new TMP-N specific primers were generated and used to successfully amplify and sequence the entire 2,500-bp TMP gene and surrounding region from all four previously negative samples suspected of being N subtypes.

This new subset of KSHV TMP genes proved to fall into an intermediate position between the TMP-P and TMP-M allele patterns, although being considerably closer to the TMP-P pattern (Fig. 5). At the nucleotide level, the 2,080-bp SAKS23 coding region DNA including the introns shows 17% differences from the BCBL-R gene, whereas at the protein level, the novel N-associated TMP gene displays 29% amino acid differences from the prototype TMP-P protein (BCBL-R) and 69% differences from the prototype TMP-M protein (HBL6). Obviously, this is vastly greater than the 0.5% to 2% nucleotide differences for B and D subtypes relative to the prototype A/C subtypes for both TMP-P and TMP-M. Therefore, we consider that TMP-N represents a third distinctive allele with evolutionary hierarchal stature equivalent to that of the TMP-P and TMP-M alleles themselves.

Interestingly, the major structural difference in the TMP-N protein compared to both TMP-P and TMP-M is the insertion of an extra nine amino acids after codon 258 in exon 5, which lies within the fifth extracellular loop. Remarkably, this is also almost exactly at the same location as the novel nine-amino acid insert within the subtype-D TMP-P gene (TKS10, ZKS3, and ZKS4). Furthermore, there is nucleotide homology at 20 out of 27 bp between these two inserts (75% identity), and both encode an NXT/S *N* -glycosylation motif at exactly the same position. This similarity indicates that the extra domain evidently must represent the primordial situation, with both the M and all other P samples having suffered independent deletions of this NXT/S motif. Each TMP allele has only one other potential NXT/S motif, which maps within extracellular loop 2 for TMP-P and TMP-N and within extracellular loop 3 for TMP-M.

Overall, the 12 transmembrane domains of the three TMP allele proteins show considerably less divergence than the intervening loop domains, and, most importantly, several blocks of amino acids within the cytoplasmic tail of TMP (encoded entirely by exon 8) are virtually completely conserved in all three versions, including regions implicated previously as SH2 or SH3 tyrosine kinase interaction motifs such as YEEVL and YASIL, or the two potential TRAF interaction motifs QSGIS and TQPTD, although a third, SPQPD, is absent in TMP-M (Glenn et al. 1999). Curiously, like both the subtype-B and -D versions of TMP-P, TMP-N also contains a 10-bp insertion at the beginning of intron 1 relative to the TMP-$P_{A/C}$ version, and the sequence here is only 2 bp different from that in the TMP-P_B versions. Two internal 23-bp and 43-bp segments of TMP-N introns 1 and 2 are also very different from the TMP-P versions (14- and 27-bp deletions or mismatches, respectively), and there are also five other 1-bp insertions or deletions spread across introns 4, 5, 6, and 7.

6.3
Extension into the ORF75E Locus Reveals Additional RHS Chimeric Junctions

Data for three of the N-subtype samples (SAKS23, SAKS25, and HKS35) were extended to cover all of the adjacent expanded ORF75E locus (Chang et al. 1994; Zong et al. 1997), where they differ at 18 polymorphic positions from the BCBL-R (A/C) prototype, although in this segment they also more closely resemble several other patterns, including K/M (4 differences), Q (4 differences), and R (5 differences). In comparison, the D, B, and F patterns for ORF75E differ from A/C at only 6, 9, and 8 positions, respectively. One important additional sample ZKS6 from a Mozambique San bushman patient was also judged to be N subtype at this locus (Fig. 7).

At least three additional chimeric boundaries are detectable within the ORF75E locus in other African genomes. The most common one occurs between positions 159 and 237 in 10 samples (nine from Uganda) that we refer to as the R/A subtype (Fig. 8). All were A[B] in ORF75 and also have common patterns in T0.7 and LANA. Eight were sequenced for TMP, and all were judged to be P[AB1]. Therefore, the size of the RHS Eurasian-derived A/C-like segment present in nearly half of the African chimeric genomes is only 2.6 kb. Interestingly, the second type of chimera here (R/M) also involves almost exactly the same location, but this time joining an R pattern on the left with a K/M pattern on the right in SAKS26 (Fig. 7) and HKS67 (not shown). The third obvious chimeric junction in this region involves an A/C-like pattern to the left of position 528 joined to a B3 pattern on the right in all four American B genomes (JKS15, OKS7 and OKS8, and VG1) (Fig. 7). HKS54 and HKS56 are also very similar, but with slightly different boundaries (A/G and A/B6) (Fig. 8).

7 Differential Evolutionary History of Segments Along the KSHV Genome

7.1 Age of Divergence and Origin of the Three TMP Alleles

In attempting to estimate the age of the evolutionary divergence of TMP-P, TMP-N, and TMP-M alleles from each other, we have based our calculations on the presumption that the B subtypes of both TMP-P and TMP-M (which show between 0.9% and 1.2% nucleotide divergence each from the A/C forms) are judged to have branched away from their common modern human ancestor close to 100,000 years ago. Intriguingly, although divergence of the D/E (Pacific Rim) version of TMP-M in TKS13 (0.85%) fits well with the notion of being 70,000 years separated from the A/C and B versions, the particular TMP-P allele present in TKS10, ZKS3, and ZKS4 (2.3% diverged) is obviously much older than the B version and is projected to have diverged from both the A/C and B versions between 200,000 and 250,000 years ago (Figs. 1 and 5). An extrapolation for 21.3% nucleotide divergence would suggest an evolutionary distance of 2 to 4 million years for the TMP-N allele from TMP-P, with a further extrapolation to at least 10 to 15 million years for TMP-M from TMP-P. Note that the equivalent gene (R15) in the RRV rhesus rhadinovirus (the RV2 prototype herpesvirus), although showing a strong structural resemblance with 12 transmembrane domains and eight matching coding exons, has no measurable residual amino acid homology to any of these KSHV TMP alleles (Alexander et al. 2000; Searles et al. 1999). The original divergence of rhesus monkeys from great apes and humans is expected to be on the order of 20 to 25 million years ago, but because KSHV fits into the RV1 rather than RV2 lineage of Old World primate rhadinoviruses (Rose 2003; Rose et al. 1997), the RRV difference is probably misleadingly large and that in rhesus RFMHV (an RV1 class) virus would likely provide a better yardstick. In the future, sequence data for the TMP genes from the three distinct KSHV-like viruses known in chimpanzees (Lacoste et al. 2000c), which would be expected to have diverged from the equivalent human and prehuman versions only 5 to 6 million years ago, will go a long way toward clarifying the exact age of divergence of the three TMP alleles found in modern humans.

Whether the events that generated these chimeras occurred between multiple distinct viruses all infecting the same prehuman hosts or instead were derived by recombination with exotic nonhuman but nevertheless "hominid" lineages remains to be elucidated. However, it is very clear from the easily recognized recombination junctions found near the N-terminal regions of ORF75 that both the TMP-N and TMP-M alleles themselves are the only segments of modern-day KSHV genomes that diverged more than 1 million years ago, whereas all of the genomes carrying them that have been detected so far represent chimeras containing these relics of anciently diverged viruses recombined into predominantly modern lineage human KSHV genomes. Nevertheless, the modern KSHV genomes that they originally recombined with (as represented by the adjacent constant regions in the T0.7, LANA, ORF75E, and UPS75 loci, etc.) are also both very distinctive from and significantly older than the standard subtype A/C, B, and D/E level divergence seen in the VIP gene at the LHS and in the central constant segment loci of all other sequenced KSHV genomes. Recombination with, for example, *Homo neanderthalis*, *Homo erectus*, or *Homo robustus* versions of KSHV would be the type of scenario that we envisage may have introduced these small segments of now presumably extinct KSHV-like viruses into chimeric modern human genomes.

However, we cannot exclude the alternative possibility that the complete highly diverged original M genomes that evolved separately in prehumans may still exist in modern *Homo sapiens* but have not yet been detected, perhaps because they are not associated with KS. Consider that (1) chimpanzees are known to have three distinct rhadinoviruses, two in the RV2 RRV-like KSHV-like lineage and one in the RV1 lineage (Lacoste et al. 2001); (2) all three major branches of modern humans (Eurasian, African, and south Asia-Pacific Rim) have their own distinctive versions of both the hugely diverged TMP-P and TMP-M alleles as either complete P or chimeric P/M KSHV genomes; and (3) at least one of the oldest groups of humans in southern Africa has a representative of yet a third rhadinovirus TMP allele as P/N chimeras. Therefore, the possibility that there are or were three distinct human RV1 lineage rhadinovirus species cannot be ignored.

7.2 Timing of the P/M Chimeric Events

Under either scenario, a question also arises about whether the final recombination event(s) that created the P/M chimeras occurred just once before modern humans came out of Africa or occurred on three separate occasions, once in the African, once in the Pacific Rim, and once in the Eurasian lineage. For the first option to be correct, the small founding human population that initially successfully exited the African continent would have had to have brought two viruses with it, one P and the other P/M, both of which then independently diverged within the three main branches. However, the fact that the three different chimeras each have different junction boundaries clearly favors the alternative scenario in which just a single *Homo sapiens* P genome came out of Africa along with the founding modern humans, but that each of the three branches independently recombined with pre-existing Neanderthal version P/M chimeras that had also diverged into three separate lineages in the three different continental subgroups. In fact, we probably also have to conclude that this type of event occurred at least twice in Africa to create both the relatively common B3/M chimera and the rare B6/M chimera found in HKS56. In addition, a secondary transfer event must have also introduced the Eurasian type M′[A] exotic TMP allele into a B/R genome in Africa (SAKS26 and FTKS2).

We originally favored the scenario that the source (in terms of both time and place) for the origin of the KSHV P/M chimeric genomes was the still overlapping *Homo sapiens* and Neanderthal populations in the Middle East as recently as 25,000 years ago, but the new evidence both here and from Kakoola et al. (Kakoola et al. 2001) for distinct African and now also south Asia-Pacific Rim versions of both the TMP-P and TMP-M alleles instead implies an alternative model in which the original TMP-M allele (probably already in the form of a P/M chimera) must have already been present before the migrations out of Africa. However, whether it was present within modern

Homo sapiens at that time or rather just in an older "humanoid" subgroup is a critical point. If the former, this second type of KSHV virus would have had to be retained in all three branches during the same founder bottleneck events that also led to the divergence of the P versions of the VIP gene into just one main subtype (A/C, B, or D/ E) within each branch. This seems doubtful. Therefore, a more complex version of the original model suggested above, in which all three subsets of TMP-M alleles were instead acquired independently by multiple but rare transfers into modern human P genomes from already existing older and already diverged "Neanderthal" chimeric P/M genomes that occurred after the primary modern *Homo sapiens* migrations out of Africa, seems to be a much more likely prospect. This model also suggests that some form of selective advantage for these particular chimeric recombination products may have been involved (within modern humans at least). Furthermore, the unexpectedly high divergence of the south Asian-Pacific Rim D version of TMP-P, but not of the E version of TMP-M, provides an additional complication to the single parental "out of Africa" concept that needs to be considered in any all encompassing model. However, it could be entirely consistent with divergence of the parental "Neanderthal" P/M genomes into three branches with a temporal pattern different from that for modern humans.

7.3
Source of the Exotic TMP-N Alleles

Presumably the rare P/N genomes would have to represent another example of a similar scenario in Africa, although presumably in this case with a different prehuman KSHV-like virus having a TMP gene allele that had diverged at an intermediate temporal stage between the P and M allele versions. However, it is still necessary to evaluate the full length of the central segments of all of these genomes more thoroughly before any compelling theories can be developed about the source, age, origin, and history of the original parental N-class KSHV genome. Nevertheless, it is probably highly significant that four of the five examples of genomes with both N-subtype TMP alleles and large N-associated adjacent internal segments came from South African sources (and the other one was from Uganda). We have also observed a number of additional South African KS genomes (but rarely Ugandan ones) that have partial N-like constant region segments, but without the associated N-allele TMP genes (Zong J-C and Hayward, G.S., unpublished data). Note that many of these South African patients came from a mining area where there is significant mixed heritage involving Zhosa or Khoisan ancestry. Furthermore, ZKS6, which contains an N-subtype ORF75E gene at least, came from a patient of known San bushman ancestry who emigrated to South Africa and then to New Zealand from Mozambique. Similarly, Whitby et al. (Whitby et al., 2004) have recently described a San bushman KSHV sample from Botswana (SAN1) that also contains an N-subtype ORF75 gene. These observations strongly hint that both the TMP-N allele and the N-subtype ORF75 gene may be indigenous in Zhosa or Khoisan populations (and may never have left Africa).

7.4 Only Modern Versions of the VIP Genes Have Been Detected

In the case of the VIP genes on the LHS of modern human KSHV genomes, there seems to be a very different evolutionary dynamic in action. Two overall general conclusions can be made. First, only modern P forms of the VIP gene are found, which split relatively recently into the A/C, B, and D/E branches corresponding to the diasporas created by migrations out of East Africa. Second, in sub-Saharan Africa, the VIP genes are essentially randomized and unlinked relative to their associated adjacent constant region subgroup patterns. In distinct contrast to the retention of alternative TMP alleles that are ancient chimeric relics of other viruses on the RHS, no VIP gene subtypes older than the estimated 100,000 years divergence created by the migrations into and out of Africa have been detected. Admittedly, we probably would not be able to detect true M or N versions of the VIP gene directly with the P-subtype primers that we use. However, no KSHV genomes that are detectable with ORF26, T0.7, or UPS75 locus primers (which do amplify all N, M, B, R, and Q constant region variants) have ever proved not to have a detectable linked modern P-subtype VIP gene. Furthermore, all VIP genes found so far fit into the B, A/C, D/E, or F/G branches of the P subtype. Therefore, on the basis of the temporal yardstick of 40,000 years divergence from each other for the known aboriginal Hwalien (Taiwan), Australian, Polynesian, Japanese (Ainu), and Amerindian versions within the D/E branch, we estimate that all known versions of VIP diverged no more than 100,000 years ago. We envisage that the B branch separated from the common ancestor of all of the others approximately 100,000 years ago, then the D/E branch diverged from the A/C and F/G line 70,000 years ago, and finally the A/C branch diverged from the F/G branch about 40,000 to 50,000 years ago. Subsequently, the Eurasian A plus C branches likely diverged from each other no more than 30,000 years ago, with the four main subclusters (C′, C″, A′, A″) each having branched off 15,000 years ago. Similarly, in sub-Saharan Africa, the original divergence of all known B-subtype VIP genes into the B1 and B2 variants appears to have occurred between 25,000 and 30,000 years ago, but in contrast the entire A5 cluster evidently originated no more than 4,000 years ago.

7.5 Source and Rapid Penetrance of the Anomalous A5 VIP Subtype Within Sub-Saharan Africa

How then did the "European-like" A5 VIP genes become distributed seemingly uniformly throughout central, eastern and southern Africa, as well as become associated almost randomly with genomes that contain B, R, Q, or N constant regions? The interpretation partially depends on the age of divergence of the constant region subtypes themselves, which is very difficult to estimate and clearly varies between gene loci with a pronounced tendency to be greater toward the RHS than on the LHS (Zong, J-C and Hayward, G.S., unpublished data). At nearly all constant region loci, we recognize five distinctive African subtype clusters (referred to as B, R, Q, N, and M′) that are at least as far diverged from the A/C cluster as are the Bs, as well as F-, G-, D/E-, J-, and K-subtype clusters that are intermediate between the A/C and B subtypes. Notably, B, R, Q, and N are all unique to sub-Saharan Africa. Consequently, we judge that the R, Q, N, and M subtypes of constant region sequences on the RHS all diverged from a common ancestor with the A/C and B subtypes well before the "out of Africa" migrations, whereas on the LHS some may have diverged contemporaneously with B from A/C. Overall, it is very obvious that there have been no VIP genes yet detected that have the equivalent patterns of divergence expected for the older R, Q, N, and M versions of VIP. Within sub-Saharan Africa, they have all evidently been lost and replaced by relatively modern B1, B2, or A5 VIP genes.

Evidently, a prototype A5 VIP gene must have been originally introduced from Europe or north Africa (probably just once) into presumably a B-subtype genome in the ancestors of the Bantu people at Bok in Nigeria before their expansion across most of the continent beginning about 4,000 years ago. Recombination into the genomes of other ethnic African populations who were displaced or absorbed would account for some spread into the hybrid B/R and B/Q genomes that are now most common among the predominantly Bantu samples that we analyzed in Uganda. However, the Bantu expansion did not proceed beyond the Cook River in South Africa before Europeans arrived several hundred years ago. Yet many of the South African genomes with N-subtype segments, which are presumably of Zhosa or Khoisan origin, also have A5 VIP genes! Furthermore, even the B-subtype VIP genes from South Africa are very little different from those found in Uganda or elsewhere in Africa, in contrast to the often dramatic differences in their constant region subtype patterns (i.e., including Q, R, and N subtypes). We propose that the most likely explanation is a very rapid, and recent, "aggressive" spread through recombination events, which are also presumably connected to some strong selective advantage in combining a Eurasian A5-subtype VIP gene with otherwise African-specific B-, Q-, R-, or N-subtype constant region genomes. Unfortunately, there is too little divergence in the ORF4 region adjacent to VIP to be able to detect chimeric boundaries there or to judge whether or not the A5/B LHS segment was subsequently introduced into R, Q, and N genomes on just one or on multiple occasions (although we assume the latter).

8 Overall Significance

8.1 Precedents From Other Herpesviruses

There are several obvious precedents for multiple distinct herpesvirus species from the same virus genus coexisting within the same host species. For example, human HSV-1 and HSV-2 have an average protein level divergence of approximately 30%, which is similar to the average protein level divergence of human and chimpanzee CMVs (Davison et al. 2003). HSV-1 and HSV-2 have different preferential sites for establishment of latency in their human hosts and evidently do not successfully recombine in vivo, but can do so relatively easily in coinfected cell cultures under laboratory conditions. Similarly, humans harbor three distinct roseoloviruses with similar but distinct biological properties, HHV6A, HHV6B, and HHV7. The overall divergence level of HHV6 and HHV7 is similar to that between HSV-1 and HSV-2, or between human and chimpanzee CMVs, but although the level of divergence of several genes in HHV6A and HHV6B reaches 15% to 25% at the protein level, most of the rest of their genomes differ by no more than 5% to 10% at the nucleotide level. Again, no known chimeras have been detected, but most individuals harbor HHV7 plus one or both of the HHV6 variants simultaneously.

Finally, the human lymphocryptovirus EBV has genomes that fall into two subtypes, often referred to as EBV-1 and EBV-2 or, preferably, EBV-A and EBV-B. Here the situation more closely resembles that in KSHV, with one gene in particular, EBNA2, occurring as two highly dimorphic alleles known as EBNA2A and EBNA2B that differ by approximately 45% at the amino acid level (Dambaugh et al. 1984), compared to a 55% amino acid divergence for both from the equivalent gene in the baboon lymphocryptovirus, H. papio or HPV1 (Ling et al. 1993). Several of the other EBV latency genes also split into A and B subtypes but show only 3% to 5% amino acid divergence, and, except for the somewhat hypervariable LMP1 gene, the differences elsewhere are likely to be no more than a few percent at the nucleotide level.

Therefore, similar to the situation with the KSHV TMP alleles, one of the two forms of human EBV genomes evidently carries a chimeric segment consisting of just the second allelic type of EBNA-2 gene, which must have diverged perhaps 15 to 20 million years ago, and would seem likely to have been derived by an exotic cross-species recombination event into a modern human EBV genome that was slightly different from the current predominant version of human EBV-A. Just as with KSHV, some relatively rare recombinant chimeras containing segments of both EBV-A and EBV-B are known, but the two EBNA2 alleles are both found spread throughout the Eurasian, African, and south Asia-Pacific Rim (Papua New Guinea) branches of modern humans (Aitken et al. 1994). There also appear to be distinctive Eurasian, African, and south Asian-Pacific versions of at least the EBNA2A allele (Shim et al. 1998).

Although it seems extremely unlikely that a complete version of the parental EBV-B-like virus that was the source of the second human EBNA2 allele still exists in modern humans, it does seem plausible that either a complete or a chimeric version of it may have existed in some other now extinct branch of prehumans, such as Neanderthals, and was then transferred probably only once early during the period when modern humans overlapped with them, originally perhaps 100,000 to 200,000 years ago in Africa or even as recently as 25,000 to 30,000 years ago within the Middle East or southern Europe.

8.2 Contribution of Chimeric Fixation to Herpesvirus Genomic Evolution

Irrespective of their precise origins, the presence of these chimeric genomes and alternative alleles of key genes in both KSHV and EBV probably reflect a somewhat unexpected but normal mechanism by which herpesvirus genomes evolve. There are also three radically different alleles of the STP/TIP genes in herpesvirus saimiri (HVS) that may represent a similar situation. Even although herpesviruses are not supposed to successfully cross between species barriers as whole genomes, our data here provide evidence that partial chimeric versions of exotic viral cross-species recombination events probably do occur and do survive successfully at least temporarily (irrespective of the issue of whether a cross-species host transfer is also involved). Presumably, for closely related viral sequences with greater than 85% nucleotide identity, unless there is a selective advantage, the rare novel form will tend to be diluted out and lost by homologous recombination, as we see occurring with the progressively shorter segments of the M-associated K/M constant region sequences found in Eurasian KSHV genomes that retain TMP-M (Kakoola et al. 2001; Poole et al. 1999; Zong et al. 2002). However, for genes that are sufficiently diverged already to be unable to undergo homologous recombination, that is, greater than 15% differences at the nucleotide level, such as TMP-P, -M, and -N or EBNA2A and 2B, they would not become scrambled or diluted out, especially if there is a positive selective biological effect, and can become fixed in the population, initially just as alternative alleles but eventually perhaps after certain bottlenecks or founder effects narrow down the range of variants in a subpopulation, they could even be destined to take over as the only extant version for a particular virus species.

Acknowledgements These studies were supported by NIH Research Grants R01-CA-73585 and P01-CA-81400 awarded to G.S.H. We thank the many colleagues who provided anonymous DNA samples or clinical paraffin sections for analysis that made this work possible. Many of them were acknowledged previously (32, 44, 45), but we especially want to thank our colleagues Henry Kajumbula, Edward Katange-Mbide, and William Boto from the Uganda Cancer Institute, University of Kampala, Uganda, and Lee Alagiozioglou, Lynn Moris, and Fred Sitas from The National Institute of Virology, Sandringhan, South Africa, for the Ugandan HKS and South African SAKS collections, respectively, for which the primary data reports have not yet been published. We also acknowledge the major contribution of Peter S. Kim to the identification of the TMP-N allele.

References

Aitken C, Sengupta SK, Aedes C, Moss DJ, and Sculley TB (1994). Heterogeneity within the Epstein-Barr virus nuclear antigen 2 gene in different strains of Epstein-Barr virus. J Gen Virol *75*, 95–100

Alagiozoglou L, Sitas F, and Morris L (2000). Phylogenetic analysis of HHV-8 in South Africa and identification of a novel subgroup. J Gen Virol *81*, 2029–2038

Alexander L, Denekamp L, Knapp A, Auerbach MR, Damania B, and Desrosiers RC (2000). The primary sequence of rhesus monkey rhadinovirus isolate 26–95: sequence similarities to Kaposi's sarcoma-associated herpesvirus and rhesus monkey rhadinovirus isolate 17577. J Virol *74*, 3388–3398

Biggar R, Whitby D, Marshall V, Linhares AC, and Black F (2000). Human Herpesvirus 8 in Brazilian Amerindians: A hyperendemic population with a new subtype. J Infect Dis *181*, 1562–1568

Brinkmann MM, Glenn M, Rainbow L, Kieser A, Henke-Gendo C, and Schulz TF (2003). Activation of mitogen-activated protein kinase and NF-κB pathways by a Kaposi's sarcoma-associated herpesvirus K15 membrane protein. J Virol *77*, 9346–9358

Chang Y, Cesarman E, Pessin MS, Lee F, Culpepper J, Knowles DM, and Moore PS (1994). Identification of herpesvirus-like DNA sequences in AIDS-associated Kaposi's sarcoma. Science *266*, 1865–1869

Choi J-K, Lee B-S, Shim SN, Li M, and Jung JU (2000). Identification of the novel K15 gene at the rightmost end of the Kaposi's sarcoma-associated herpesvirus genome. J Virol *74*, 436–446

Chokunonga E, Levy LM, Bassett MT, Borok MZ, Mauchaza BG, Chirenje MZ, and Parkin DM (1999). AIDS and cancer in Africa: the evolving epidemic in Zimbabwe. AIDS *13*, 2583–2588

Cook PM, Whitby D, Calabro ML, Luppi M, Kakoola DN, Hjalgrim H, Ariyoshi K, Ensoli B, Davison AJ, and Schulz TF (1999). Variability and evolution of Kaposi's sarcoma-associated herpesvirus in Europe and Africa. International collaborative group. AIDS, 1165–1176

Damania B, and Jung JU (2001). Comparative analysis of the transforming mechanisms of Epstein-Barr virus, Kaposi's sarcoma-associated herpesvirus, and *Herpesvirus saimiri*. Adv Cancer Res *80*, 51–82

Dambaugh T, Hennessey K, Chamnankit L, and Kieff E (1984). U2 region of Epstein-Barr virus DNA may encode Epstein-Barr virus nuclear antigen 2. Proc Natl Acad Sci USA *81*, 7632–7636

Davison AJ, Dolan A, Akter P, Addison C, Dargan DJ, Alcendor DJ, McGeoch DJ, and Hayward GS (2003). The human cytomegalovirus genome revisited: comparison with the chimpanzee cytomegalovirus genome. J Gen Virol *84*, 17–28

Dedicoat M, and Newton R (2003). Review of the distribution of Kaposi's sarcoma-associated herpesvirus (KSHV) in Africa in relation to the incidence of Kaposi's sarcoma. Br J Cancer *88*, 1–3

Fouchard N, Lacoste V, Couppie P, Develoux M, Mauclere P, Michel P, Herve V, Pradinaud R, Bestetti G, Huerre M, et al. (2000). Detection and genetic polymorphism of human herpesvirus Type 8 in endemic or epidemic Kaposi's Sarcoma from West and Central Africa, and South America. Int J Cancer *85*, 166–170

Glenn M, Rainbow L, Aurade F, Davison A, and Schulz TF (1999). Identification of a spliced gene from Kaposi's sarcoma-associated herpesvirus encoding a protein with similarities to latent membrane proteins 1 and 2A of Epstein-Barr virus. J Virol *73*, 6953–6963

Hayward GS (1999). KSHV Strains: The origin and global spread of the virus. (Review) Semin Cancer Biol, 187–199

Kadyrova E, Lacoste V, Duprez R, Pozharissky K, Molochkov V, Huerre M, Gurtsevitch V, and Gessain A (2003). Molecular epidemiology of Kaposi's sarcoma-associated herpesvirus/human herpesvirus 8 strains from Russian patients with classic, posttransplant, and AIDS-associated Kaposi's sarcoma. J Med Virol *71*, 548–556

Kajumbula H, Wallace RG, Zong JC, Hokello J, Sussman N, Simms S, Rockwell RF, Pozos R, Hayward GS, and Boto W (2006). Ugandan Kaposi's sarcoma-associated herpesvirus phylogeny: Evidence for cross-ethnic transmission of viral subtypes. Intervirology *49*, 133–143

Kakoola DN, Sheldon J, Byabazaire N, Bowden RJ, Katongole-Mbidde E, Schulz TF, and Davison AJ (2001). Recombination in human herpesvirus-8 strains from Uganda and evolution of the K15 gene. J Gen Virol *82*, 2393–2404

Kasolo FC, Mpabalwani E, and Gompels UA (1997). Infection with AIDS-related herpesviruses in human immunodeficiency virus-negative infants and endemic childhood Kaposi's sarcoma in Africa. J Gen Virol *78*, 847–856

Lacoste V, Judde JG, Briere J, Tulliez M, Garin B, Kassa-Kelembho E, Morvan J, Couppie P, Clyti E, Forteza Vila J, *et al.* (2000a). Molecular epidemiology of human herpesvirus 8 in Africa: both B and A5 K1 genotypes, as well as the M and P genotypes of K14.1/K15 loci, are frequent and widespread. Virology *278*, 60–74

Lacoste V, Kadyrova E, Chistiakova I, Gurtsevitch V, Judde JG, and Gessain A (2000b). Molecular characterization of Kaposi's sarcoma-associated herpesvirus/human herpesvirus-8 strains from Russia. J Gen Virol *81* Pt 5, 1217–1222

Lacoste V, Mauclere P, Dubreuil G, Lewis J, Georges-Courbot MC, and Gessain A (2000c). KSHV-like herpesviruses in chimps and gorillas. Nature *407*, 151–152

Lacoste V, Mauclere P, Dubreuil G, Lewis J, Georges-Courbot MC, and Gessain A (2001). A novel gamma 2-herpesvirus of the Rhadinovirus 2 lineage in chimpanzees. Genome Res *11*, 1511–1519

Lagunoff M, and Ganem D (1997). The structure and coding organization of the genomic termini of Kaposi's Sarcoma-associated herpesvirus (Human Herpesvirus 8). Virology *236*, 147–154

Lee BS, Connole M, Tang Z, Harris NL, and Jung JU (2003). Structural analysis of the Kaposi's sarcoma-associated herpesvirus K1 protein. J Virol *77*, 8072–8086

Lee BS, Lee SH, Feng P, Chang H, Cho NH, and Jung JU (2005). Characterization of the Kaposi's sarcoma-associated herpesvirus K1 signalosome. J Virol *79*, 12173–12184

Ling PD, Ryon JJ, and Hayward SD (1993). EBNA-2 of herpesvirus papio diverges significantly from the type A and type B EBNA-2 proteins of Epstein-Barr virus but retains an efficient transactivation domain with a conserved hydrophobic motif. J Virol *67*, 2990–3003

Meng YX, Sata T, Stamey FR, Voevodin A, Katano H, Koizumi H, Deleon M, De Cristofano MA, Galimberti R, and Pellett PE (2001). Molecular characterization of strains of human herpesvirus 8 from Japan, Argentina and Kuwait. J Gen Virol *82*, 499–506

Meng YX, Spira TJ, Bhat GJ, Birch CJ, Druce JD, Edlin BR, Edwards R, Gunthel C, Newton R, Stamey FR, *et al.* (1999). Individuals from North America, Australasia, and Africa are infected with four different genotypes of human herpesvirus 8. Virology *261*, 106–119

Nicholas J, Zong J-C, Alcendor DJ, Ciufo DM, Poole LJ, Sarisky RT, Chiou CJ, Zhang X, Wan X, Guo H-G, *et al.* (1998). Novel organizational features, captured cellular genes and strain variability within the genome of KSHV/HHV8. J Natl Cancer Inst Monogr *23*, 79–88

Poole LJ, Zong J-C, Ciufo DM, Alcendor DJ, Cannon JS, Ambinder R, Orenstein J, Reitz MS, and Hayward GS (1999). Comparison of genetic variability at multiple loci across the genomes of the major subgroups of Kaposi's sarcoma associated herpesvirus (HHV8) reveals evidence for recombination and for two distinct types of ORF-K15 alleles at the right hand end. J Virol *73*, 6646–6660

Rose TM, Ryan, J. T., Schultz, E.R., Raden, B.W., Tsai, C-C (2003). Analysis of 4.3 kilobases of divergent locus B of macaque retroperitoneal fibromatosis-associated herpesvirus reveals a close similarity in gene sequence and genome organization to Kaposi's sarcoma-associated herpesvirus. JVirol *77*, 5084–5097

Rose TM, Strand KB, Schultz ER, Schaefer G, G.W. Rankin J, Thouless ME, Tsai C-C, and Bosch ML (1997). Identification of two homologs of the Kaposi's sarcoma-associated herpesvirus (human herpesvirus 8) in retroperitoneal fibromatosis of different macaque species. J Virol *71*, 4138–4144

Russo JJ, Bohenzky RA, Chien M-C, Chen J, Yan M, Maddalena D, Parry JP, Peruzzi D, Edelman IS, Chang Y, and Moore PS (1996). Nucleotide sequence of the Kaposi sarcoma-associated herpesvirus (HHV8). Proc Natl Acad Sci USA *93*, 14862–14867

Searles RP, Bergquam EP, Axthelm MK, and Wong SW (1999). Sequence and genomic analysis of a rhesus macaque rhadinovirus with similarity to Kaposi's sarcoma-associated herpesvirus/human herpesvirus 8. J Virol *73*, 3040–3053

Sharp TV, Wang HW, Koumi A, Hollyman D, Endo Y, Ye H, Du MQ, and Boshoff C (2002). K15 protein of Kaposi's sarcoma-associated herpesvirus is latently expressed and binds to HAX-1, a protein with antiapoptotic function. J Virol *76*, 802–816

Shim YS, Kim CW, and Lee WK (1998). Sequence variation of EBNA2 of Epstein-Barr virus isolates from Korea. Mol Cells *8*, 226–232

Stebbing J, Bourboulia D, Johnson M, Henderson S, Williams I, Wilder N, Tyrer M, Youle M, Imami N, Kobu T, *et al.* (2003). Kaposi's sarcoma-associated herpesvirus cytotoxic T lymphocytes recognize and target Darwinian positively selected autologous K1 epitopes. J Virol *77*, 4306–4314

Treurnicht FK, Engelbrecht S, Taylor MB, Schneider JW, and van Rensburg EJ (2002). HHV-8 subtypes in South Africa: identification of a case suggesting a novel B variant. J Med Virol *66*, 235–240

Wang L, Wakisaka N, Tomlinson CC, DeWire SM, Krall S, Pagano JS, and Damania B (2004). The Kaposi's sarcoma-associated herpesvirus (KSHV/HHV-8) K1 protein induces expression of angiogenic and invasion factors. Cancer Res *64*, 2774–2781

Whitby D, Marshall VA, Bagni RK, Wang CD, Gamache CJ, Guzman JR, Kron M, Ebbesen P, and Biggar RJ (2004). Genotypic characterization of Kaposi's sarcoma-associated herpesvirus in asymptomatic infected subjects from isolated populations. J Gen Virol *85*, 155–163

Zhang YJ, Davis TL, Wang XP, Deng JH, Baillargeon J, Yeh IT, Jenson HB, and Gao SJ (2001). Distinct distribution of rare US genotypes of Kaposi's sarcoma- associated herpesvirus (KSHV) in South Texas: implications for KSHV epidemiology. J Infect Dis *183*, 125–129

Zong J-C, Ciufo DM, Alcendor DJ, Wan X, Nicholas J, Browning P, Rady P, Tyring SK, Orenstein J, Rabkin C, *et al.* (1999). High level variability in the ORF-K1 membrane protein gene at the left end of the Kaposi's sarcoma associated herpesvirus (HHV8) genome defines four major virus subtypes and multiple clades in different human populations. J Virol *73*, 4156–4170

Zong JC, Ciufo DM, Viscidi R, Alagiozoglou L, Tyring S, Rady P, Orenstein J, Boto W, Kajumbula H, Romano N, *et al.* (2002). Genotypic analysis at multiple loci across Kaposi's sarcoma herpesvirus (KSHV) DNA molecules: clustering patterns, novel variants and chimerism. J Clin Virol *23*, 119–148

Zong J-C, Metroka C, Reitz MS, Nicholas J, and Hayward GS (1997). Strain variability among Kaposi sarcoma associated herpesvirus (HHV8) genomes: Evidence that a large cohort of U.S.A. AIDS patients may have been infected by a single common isolate. J Virol *71*, 2505–2511

CTMI (2006) 312:43–69

Rhesus Monkey Rhadinovirus: A Model for the Study of KSHV

C. M. O'Connor · D. H. Kedes (✉)

Departments of Microbiology and Internal Medicine, Division of Infectious Diseases and Myles H. Thaler Center for AIDS and Human Retrovirus Research, University of Virginia, Charlottesville, VA 22901, USA
kedes@virginia.edu

Abstract Rhesus monkey rhadinovirus (RRV) is one of the closest phylogenetic relatives to the human pathogen Kaposi sarcoma-associated herpesvirus (KSHV)—a gamma-2 herpesvirus and the etiologic agent of three malignancies associated with immunosuppression. In contrast to KSHV, RRV displays robust lytic-phase growth in culture, replicating to high titer, and therefore holds promise as an effective model for studying primate gammaherpesvirus lytic gene transcription as well as virion structure, assembly, and proteomics. More recently, investigators have devised complementary latent systems of RRV infection, thereby also enabling the characterization of the more restricted latent transcriptional program. Another benefit of working with RRV as a primate gammaherpesvirus model is that its efficient lytic growth makes genetic manipulation easier than that in its human counterpart. Exploiting this quality, laboratories have already begun to generate mutant RRV, setting the stage for future work investigating the function of individual viral genes. Finally, rhesus macaques support experimental infection with RRV, providing a natural in vivo model of infection, while similar nonhuman systems have remained resistant to prolonged KSHV infection. Recently, dual infection with RRV and a strain of simian immunodeficiency virus (SIV) has led to a lymphoproliferative disorder (LPD) reminiscent of multicentric Castleman disease (MCD)—a clinical manifestation of KSHV infection in a subset of immunosuppressed patients. RRV, in short, shows a high degree of homology with KSHV yet is more amenable to experimental manipulation both in vitro and in vivo. Taken together, these qualities ensure its current position as one of the most relevant viral models of KSHV biology and infection.

1 Introduction

After the 1994 Moore and Chang discovery of Kaposi sarcoma-associated herpesvirus (KSHV) (Chang et al. 1994) and the subsequent series of investigations linking this virus to the human diseases Kaposi sarcoma (KS) (reviewed in Boshoff and Chang 2001), primary effusion lymphoma (PEL), and multicentric Castleman disease (MCD) (reviewed in Boshoff and Weiss 1998; Moore and Chang 2001), two separate groups isolated a homologous herpesvirus from rhesus macaques (*Macaca mulatta*). The first group was led by R. C. Desrosiers at the New England Primate Research Center (NEPRC) (Desrosiers et al. 1997) and the second by S. Wong at the Oregon Regional Primate Research Center (ORPRC) (Wong et al. 1999). Desrosiers' laboratory designated this new virus rhesus monkey rhadinovirus (RRV). Sequence analysis of the two separate isolates (Alexander et al. 2000; Searles et al. 1999) confirmed their classification as gamma-2-herpesviruses and demonstrated that the two were highly homologous to each other as well as to the human pathogen (Alexander et al. 2000; Searles et al. 1999).

The Desrosiers isolate, RRV 26-95H, originated from a colony of healthy monkeys at the NEPRC. The investigators originally noted that coculturing peripheral blood mononuclear cells (PBMC) from rhesus monkeys with primary rhesus fibroblasts (RhF) led to a cytopathic effect (CPE) in the latter cells. The investigators subsequently transfected primary RhF with RNase-treated virion DNA and propagated the virus on these same cells. Finally, they demonstrated that infection of the primary RhF with column-purified virus also led to similar CPE (Desrosiers et al. 1997).

In contrast to the discovery of the NEPRC isolate from healthy animals, the Wong group isolated RRV (isolate 17577 at the ORPRC) from SIV-infected macaques that had developed a lymphoproliferative disorder (LPD), characterized by splenomegaly, hepatomegaly, angiofollicular lymphadenopathy, and hypergammaglobulinemia, all of which are also clinical manifestations of MCD in humans (Bergquam et al. 1999). This group not only demonstrated that healthy animals in the same colony harbored a natural infection within circulating B lymphocytes (Bergquam et al. 1999) but also noted that coinfection of healthy animals with SIVmac239 and RRV-17577 led to the development of LPD (Wong et al. 1999). Mansfield et al. experimentally infected macaques with RRV both alone and in the context of a different SIV strain, SIVmac251. This coinfection resulted in no specific disease, although the investigators did find that the RRV 26-95H displayed a preferential tropism for $CD20^+$ B lymphocytes (Mansfield et al. 1999). The reason for the different outcomes in these two sets of animal experiments remains unclear.

In the same year as the initial discovery of RRV, Rose et al. identified two additional herpesviruses in fixed tissue samples from macaques with simian retroperitoneal fibromatosis (RF) (Rose et al. 1997), a rare vascular tumor similar to KS (Giddens 1979). These investigators employed degenerative PCR primers to the highly conserved herpesvirus DNA polymerase genes to screen the RF tissue for a KSHV-related homologue. The RF tissue from both *Macaca nemestrina* and *Macaca mulatta* supported amplification of this region, and the investigators designated the predicted source viruses as RF-associated herpesvirus $(RFHV)_{Mn}$ and $RFHV_{Mm}$, respectively (Rose et al. 1997). Subsequent sequence analysis of both isolates revealed an even closer homology to KSHV than RRV (Rose et al. 2003, 1997; Schultz et al. 2000; Strand et al. 2000). Rose and colleagues cloned over 4 kb of the $RFHV_{Mn}$ and $RFHV_{Mm}$ divergent locus B (DL-B) (Rose et al. 2003), a region of the genome that is 3′ to the open reading frames (ORFs) encoding gB (ORF8) and the DNA polymerase (ORF9), both of which were sequenced previously by Schultz et al. (Schultz et al. 2000). Within the DL-B region these investigators identified viral homologues of cellular interleukin-6 (IL-6), dihydrofolate reductase (DHFR), and thymidylate synthase (TS), as well as the homologues of KSHV ORF10 and K3 (vMIR1) (Rose et al. 2003). Unfortunately, limited availability of well-preserved RF tissue and spontaneous disappearance of this tumor from the macaque colonies have prevented the isolation or growth of $RFHV_{Mn}$ and $RFHV_{Mm}$ in tissue culture (Rose et al. 2003). The remainder of this review, therefore, will focus on RRV.

2 RRV Genome

2.1 Genomic Organization

Analyses of the genomes of the two RRV isolates have demonstrated a high degree of similarity to each other, as well as to KSHV (Alexander et al. 2000). Nearly all of the genes in RRV have a homologue in KSHV (Fig. 1), although the monkey and the human viruses each have a small subset of

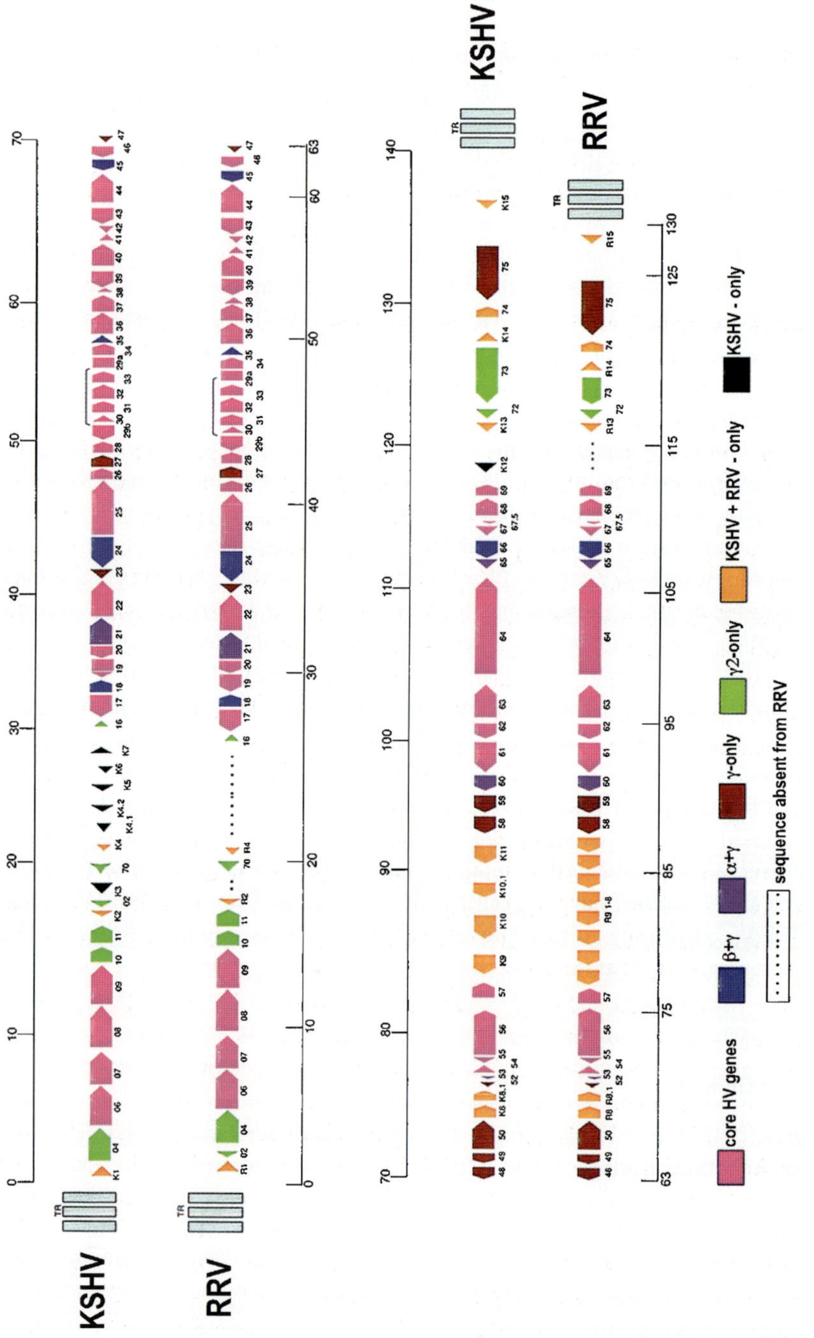
KSHV
RRV
TR
core HV genes
β+γ
α+γ
γ-only
γ2-only
KSHV + RRV - only
KSHV - only
sequence absent from RRV

Fig. 1 Alignment of RRV and KSHV genomes. The ORFs are colored according to the code at the *bottom* of the figure. The *blunted and arrowhead ends* of each ORF signify 5′ and 3′ directionality of genes, respectively. *Numbers* designate each ORF and those preceded with *K* and *R* indicate those unique to RRV and KSHV, respectively. *Lines* connecting two ORFs designate spliced genes, and *TR* indicates terminal repeats. The *demarcated lines* specify the approximate length in kbp. (Adapted with permission from Alexander et al. 2000)

species-specific genes (see below). Sequence analysis of RRV17577 long unique region (LUR) and the characterization of the terminal repeats (TR) revealed that the genome was approximately 130,000 bp and had a G+C content of 52.2%. Sequence scanning predicted 79 ORFs, and 67 of these were similar to those of KSHV and herpesvirus saimiri (HVS). The same group also found eight genes encoded only by RRV and KSHV and four that are unique to RRV (Searles et al. 1999). Shortly thereafter, investigators at the NEPRC also sequenced the LUR of RRV isolate 26-95 (Alexander et al. 2000) and compared it to the published sequence for RRV17577 (Searles et al. 1999). The RRV26-95 LUR consists of 130,733 bp, and sequence analysis predicted an additional five ORFs for a total of 84. The overall sequence and genome architecture was nearly identical to the RRV17577 isolate and, again, highly similar to the KSHV LUR. Specifically, all of the ORFs of RRV26-95 have a homologue in KSHV, but the slightly shorter monkey virus genome lacks homologues of K3, K5 (vMIR2), K7 (viral inihibitor of apoptosis vIAP), and K12 (kaposin)—see below. In addition, RRV differs from KSHV in the number of ORFs encoding the macrophage inflammatory protein (MIP)-1 (RRV has 3 and KSHV 4) and viral interferon regulatory factors (vIRFs) (RRV has 1 and KSHV 8). Finally, the ORF encoding a viral homologue of cellular dihydrofolate reductase (DHFR) is located near the left end (5′) of the genome in RRV26-95, as it is in HVS, whereas this ORF is displaced over 16 kbp rightward in KSHV and RFHV (Rose et al. 2003). The RRV26-95 DHFR ORF also retains a greater similarity to the HVS (74% similarity) than to the KSHV (55% similarity) DHFR homologue (Alexander et al. 2000).

The ORFs identified for RRV26-95 all have sequences similar to the ORFs of RRV17577, with a few minor exceptions. Sequences mapping to ORF67.5 of RRV26-95 are present in RRV17577 but were not originally designated as an ORF. Further, ORF R15 in RRV17577 is the sequence and positional homologue of R14 of RRV26-95 and K14 of KSHV (Alexander et al. 2000; Searles et al. 1999). Finally, meaningful direct comparisons between R15 and K15 await further characterization of the RRV protein products. Table 1 gives a detailed comparison among the two RRV isolates and KSHV. Together, these

Table 1 Comparison of the predicted proteins encoded by RRV isolates 26-95 and 17577 and KSHV (adapted with permission from Alexander, Denekamp et al. 2000)

	RRV26-95	RRV17577			KSHV			
ORF	Size (aa)	Size (aa)	% Identity	% Similarity	Size (aa)	% Identity	% Similarity	Putative function
R1	423	423	98.1	98.6	289	28	33	
2	188	188	99.9	99.5	210	45.5	54.4	DHFR
4	395	645	64.5	72.2	550	35.7	42.3	*CP
6	1132	1132	100	100	1133	63.3	71.3	ssDBP
7	686	686	99.6	99.6	695	52.2	60.8	TP
8	829	829	96.7	97.1	845	66.4	73.9	gB
9	1014	1014	100	100	1012	67.4	75.5	POL
10	416	416	100	100	418	35.3	44	dUTPase-related protein
11	409	409	100	100	407	33	42.8	dUTPase-related protein
R2	207	207	100	100	204	22.6	30.6	vIL-6
70	333	333	98.8	99.1	337	65.8	71.8	TS
R4	118	115	94.8	95.6	94	32.6	42.4	vMIP
16	187	187	100	100	175	46	58	vBcl-2
17	536	536	99.4	99.6	553	45.8	52.7	SCAF/PRO
18	257	299	100	100	257	58	68.1	Unknown
19	547	547	99.3	99.3	549	52.9	61.4	Packaging protein
20	350	350	99.4	99.4	320	46.9	53.9	Unknown
21	557	557	98.9	99.1	580	46.4	54.9	TK
22	726	704	74.2	79.4	730	41.2	50.8	gH
23	402	402	99.8	99.8	404	49	57.5	Egress protein
24	732	732	99.6	99.6	752	59.1	66.7	Unknown
25	1378	1378	99.6	99.6	1376	72.3	79.8	MCP
26	305	307	98.4	98.4	305	64.3	71.8	TRI-2
27	269	269	100	100	290	27.8	37.2	Unknown
28	91	91	100	100	102	26.5	30.1	gp150
29b	348	348	99.4	99.4	351	66.4	77.8	Packaging protein

Table 1 (continued)

	RRV26-95	RRV17577			KSHV			
ORF	Size (aa)	Size (aa)	% Identity	% Similarity	Size (aa)	% Identity	% Similarity	Putative function
30	76	76	98.7	98.7	77	38.2	51.3	Unknown
31	217	217	100	100	224	46.1	57.3	Unknown
32	464	464	100	100	454	42.9	51.2	Transport protein
33	336	336	100	100	312	42.7	52.8	MyrPBP
29a	327	327	100	100	312	61.2	66.7	Packaging protein
34	327	327	99.7	99.7	327	49.4	59.8	Unknown
35	149	149	100	100	151	37.2	49.3	Unknown
36	435	435	100	100	444	48.7	59.1	vPK
37	472	480	100	100	486	63.9	72.6	SOX
38	69	69	100	100	61	45	56.7	MyrP
39	378	378	100	100	399	59.5	73.3	gM
40	468	468	99.8	99.8	457	33	42.5	Helicase-primase
41	203	203	99.5	99.5	205	28.7	37.9	Helicase-primase
42	272	272	99.6	99.6	278	48	58.7	
43	576	576	100	100	605	62	70.1	PORT
44	790	790	100	100	788	66	74	Helicase-primase
45	353	352	99.4	99.4	407	37	43.6	vIRF-7BP
46	230	255	83.5	85.2	255	56.8	68.6	UDG
47	163	169	56.4	62	167	31.2	38.8	gL
48	389	389	100	100	402	29.6	39.1	Unknown
49	301	301	99.7	99.7	302	53.8	65.8	Unknown
50	514	514	99.7	100	631	44.9	54.7	RTA
R8								
R8.1								*8.1
52	139	139	100	100	131	47.7	60.8	Unknown
53	104	104	100	100	110	49	53.8	gN
54	290	290	100	100	318	43.1	51	dUTPase
55	210	210	100	100	227	55.2	62.8	PalmP

Table 1 (continued)

	RRV26-95	RRV17577			KSHV			
ORF	Size (aa)	Size (aa)	% Identity	% Similarity	Size (aa)	% Identity	% Similarity	Putative function
56	828	828	100	100	843	52.7	61.8	DNA replication protein
57	442	442	99.8	100	275	47.1	61.4	MTA
R9.1	415	415	98.3	98.8	449	28.7	60.6	vIRF
R9.2	415	415	99.5	99.5	449	23.2	37.2	vIRF
R9.3	351	351	100	100	449	24.3	31.9	vIRF
R9.4	361	253	99.6	99.6	449	25.5	34.2	vIRF
R9.5	385	385	100	100	449	29.5	38.4	vIRF
R9.6	390	290	100	100	449	22.9	33.4	vIRF
R9.7	355	355	99.4	99.4	449	25	34.6	vIRF
R9.8	364	364	100	100	449	26.8	33.6	vIRF
58	360	360	99.7	100	357	38.9	46.2	EpiL
59	394	394	100	100	396	51.8	60.2	PF
60	304	314	100	100	305	70.1	78.3	RNR-S
61	788	788	99	99.2	792	61.6	69.6	RNR-L
62	331	331	99.4	99.4	331	57.6	65.4	TRI-1
63	939	939	99.8	99.8	927	43.5	52.9	LTPBP
64	2548	2548	99.6	99.6	2635	41.3	51.2	LTP
65	169	169	100	100	170	45.5	55.8	SCIP
66	448	448	99.8	99.8	429	47.3	52.7	Egress protein
67	222	224	87.8	87.8	271	67.1	71.2	Unknown
67.5	86	86	98.8	98.8	80	62.5	71.2	Unknown
68	457	457	100	100	545	49.1	58.3	Unknown
69	297	297	100	100	225	65.6	73.1	Unknown
R13	174	174	99.4	99.4	188	33.1	40.1	FLIP
72	254	254	99.6	99.6	257	39.4	50.2	vCyclin D
73	448	447	99.8	97.8	1162	18.3	34.9	LANA
R14	253	253	100	100	348	32.8	37.6	vOx2
74	342	342	100	100	342	42	54.7	vGPCR
75	1298	1298	99.8	99.9	1296	44.9	53.2	vFGARAT
R15					490			LAMP

* Protein named accordingly: R, for RRV; K, for KSHV

data demonstrate that the independent RRV isolates are the same virus species and that both are highly related to KSHV in both structural organization of the genome and overall sequence (Alexander et al. 2000; Searles et al. 1999).

2.2 Genomic Structure—Conservation with KSHV

The 130.7-kbp LUR of RRV with its 84 predicted ORFs, is approximately colinear with KSHV (Fig. 1). As we mentioned above, four genes are unique to KSHV and absent in the RRV genome. These include K3 and K5 [bovine herpesvirus immediate early protein 4 (IEP4) homologues involved in immune modulation], K7 (vIAP), and K12 (kaposin, a transforming gene). The absence of a homologue to K12 is particularly noteworthy, as this is the most abundantly transcribed region of the genome in KSHV latent infection and, thus, may be a potentially important clue to understanding the differences in the biology and/or pathogenesis of the two viruses. Interestingly, adjacent to the K12 region of the KSHV genome are two G+C-rich direct repeating units, DR-1 and DR-2. Although RRV lacks the K12 homologue, it has retained the G+C-rich repeating elements in positions that are similar to those of DR-1 and DR-2 within the KSHV genome, suggesting that the monkey virus may have lost this gene during evolution from a common herpesvirus ancestor (Alexander et al. 2000). The eight genes specific to RRV and KSHV are the homologues to interleukin-6 (vIL-6), macrophage inflammatory proteins (vMIPs), interferon regulatory factors (vIRFs), Fas-associating protein with death domain-like interleukin-1b converting enzyme (FLICE) inhibitory protein (vFLIP), vOx-2, and vIL-8 receptor (Alexander et al. 2000) In addition, although RRV's vIRFs show little similarity to those of KSHV, the RRV vIRFs have a high degree of similarity to each other, suggesting an amplification event in which the four ancestral vIRFs duplicated, yielding the eight now present in the genome (Alexander et al. 2000; Searles et al. 1999). The vIRFs encoded by R6, R7, R8, and R9 are most similar to those encoded by R10, R11, R12, and R13, respectively. R6, R7, R8, R10, and R11 are similar to K9, but only R10 has higher than 30% similarity to K9. Thus, Searles et al. suggest that a possible duplication event occurred, such that the four vIRFs of KSHV were duplicated en masse, creating the eight now present in the RRV genome (Searles et al. 1999).

RRV and KSHV also share a high degree of homology in nucleotide sequence as well as the G+C distribution at the origin of lytic replication (OriLyt) (Pari et al., 2001), although the location and number of OriLyts in RRV differ from those in KSHV (AuCoin et al. 2002). The RRV OriLyt maps between ORF69 and ORF71 and has an A+T region that is followed by a short, downstream G+C-rich sequence (Pari et al. 2001). KSHV has a left- and a right-end

OriLyt, with the latter showing the most similarity to the OriLyts of RRV (AuCoin et al. 2002).

3
RRV Transcriptional Program and Characterization of Individual Proteins

Recent reports on the transcription program of RRV also strengthen the argument for using RRV as a model system for the study of KSHV. After the cloning of full-length cDNAs for replication and transcription activator (RTA), investigators showed that RTA is a strong transcriptional transactivator of early RRV promoters, including those of R8, ORF57, and gB. Because KSHV RTA activates the promoters of the KSHV homologues of these same three genes, this work helped demonstrate the conservation of RTA function between RRV and KSHV (DeWire et al. 2002). In further support of using RRV to study KSHV transcription, Dittmer and colleagues recently described the overall transcription program of RRV after de novo infection (Dittmer et al., 2005). Using real-time reverse transcriptase (RT)-based polymerase chain reaction (PCR) techniques to profile the whole genome of RRV after infection of RhFs, they found that the lytic transcription program of RRV closely resembles that of KSHV after the induction of lytic reactivation (Dittmer et al. 2005). However, in contrast to KSHV culture systems that are marked by inefficient reactivation from a latent infection, de novo infection of RhFs with RRV results in highly efficient spontaneous lytic replication. Thus, RRV affords a powerful advantage for studying primate gammaherpesvirus lytic replication.

Although they share limited sequence homology, two transmembrane signaling molecules, R1 and K1, are unique to RRV and KSHV, respectively, and both contribute to cell growth and transformation (reviewed in Damania et al. 1999a), although the mechanism(s) by which this occurs remains poorly understood. Both K1 and R1 oligomerize through disulfide bonding (Damania et al. 1999b; Lee et al. 1998), transmit extracellular signals into the cytoplasm through ITAM motifs, and also share corresponding positions, located at the leftmost end of their respective genomes. Recently, L. Wang and colleagues demonstrated that K1 expression results in vascular endothelial growth factor (VEGF) promoter activation, leading to the expression and secretion of this growth factor, and also induces matrix metalloproteinase-9 (MMP-9) expression in endothelial cells. These results suggest that K1 may play a role in KSHV pathogenesis by inducing the expression of these proteins involved in both angiogenesis and invasion (Wang et al. 2004). Finally, using a recently described latent system of infection (see below), investigators have found that the RRV latency-associated nuclear antigen (LANA), significantly shorter than its KSHV counterpart, is also expressed during the lytic phase of infection and that RRV LANA inhibits lytic replication through the repression of RTA transcription (DeWire and Damania 2005). KSHV LANA, more intensely studied than its RRV counterpart, has multiple functions, including a role in tethering the viral genome to host cell chromosomes during latency (Cotter and Robertson 1999; Cotter et al. 2001; Garber et al. 2001; Grundhoff and Ganem 2003; Hu et al. 2002). Additionally, KSHV LANA regulates viral and cellular promoters, including its own (Jeong et al. 2001), acting as either an activator or a repressor of transcription (Schwam et al. 2000). The full extent of the parallels between the LANA homologues of KSHV and RRV awaits further investigation.

4
In Vivo and In Vitro Infection with RRV

RRV serves as a useful model system for the study of KSHV both in vivo and in vitro. In contrast, KSHV appears unable to persistently infect either mice or rhesus macaques—a major limitation in the study of the human virus (Dittmer et al. 1999; Renne et al. 2004). As described above, the natural host of RRV, rhesus macaques, can serve as a primate model system to study RRV biology and pathogenesis following de novo infection. Even for in vitro work RRV offers some advantages over, and can complement direct work with KSHV. Specifically, one of the major difficulties facing KSHV investigators interested in studying the viral transcription program is that only a small percentage (25%–30% or less) (Chan et al. 1998) of latently infected PEL cells (a B cell line) in culture respond to chemical induction and support lytic reactivation. Furthermore, of these only a fraction (approximately 25%) complete the lytic cycle and actually produce virus (Renne et al. 1996). More recently, transduction of the lytic "switch" gene, ORF50 (encoding RTA), into these cells led to higher proportions of the culture initiating the lytic cycle (Bechtel et al. 2003). Nevertheless, with KSHV culture systems, the background of simultaneous latent gene expression and frequently incomplete lytic reactivation complicates the study of the lytic gene transcription program. These in vitro KSHV systems are likewise not ideal for latent studies, because the converse complication exists with low but significant levels of spontaneous lytic reactivation remaining within 2%–5% of cells in the absence of any exogenous stimulation (Miller et al. 1997; Zhong et al. 1996). Although investigators have had some success at KSHV de novo infection of endothelial cells, the inability to reactivate efficiently the resultant latent infection and obtain high titers or to passage the virus over extended periods after de novo infection remains a major impediment to the development of a tractable genetic system similar to those that exist for other herpesviruses such as herpes simplex virus-1 (HSV-1), for example (Dezube et al. 2002; Lagunoff et al. 2002; Renne et al. 1998; Sakurada et al. 2001).

Infection of RhFs with RRV, in contrast, yields high titers of virus (e.g., 10^6 PFU/ml) in culture media within 6 days after infection (O'Connor et al. 2003). Further, the kinetics of RRV gene expression after infection of RhFs resembles closely that of KSHV lytic reactivation (DeWire et al. 2002). Two separate RhF cell lines are available, and both are permissive to RRV infection. In both cases, the cells are immortalized with telomerase, facilitating serial passaging (DeWire et al. 2002; Kirchoff et al. 2002). Unlike the KSHV in vitro system, infection of the RhFs with RRV is in the context of de novo infection (Alexander et al. 2000). The RRV in vitro system also allows researchers to perform traditional plaque assays (DeWire, 2003), a technique, that has eluded current KSHV in vitro systems.

Recently, DeWire and Damania described a latent system of RRV infection (DeWire and Damania 2005). After infection, both BJAB and 293-HEK cells harbor latent RRV and the virus can be reactivated with the addition of TPA. Additionally, transient transfection of RRV ORF50, which encodes the lytic switch protein RTA, into the BJAB cell line results in approximately four-fold reactivation of the viral lytic life cycle (DeWire and Damania 2005). Both the lytic and latent in vitro systems of RRV infection provide useful tools for the study of primate gammaherpesviruses in culture.

Additionally, naive rhesus monkeys are susceptible to experimental infection with RRV (Mansfield et al. 1999; Wong et al. 1999), and the manifestations of infection approximate many of the clinical diseases of KSHV infection in humans. Infection of RRV-naive, SIV-positive macaques with RRV-17577, for example, led to the development of a hyperplasic B lymphocytic LPD, resembling MCD observed in patients dually infected with KSHV and HIV (Wong et al. 1999). In addition, experimental infection of immunocompetent, SIV-negative animals coinfected with RRV 26-95 resulted in clinical lymphadenopathy characterized initially by paracortical hyperplasia and vascular hypertrophy/hyperplasia that was subsequently replaced by follicular hyperplasia (Mansfield et al. 1999). Similarly, HIV-negative MCD patients who harbor KSHV also present with B cell proliferation and angioimmunoblastic lymphadenopathy. In this same study, three of the four animals developed arteriopathy that resembled the vascular endothelial lesions observed in KS patients, and therefore the authors suggest a possible role for RRV in development of this lesion (Mansfield et al. 1999). Because small-animal models capable of long-term propagation of KSHV are still only in initial stages of development (Parsons et al. 2006), the in vivo system of RRV is an attractive model system for the study of these two closely related viruses.

5
Structural Conservation Between RRV and KSHV Capsids

KSHV lytic replication leads to the production of three (denoted A, B, and C) biochemically and structurally distinct capsid species (Nealon et al. 2001), similar to those in other herpesviruses (reviewed in Homa and Brown, 1997). A capsids are empty icosahedral structures (Fig. 2A), B capsids contain an inner ringlike structure (Fig. 2B), presumably the viral scaffolding (SCAF), and C capsids contain the viral DNA (Fig. 2C). The C capsids most likely mature to virions, after gaining tegument and envelope proteins. Because the KSHV culture system does not permit efficient, productive lytic reactivation, the yield of these capsid species is also low. Among the KSHV capsids that lytic reactivation does produce, the C capsids are most poorly represented, whereas A- and B-type capsids predominate (Fig. 3) (Nealon et al. 2001). For these reasons, studying the capsid assembly and maturation of KSHV capsids into virions is particularly challenging, and, as we describe below, RRV culture systems offer an attractive alternative.

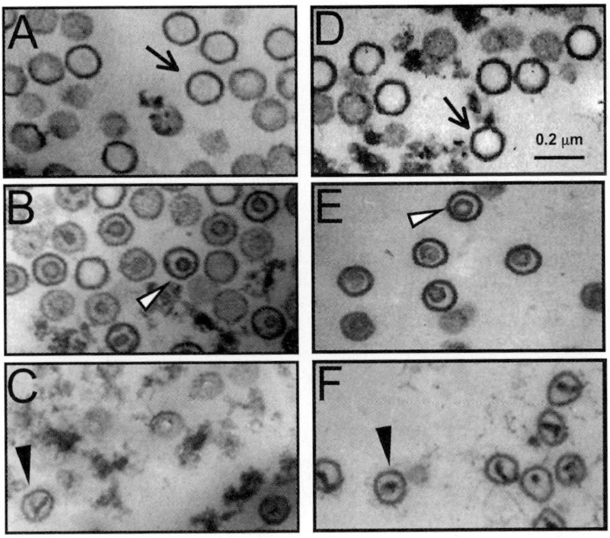

Fig. 2A–F TEMs of KSHV and RRV capsids. **A–C** KSHV capsids. **D–F** RRV capsids. **A, D** A capsids (*arrows*) are empty icosahedral structures. **B, E** B capsids (*white arrowhead*) have an inner ringlike structure, likely composed of SCAF. **C, F** C capsids (*black arrowheads*) contain a dense core consistent with encapsidated viral DNA. *Bar* indicates 0.2 μm. (Adapted with permission from Nealon et al. 2001 and O'Connor et al. 2003)

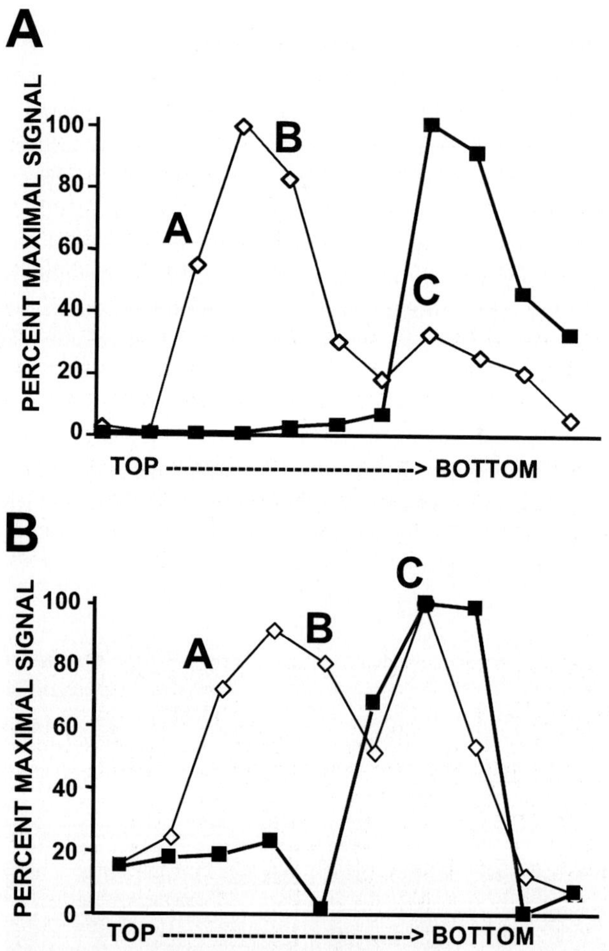

Fig. 3A, B Encapsidated viral DNA cosediments with C capsids—the predominant species in RRV. The MCP profiles (*open diamonds*) across fractions from sucrose gradients reveal distinct sedimentation velocities of the three different KSHV and RRV capsid species (identified by TEM, see Fig. 2; the letters *A*, *B*, and *C* designate the fractions containing the each of the isolated populations, respectively). The viral DNA profile (*filled squares*) from these same fractions confirms the location of C capsids. **A** KSHV A- and B-type capsids predominate (measured by intensity of Coomassie-stained MCP), whereas the C capsids are in low abundance. **B** In contrast, RRV C capsids are the most abundant capsid type, corresponding with the peak of viral DNA. (Adapted with permission from Nealon et al. 2001 and O'Connor et al. 2003)

De novo RRV infection of RhFs is, in contrast to KSHV reactivation, highly efficient in vitro (see above). Like KSHV, RRV lytic replication results in the synthesis of the same three capsid types, the A, B, and C capsids (Fig. 2D–F) (O'Connor et al. 2003). The RRV capsids are comprised of proteins similar to those of the KSHV capsids. All three capsid species in the two viruses contain the major capsid protein (MCP/ORF25), the triplex proteins TRI-1 (ORF62) and TRI-2 (ORF26), and the small capsomer interacting protein (SCIP/ORF65). Additionally, the B capsids contain the scaffolding protein (SCAF), and the C capsids, instead, contain the viral DNA (Fig. 2E and F). Structural analyses of these particles revealed that they were highly similar to those of KSHV, in both transmission electron (Fig. 2) (O'Connor et al., 2003) and cryoelectron (cryoEM) microscopic and three-dimensional (3D) reconstruction studies (Fig. 4) (Yu et al. 2003). Furthermore, the elevated levels of RRV capsids have led to high-resolution reconstructions for the three RRV capsid types, representing the first reconstruction of the A, B, and C capsids for gammaherpesviruses (Yu et al. 2003). In contrast to patterns of capsid production during KSHV replication, RRV replication results in high

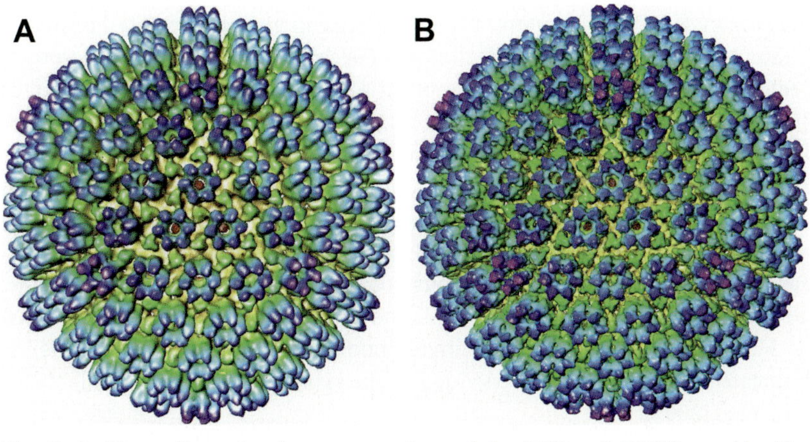

Fig. 4A, B Three-dimensional reconstructions of the RRV and KSHV capsids. The KSHV (**A**) and RRV (**B**) capsids are reconstructed to 15-Å and 24-Å resolution, respectively, and viewed along the icosahedral threefold axis. The maps are color coded according to the particle radius, such that the upper domains of the pentons and hexons are in *blue* (between radii of 570 and 640 Å), the connecting triplexes are in *green* (between radii of 510 and 560 Å), and the shell is in *yellow* (between radii of 460 and 510 Å). The higher resolution of the RRV compared to the KSHV capsid results from the integration of greater numbers of more highly purified RRV particles and allows better definition of the individual protein components. (Figure provided by Z. Hong Zhou and Xuekui Yu and modified, with permission, from Wu et al. 2000 and Yu et al. 2003)

proportions of the genome-filled C capsids (Fig. 3) (O'Connor 2005). The most abundant capsid species produced after KSHV lytic reactivation is the A capsid (Fig. 3A) (Nealon et al. 2001). In contrast, the genome-filled C-type capsid is the most prevalent capsid type produced after de novo infection of RhF with RRV (Fig. 3B) (O'Connor 2005). This difference in the production of the genome-containing C capsids provides the first potential explanation for the differences in replication efficiency (production of mature virions) between the two viruses.

6 Structural Conservation Between RRV and KSHV Virions

Accurately determining the protein composition of mature herpesvirus particles is essential to understanding not only their assembly but also early events following infection. Further, the ways in which KSHV and RRV rapidly initiate changes to the host cell are most likely distinct from other herpesviruses because they also differ in their modes of transmission, host cell tropism, balance of latent and lytic replication, and pathogenesis. Recently, two groups working on KSHV have reported their findings on virion-associated proteins with mass spectrometric (MS) techniques (Bechtel et al. 2005; Zhu et al. 2005). Bechtel et al. induced KSHV lytic replication of five liters of latently infected BCBL-1 cells, treated gradient-purified virus with trypsin, and separated the viral proteins by SDS-PAGE. By excising visibly stained gel slices, the investigators identified a total of 18 proteins, 10 of which are encoded by the virus (Table 2) (Bechtel et al. 2005). In a similar study by Zhu et al., KSHV lytic replication was reactivated in BCBL-1 cells, and the virus was subsequently purified by both step- and continuous gradient centrifugation. After separation of the viral proteins by SDS-PAGE, the investigators excised visible bands for MS analysis. In this study, the investigators reported a total of 26 proteins associated with the virus (24 encoded by the virus and 2 by the cell) (Table 2) (Zhu et al. 2005).

Taking advantage of the high yield of virus from RRV infection of RhF in culture, our laboratory has recently optimized the purification methods used to isolate both virions and capsids. In contrast to KSHV culture systems, much smaller (<1 or 2 liters) preparations of RRV-infected media are able to provide sufficient material to allow similar MS analyses (O'Connor and Kedes 2006). In our work, we treated virions with one of two proteases before gradient centrifugation, eliminating any contaminating cellular or viral proteins stuck nonspecifically to the outside of the particles and thus ensuring the purity of the particles. We used both velocity sedimentation and density

Table 2 Virion proteins of RRV and KSHV

Name	Abbrev.	RRV	KSHV	EBV	MHV-68	HSV-1	VZV	HCMV	HHV6/7	RRV	KSHV[1]	KSHV[2]	MHV-68[3]	EBV[4]	HCMV[5]
Capsid															
Major capsid protein	MCP	ORF25	ORF25	BcLF1	ORF25	UL19	40	UL86	U57	+	+	+	+	+	+
Triplex-1	TRI-1	ORF62	ORF62	BORF1	ORF62	UL38	20	UL46	U29	+	+	+	+	+	+
Triplex-2	TRI-2	ORF26	ORF26	BDLF1	ORF26	UL18	41	UL85	U56	+	+	+	+	+	+
Small capsomer interacting protein	SCIP	ORF65	ORF65	BFRF3	ORF65	UL35	23	UL48.5	U32	+	–	+	+	+	+
Portal protein	PORT	ORF43	ORF43	BBRF1	ORF43	UL6	54	UL104	U76	+	–	–	–	+	–
Packaging protein	–	ORF19	ORF19	BVRF1	ORF19	UL25	34	UL77	U50	+	–	–	–	+	–
Scaffolding protein	SCAF	ORF17.5	ORF17.5	BdRF2	ORF17.5	UL26.5	33.5	UL80.5	U53.5	+	–	+	–	+	–
Protease	PRO	ORF17	ORF17	BVRF2	ORF17	UL26	33	UL80	U53	+	–	–	–	–	+
Tegument															
DNA binding protein	ssDBP	ORF6	ORF6	BALF2	ORF6	UL29	29	UL57	U41	–	–	+	–	+	+
Transport protein	TP	ORF7	ORF7	BALF3	ORF7	UL28	30	UL56	U40	–	–	+	–	–	–
DNA polymerase	POL	ORF9	ORF9	BALF5	ORF9	UL30	28	UL54	U38	–	–	–	–	–	+

Table 2 (continued)

Name	Abbrev.	RRV	KSHV	EBV	MHV-68	HSV-1	VZV	HCMV	HHV6/7	RRV	KSHV[1]	KSHV[2]	MHV-68[3]	EBV[4]	HCMV[5]
dUTPase-elated protein	–	ORF11	ORF11	LF2	ORF11	γ	γ	γ	γ	–	–	+	–	γ	γ
Unknown	–	ORF20	ORF20	BXRF1	ORF20	UL24	35	UL76	U49	–	–	–	+	–	–
Thymidine kinase	TK	ORF21	ORF21	BXLF1	ORF21	UL23	36	n/a	n/a	+	+	+	–	+	n/a
Egress protein	–	ORF23	ORF23	BTRF1	ORF23	UL21	38	UL88	U59	+	–	–	–	low	+
Unknown	–	ORF24	ORF24	BcRF1	ORF24	n/a	n/a	UL87	U58	–	+	–	+	–	–
Unknown	–	ORF27	ORF27	BDLF2	ORF27	γ	γ	γ	γ	+	–	+	–	+	γ
Packaging protein	–	ORF29	ORF29	BGRF1/ BDRF1	ORF29	UL15	42/ 45	UL89	U66	–	–	–	+	–	+
Transport protein	–	ORF32	ORF32	BGLF1	ORF32	UL17	43	UL93	U64	+	–	–	–	+	+
Myristoylated protein binding protein	MyrPBP	ORF33	ORF33	BGLF2	ORF33	UL16	44	UL94	U65	+	+	+	–	+	+
Unknown	–	ORF35	ORF35	BGLF3.5	ORF35	UL14	46	UL95	U67	+	–	–	–	low	–
Virion protein kinase	vPK	ORF36	ORF36	BGLF4	ORF36	UL13	47	UL97	U69	+	–	–	–	+	+
Shutoff exonuclease	SOX	ORF37	ORF37	BGLF5	ORF37	UL12	48	UL98	U70	–	–	–	–	–	–

Table 2 (continued)

Name	Abbrev.	RRV	KSHV	EBV	MHV-68	HSV-1	VZV	HCMV	HHV6/7	RRV	KSHV[1]	KSHV[2]	MHV-68[3]	EBV[4]	HCMV[5]
Myristoylated	MyrP	ORF38	ORF38	BBLF1	ORF38	UL11	49	UL99	U71	+	–	–	–	+	+
protein	–	ORF42	ORF42	BBRF2	ORF42	UL7	53	UL103	U75	+	–	–	–	low	+
vIRF-7 binding protein	vIRF-7BP	ORF45	ORF45	BKRF4	ORF45	γ	γ	γ	γ	+	–	+	+	+	γ
Unknown	–	ORF48	ORF48	BRRF2	ORF48	γ	γ	γ	γ	–	–	–	+	+	γ
Unknown	–	ORF49	ORF49	BRRF1	ORF49	γ	γ	γ	γ	+	–	–	–	–	γ
Unknown	–	ORF52	ORF52	BLRF2	ORF52	γ	γ	γ	γ	+	–	+	+	+	γ
dUTPase	dUTPase	ORF54	ORF54	BLLF3	ORF54	UL50	8	UL72	U45	+	–	–	–	low	+
Palmitoylated protein	PalmP	ORF55	ORF55	BSRF1	ORF55	UL51	7	UL71	U44	+	–	–	–	+	+
mRNA transcript accumulation function	MTA	ORF57	ORF57	BSLF1/ MLF1	ORF57	UL54	4	UL69	U42	–	–	–	–	–	+
Processivity factor	PF	ORF59	ORF59	BMRF1	ORF59	UL42	16	UL44	U27	–	–	–	–	+	+
Small ribonucleotide reductase	RNR-S	ORF60	ORF60	BaRF1	ORF60	UL40	18	n/a	n/a	–	–	–	–	low	n/a
Large ribonucleotide reductase	RNR-L	ORF61	ORF61	BORF2	ORF61	UL39	19	UL45	U28	–	–	–	–	+	+

Table 2 (continued)

Name	Abbrev.	RRV	KSHV	EBV	MHV-68	HSV-1	VZV	HCMV	HHV6/7	RRV	KSHV[1]	KSHV[2]	MHV-68[3]	EBV[4]	HCMV[5]
Large tegument protein binding protein	LTPBP	ORF63	ORF63	BOLF1	ORF63	UL37	21	UL47	U30	+	+	+	–	+	+
Large tegument protein	LTP	ORF64	ORF64	BPLF1	ORF64	UL36	22	UL48	U31	+	–	+	–	–	+
Egress protein	–	ORF66	ORF66	BFRF2	ORF66	UL34	24	UL50	U34	–	–	–	–	low	+
Unknown	–	ORF67	ORF67	BFRF1A	ORF67	UL33	25	UL51	U35	–	–	–	–	–	+
Unknown	–	ORF68	ORF68	BFLF1	ORF68	UL32	26	UL52	U36	–	–	+	–	–	–
FGARAT homolog	vFGA-RAT	ORF75	ORF75	BNRF1	ORF75	γ	γ	γ	γ	+	+	+	+	+	γ
Unknown	–	γ1	γ1	BLLF1	γ1	γ1	γ1	γ1	γ1	γ1	γ1	γ1	γ1	+	γ1
Unknown	–	γ1	γ1	BZLF2	γ1	γ1	γ1	γ1	γ1	γ1	γ1	γ1	γ1	+	γ1
Unknown	–	γ1	γ1	BILF2	γ1	γ1	γ1	γ1	γ1	γ1	γ1	γ1	γ1	+	γ1
Envelope															
Complement binding protein	*CP	ORF4	ORF4	γ2	ORF4	γ2	γ2	γ2	γ2	+	–	–	–	γ2	γ2
Glycoprotein B	gB	ORF8	ORF8	BALF4	ORF8	UL27	31	UL55	U39	+	+	+	+	+	+

Table 2 (continued)

Name	Abbrev.	RRV	KSHV	EBV	MHV-68	HSV-1	VZV	HCMV	HHV6/7	RRV	KSHV[1]	KSHV[2]	MHV-68[3]	EBV[4]	HCMV[5]
Glycoprotein	*8.1	ORF R8.1	ORF K8.1	‡	‡	‡	‡	‡	‡	+	–	+	‡	‡	‡
Glyco-protein H	gH	ORF22	ORF22	BXLF2	ORF22	UL22	37	UL75	U48	+	+	+	+	+	+
Glyco-protein 150	gp150	ORF28	ORF28	BDLF3	ORF28	γ	γ	γ	γ	+	–	+	+	+	γ
Glyco-protein M	gM	ORF39	ORF39	BBRF3	ORF39	UL10	50	UL100	U72	+	–	+	–	+	+
Glyco-protein L	gL	ORF47	ORF47	BKRF2	ORF47	UL1	69	UL115	U82	+	–	+	–	+	+
Glyco-protein N	gN	ORF53	ORF53	BLRF1	ORF53	UL49A	9A	UL73	U46	+	–	+	–	+	+
Epithelial ligand	EpiL	ORF58	ORF58	BMRF2	ORF58	γ	γ	γ	γ	+	–	–	–	+	γ

Data from [1] Bechtel, J.T., et al. (2005); [2] Zhu, F.X., et al. (2005); [3] Bortz, E., et al (2003); [4] Johannsen, E., et al. (2004); [5] Varnum, S.M., et al. (2004)
γ gammaherpesvirus-specific protein
γ1 gamma-1 herpesvirus-specific protein, and not detected in γ2
γ2 gamma-2 herpesvirus-specific protein, and not detected in γ1
* Protein named accordingly: R, for RRV; K, for KSHV
‡ Protein only encoded by RRV and KSHV
n/a Protein not encoded by the respective virus
"low" Signifies those proteins the investigators found in low abundance and did not include in their final list of EBV-associated proteins

centrifugation and analyzed the particles with two different MS techniques: multidimensional protein identification technology (MudPIT) and tandem MS of excised gel slices. We identified 33 virally encoded proteins associated within RRV. These included seven capsid proteins, 17 putative tegument proteins, and nine envelope proteins. Five of the 33 proteins are gammaherpesvirus specific, and three of these have no known function. Additionally, we identified three proteins not previously associated with a herpesvirus, and seven are new findings for a gamma-2-herpesvirus (O'Connor and Kedes 2006). In addition to the five capsid proteins we previously demonstrated as RRV capsid components (O'Connor et al. 2003), we identified the portal protein (PORT/ORF43) and the packaging protein (ORF19) (O'Connor and Kedes 2006). The 17 proteins we assigned to the tegument are tentative assignments, as their localization is based on homology with other herpesviruses, and thus require direct experimental evidence. Five of the 17 putative tegument proteins are gammaherpesvirus specific, and three of these five have no known function. Three of the nine envelope proteins are gammaherpesvirus specific, including the KSHV complement control protein (KCP) homologue encoded by ORF4. Finally, after detergent treatment of purified virions, a subset (six) proteins remain associated with the capsid, suggesting that these proteins interact more directly with and/or with higher affinity for the underlying capsid. We hypothesize that these last proteins may play a role in assembly or transport of viral or subviral particles during entry or egress (O'Connor and Kedes 2006).

Although the multiple techniques and rigorous purification methods are a potential explanation for the increased number of proteins identified in this RRV study, it is also likely that the greater abundance of purified virions played a major role. Because RRV infection of RhFs result in a greater number of particles produced, the loss we potentially incur throughout our purification protocol is not as significant as with KSHV, where the starting material is not as great. In the case of KSHV, this loss is detrimental as this, in turn, reduces the amount of protein for staining after SDS-PAGE separation. For the more abundant virion proteins, such as MCP (present in 960 copies/capsid), this is of little to no consequence, but for proteins that are present in the virions in low abundance, such as the portal protein (PORT/ORF43) present in only 12 copies/capsid, this can be a significant factor in protein detection. It is possible, therefore, that because of the low yield of KSHV virions, coupled with sample loss throughout the protocol, some of the less abundant virion-associated proteins were not detected in the KSHV studies but, in contrast, were found in the RRV investigation.

Thus the RRV system allows the detailed study of the structure and proteomic composition of gamma-2-herpesviruses. Defining the molecular composition of this primate gammaherpesvirus has begun to lay the foundation for the next stage of this research—to explore true contributions of cellular components of the mature virion and to define the functions of each of these virion proteins, examining their potential roles in initial infection, entry, assembly, and egress.

7
The Future of RRV as a Model for the Study of KSHV

With its high level of sequence and genome conservation with KSHV, efficient lytic growth in cell culture, and ability to infect naive rhesus monkeys, RRV clearly remains a highly attractive model system for studying primate gammaherpesvirus biology. Further, the RRV lytic system allows for genetic manipulation of the virus, including the production of recombinant and mutant viruses. Recently, DeWire et al. developed an RRV-GFP construct, in which the gene encoding GFP was inserted into the genome between two ORFs. This RRV-GFP and wild-type RRV replicate to similar titers and with similar kinetics (DeWire et al. 2003), providing clear support for the feasibility and usefulness of these methods in the study of the virus. It will also undoubtedly be equally amenable to siRNA and bacterial artificial chromosome (BAC) techniques of genetic manipulation that together will aid in characterizing the overall biology as well as the functions of individual proteins encoded by the virus. We anticipate that future work with RRV will continue to give insights into not only the mechanisms underlying its own replication, pathogenesis, structure and assembly, and persistence but also those of its human homologue, KSHV.

Acknowledgements The authors wish to thank Laura Adang (University of Virginia, USA) for critical review of this chapter and Z. Hong Zhou (University of Texas, Houston, USA) and Blossom Damania (University of North Carolina, USA) for granting permission to modify their figures from published work. DHK also acknowledges the following granting agencies for support: The Doris Duke Foundation and the National Cancer Institute (National Institutes of Health, USA).

References

Alexander L, Denekamp L, Knapp A, Auerbach MR, Damania B, Desrosiers RC (2000) The primary sequence of rhesus monkey rhadinovirus isolate 26-95: sequence similarities to Kaposi's sarcoma-associated herpesvirus and rhesus monkey rhadinovirus isolate 17577. J Virol 74:3388–98

AuCoin DP, Colletti KS, Xu Y, Cei SA, Pari GS (2002) Kaposi's sarcoma-associated herpesvirus (human herpesvirus 8) contains two functional lytic origins of DNA replication. J Virol 76:7890–6

Bechtel JT, Liang Y, Hvidding J, Ganem D (2003) Host range of Kaposi's sarcoma-associated herpesvirus in cultured cells. J Virol 77:6474–81

Bechtel JT, Winant RC, Ganem D (2005) Host and viral proteins in the virion of Kaposi's sarcoma-associated herpesvirus. J Virol 79:4952–64

Bergquam EP, Avery N, Shiigi SM, Axthelm MK, Wong SW (1999) Rhesus rhadinovirus establishes a latent infection in B lymphocytes in vivo. J Virol 73:7874–6

Boshoff C, Chang Y (2001) Kaposi's sarcoma-associated herpesvirus: a new DNA tumor virus. Annu Rev Med 52:453–70

Boshoff C, Weiss RA (1998) Kaposi's sarcoma-associated herpesvirus. Adv Cancer Res 75:57–86

Chan SR, Bloomer C, Chandran B (1998) Identification and characterization of human herpesvirus-8 lytic cycle-associated ORF 59 protein and the encoding cDNA by monoclonal antibody. Virology 240:118–26

Chang Y, Cesarman E, Pessin MS, Lee F, Culpepper J, Knowles DM, Moore PS (1994) Identification of herpesvirus-like DNA sequences in AIDS-associated Kaposi's sarcoma. Science 266:1865–9

Cotter MA 2nd, Robertson ES (1999) The latency-associated nuclear antigen tethers the Kaposi's sarcoma-associated herpesvirus genome to host chromosomes in body cavity-based lymphoma cells. Virology 264:254–64

Cotter MA 2nd, Subramanian C, Robertson ES (2001) The Kaposi's sarcoma-associated herpesvirus latency-associated nuclear antigen binds to specific sequences at the left end of the viral genome through its carboxy-terminus. Virology 291:241–59

Damania B, Lee H, Jung JU, . 1999a. Primate herpesviral oncogenes. Mol Cells 9:345–9

Damania B, Li M, Choi JK, Alexander L, Jung JU, Desrosiers RC, . 1999b. Identification of the R1 oncogene and its protein product from the rhadinovirus of rhesus monkeys. J Virol 73:5123–31

Desrosiers RC, Sasseville VG, Czajak SC, Zhang X, Mansfield KG, Kaur A, Johnson RP, Lackner AA, Jung JU (1997) A herpesvirus of rhesus monkeys related to the human Kaposi's sarcoma-associated herpesvirus. J Virol 71:9764–9

DeWire SM, Damania B (2005) The latency-associated nuclear antigen of rhesus monkey rhadinovirus inhibits viral replication through repression of Orf50/Rta transcriptional activation. J Virol 79:3127–38

DeWire SM, McVoy MA, Damania B (2002) Kinetics of expression of rhesus monkey rhadinovirus (RRV) and identification and characterization of a polycistronic transcript encoding the RRV Orf50/Rta, RRV R8, and R8.1 genes. J Virol 76:9819–31

DeWire SM, Money ES, Krall SP, Damania B (2003) Rhesus monkey rhadinovirus (RRV): Construction of an RRV-GFP recombinant virus and the development of assays to assess viral replication. Virology 312:122–134

Dezube BJ, Zambela M, Sage DR, Wang JF, Fingeroth JD (2002) Characterization of Kaposi sarcoma-associated herpesvirus/human herpesvirus-8 infection of human vascular endothelial cells: early events. Blood 100:888–96

Dittmer D, Stoddart C, Renne R, Linquist-Stepps V, Moreno ME, Bare C, McCune JM, Ganem D (1999) Experimental transmission of Kaposi's sarcoma-associated herpesvirus (KSHV/HHV-8) to SCID-hu Thy/Liv mice. J Exp Med 190:1857–68
Dittmer DP, Gonzalez CM, Vahrson W, DeWire SM, Hines-Boykin R, Damania B (2005) Whole-genome transcription profiling of rhesus monkey rhadinovirus. J Virol 79:8637–50
Garber AC, Shu MA, Hu J, Renne R (2001) DNA binding and modulation of gene expression by the latency-associated nuclear antigen of Kaposi's sarcoma-associated herpesvirus. J Virol 75:7882–92
Giddens WE Jr, Bielitzki JT, Morton WR, Ochs HD, Myers MS, Blakely GA, Boyce JT (1979) Idiopathic retroperitoneal fibromatosis: an enzootic disease in the pigtail monkey, *Macaca nemestrina*. Lab Invest 1979:294
Grundhoff A, Ganem D (2003) The latency-associated nuclear antigen of Kaposi's sarcoma-associated herpesvirus permits replication of terminal repeat-containing plasmids. J Virol 77:2779–83
Homa FL, Brown JC (1997) Capsid assembly and DNA packaging in herpes simplex virus. Rev Med Virol 7:107–122
Hu J, Garber AC, Renne R (2002) The latency-associated nuclear antigen of Kaposi's sarcoma-associated herpesvirus supports latent DNA replication in dividing cells. J Virol 76:11677–87
Jeong J, Papin J, Dittmer D (2001) Differential regulation of the overlapping Kaposi's sarcoma-associated herpesvirus vGCR (orf74) and LANA (orf73) promoters. J Virol 75:1798–807
Kirchoff VS, Wong SJS, Pari GS (2002) Generation of a life-expanded rhesus monkey fibroblast cell line for the growth of rhesus rhadinovirus (RRV). Arch Virol 147:321–33
Lagunoff M, Bechtel J, Venetsanakos E, Roy AM, Abbey N, Herndier B, McMahon M, Ganem D (2002) De novo infection and serial transmission of Kaposi's sarcoma-associated herpesvirus in cultured endothelial cells. J Virol 76:2440–8
Lee H, Veazey R, Williams K, Li M, Guo J, Neipel F, Fleckenstein B, Lackner A, Desrosiers RC, Jung JU (1998) Deregulation of cell growth by the K1 gene of Kaposi's sarcoma-associated herpesvirus. Nat Med 4:435–40
Mansfield KG, Westmoreland SV, DeBakker CD, Czajak S, Lackner AA, Desrosiers RC (1999) Experimental infection of rhesus and pig-tailed macaques with macaque rhadinoviruses. J Virol 73:10320–8
Miller G, Heston L, Grogan E, Gradoville L, Rigsby M, Sun R, Shedd D, Kushnaryov VM, Grossberg S, Chang Y (1997) Selective switch between latency and lytic replication of Kaposi's sarcoma herpesvirus and Epstein-Barr virus in dually infected body cavity lymphoma cells. J Virol 71:314–24
Moore PS, Chang Y (2001) Molecular virology of Kaposi's sarcoma-associated herpesvirus. Philos Trans R Soc Lond B Biol Sci 356:499–516
Nealon K, Newcomb WW, Pray TR, Craik CS, Brown JC, Kedes DH (2001) Lytic replication of Kaposi's sarcoma-associated herpesvirus results in the formation of multiple capsid species: isolation and molecular characterization of A, B, and C capsids from a gammaherpesvirus. J Virol 75:2866–78
O'Connor CM (2005) Structural and Proteomic Analyses of the Gammaherpesviruses KSHV and RRV. University of Virginia, Charlottesville, 302 pp.

O'Connor CM, Damania B, Kedes DH (2003) De novo infection with rhesus monkey rhadinovirus leads to the accumulation of multiple intranuclear capsid species during lytic replication but favors the release of genome-containing virions. J Virol 77:13439–47
O'Connor CM, Kedes DH (2006) Mass spectrometric analysis of rhesus monkey rhadinovirus. J Virol 80:1574–1583
Pari GS, AuCoin D, Colletti K, Cei SA, Kirchoff V, Wong SW (2001) Identification of the rhesus macaque rhadinovirus lytic origin of DNA replication. J Virol 75:11401–7
Parsons CH, Adang LA, Overderest J, O'Connor CM, Taylor JR, Camerini D, Kedes DH (2006) KSHV targets multiple leukocyte lineages during long-term productive infection in NOD/SCID mice. J Clin Invest 116:1963–1973
Renne R, Blackbourn D, Whitby D, Levy J, Ganem D (1998) Limited transmission of Kaposi's sarcoma-associated herpesvirus in cultured cells. J Virol 72:5182–8
Renne R, Dittmer D, Kedes D, Schmidt K, Desrosiers RC, Luciw PA, Ganem D (2004) Experimental transmission of Kaposi's sarcoma-associated herpesvirus (KSHV/HHV-8) to SIV-positive and SIV-negative rhesus macaques. J Med Primatol 33:1-9
Renne R, Zhong W, Herndier B, McGrath M, Abbey N, Kedes D, Ganem D (1996) Lytic growth of Kaposi's sarcoma-associated herpesvirus (human herpesvirus 8) in culture. Nat Med 2:342–6
Rose TM, Ryan JT, Schultz ER, Raden BW, Tsai CC (2003) Analysis of 4.3 kilobases of divergent locus B of macaque retroperitoneal fibromatosis-associated herpesvirus reveals a close similarity in gene sequence and genome organization to Kaposi's sarcoma-associated herpesvirus. J Virol 77:5084–97
Rose TM, Strand KB, Schultz ER, Schaefer G, Rankin GW Jr, Thouless ME, Tsai CC, Bosch ML (1997) Identification of two homologs of the Kaposi's sarcoma-associated herpesvirus (human herpesvirus 8) in retroperitoneal fibromatosis of different macaque species. J Virol 71:4138–44
Sakurada S, Katano H, Sata T, Ohkuni H, Watanabe T, Mori S (2001) Effective human herpesvirus 8 infection of human umbilical vein endothelial cells by cell-mediated transmission. J Virol 75:7717–22
Schultz ER, Rankin GW Jr, Blanc MP, Raden BW, Tsai CC, Rose TM (2000) Characterization of two divergent lineages of macaque rhadinoviruses related to Kaposi's sarcoma-associated herpesvirus. J Virol 74:4919–28
Schwam DR, Luciano RL, Mahajan SS, Wong L, Wilson AC (2000) Carboxy terminus of human herpesvirus 8 latency-associated nuclear antigen mediates dimerization, transcriptional repression, and targeting to nuclear bodies. J Virol 74:8532–40
Searles RP, Bergquam EP, Axthelm MK, Wong SW (1999) Sequence and genomic analysis of a Rhesus macaque rhadinovirus with similarity to Kaposi's sarcoma-associated herpesvirus/human herpesvirus 8. J Virol 73:3040–53
Strand K, Harper E, Thormahlen S, Thouless ME, Tsai C, Rose T, Bosch ML (2000) Two distinct lineages of macaque gamma herpesviruses related to the Kaposi's sarcoma associated herpesvirus. J Clin Virol 16:253–69
Wang L, Wakisaka N, Tomlinson CC, DeWire SM, Krall S, Pagano JS, Damania B (2004) The Kaposi's sarcoma-associated herpesvirus (KSHV/HHV-8) K1 protein induces expression of angiogenic and invasion factors. Cancer Res 64:2774–81

Wong SW, Bergquam EP, Swanson RM, Lee FW, Shiigi SM, Avery NA, Fanton JW, Axthelm MK (1999) Induction of B cell hyperplasia in simian immunodeficiency virus-infected rhesus macaques with the simian homologue of Kaposi's sarcoma-associated herpesvirus. J Exp Med 190:827–40

Wu L, Lo P, Yu X, Stoops JK, Forghani B, Zhou ZH (2000) Three-dimensional structure of the human herpesvirus 8 capsid. J Virol 74:9646–54

Yu XK, O'Connor CM, Atanasov I, Damania B, Kedes DH, Zhou ZH (2003) Three-dimensional structures of the A, B, and C capsids of rhesus monkey rhadinovirus: insights into gammaherpesvirus capsid assembly, maturation, and DNA packaging. J Virol 77:13182–93

Zhong W, Wang H, Herndier B, Ganem D (1996) Restricted expression of Kaposi sarcoma-associated herpesvirus (human herpesvirus 8) genes in Kaposi sarcoma. Proc Natl Acad Sci U S A 93:6641–6

Zhu FX, Chong JM, Wu L, Yuan Y (2005) Virion proteins of Kaposi's sarcoma-associated herpesvirus. J Virol 79:800–11

CTMI (2006) 312:71–100

The Rta/Orf50 Transactivator Proteins of the Gamma-Herpesviridae

M. R. Staudt · D. P. Dittmer (✉)

Department of Microbiology and Immunology and Lineberger Comprehensive Cancer Center, University of North Carolina at Chapel Hill, 804 Mary Ellen Jones Bldg, CB 7290, Chapel Hill, NC 27599, USA
ddittmer@med.unc.edu

Abstract The replication and transcription activator protein, Rta, is encoded by *Orf50* in Kaposi's sarcoma-associated herpesvirus (KSHV) and other known gammaherpesviruses including Epstein-Barr virus (EBV), rhesus rhadinovirus (RRV), herpesvirus saimiri (HVS), and murine herpesvirus 68 (MHV-68). Each Rta/Orf50 homologue of each gammaherpesvirus plays a pivotal role in the initiation of viral lytic gene expression and lytic reactivation from latency. Here we discuss the Rta/Orf50 of KSHV in comparison to the Rta/Orf50s of other gammaherpesviruses in an effort to identify structural motifs, mechanisms of action, and modulating host factors.

1
Introduction

As all members of the *Herpesviridae*, the gammaherpesviruses can establish either a latent or lytic life cycle within host cells. During latency, only a few viral genes are transcribed and the virus exists as a nonintegrated circular episome within the nucleus of the infected cell (Fakhari and Dittmer 2002; Jenner et al. 2001; Paulose-Murphy et al. 2001; Sarid et al. 1998; Zhong et al. 1996). B cell latency can be disrupted by host cell signaling, such as B cell receptor cross-linking, which leads to the sequential expression of several subsets of lytic genes: immediate-early genes (IE) that encode viral transcriptional regulators; delayed-early genes (DE) that encode proteins involved in viral DNA replication, and late genes (L) that encode viral structural proteins. Herpesvirus lytic gene expression follows this temporal and sequential cascade, ultimately resulting in the production of progeny virions and destruction of and egress from the infected host cell. The Rta/Orf50 switch protein is essential to initiate lytic reactivation of all gammaherpesviruses: Epstein-Barr virus (EBV), Kaposi's sarcoma-associated herpesvirus (KSHV), rhesus rhadinovirus (RRV), herpesvirus saimiri (HVS), and murid herpesvirus 68 (MHV-68).

2
Immediate-Early Genes

Immediate-early genes define mRNAs that are transcribed in the presence of protein synthesis inhibitors, such as cycloheximide. This applies to herpesvirus genes after de novo infection of permissive cells (Roizman 1996) but also to cellular genes after serum stimulation. Lau and Nathans identified cellular immediate-early genes (*jun*/*fos*) because they constituted the first wave of mRNAs after serum stimulation of mouse fibroblasts (Lau and Nathans 1985, 1987). jun/fos mRNA levels were induced within 10 min after addition of serum and declined shortly thereafter. By comparison, the induction of c-myc, a cellular early mRNA, was delayed. c-Myc mRNA peaked at 20–45 min, after the wave of immediate-early mRNAs subsided, and stayed induced for longer periods of time. The definition of early genes for herpesviruses is more strict: Early gene transcription is dependent on immediate-early transactivators independent of the time frame. Herpesvirus immediate-early transactivators are necessary and sufficient to initiate viral replication (McKnight et al. 1987; Triezenberg et al. 1988).

Rta/Orf50 is an immediate-early protein of rhadinoviruses. It is necessary and sufficient to drive lytic replication for KSHV, HVS, RRV, and MHV-68. Ectopic expression of Rta/Orf50 will reactivate virus from latency (sufficient); deletion of Rta/Orf50 or inhibition by a dominant-negative mutant will prevent lytic reactivation and replication (necessary). Although other rhadinovirus mRNAs are transcribed in the presence of cycloheximide (CHX) (Orf57/Mta, K8/Zta, Orf45) and are therefore considered immediate-early genes, their gene products are not sufficient to reactivate virus from latency. Whether any of these are necessary for lytic replication remains to be established.

Figure 1 shows an array analysis of RRV transcription at 6 h after de novo infection in the presence or absence of permissive fibroblasts (from Dittmer et al. 2005). Here, the levels for each viral mRNA were measured by quantitative real-time RT-PCR, and for each viral mRNA the number of cycles that were required to obtain a fixed amount of product was plotted. Rta/Orf50 is the most abundant mRNA at early times after infection and is unaffected by CHX. In contrast, the majority of RRV transcripts are not transcribed that early in the infection process (requiring more than 40 cycles of PCR to detect a signal under either condition) or are significantly inhibited by CHX. The latter includes mRNAs driven by Rta/Orf50-responsive promoters. But there are also a significant number of mRNAs that were transcribed in the presence of the protein synthesis inhibitor. By definition these are immediate-early genes, and their transcription is dependent only on preformed (i.e., immediate early) cellular regulators or RRV virion transactivators. Yet at the same time many of these genes are also Rta/Orf50-responsive. Rta/Orf50's own promoter falls into this class. In the case of the Rta/Orf50 promoter, transactivation by Rta/Orf50 protein establishes a direct positive feedback loop that locks the lytic transcription cascade into place. On the basis of extensive transcriptional profiling, we would speculate that the gamma herpesviridae evolved a more plastic, less ridged transcriptional control program than the alpha and beta herpesviridae to cope with the various signaling events, cytokine exposures, and growth stimuli in the life of a latently infected lymphocytes.

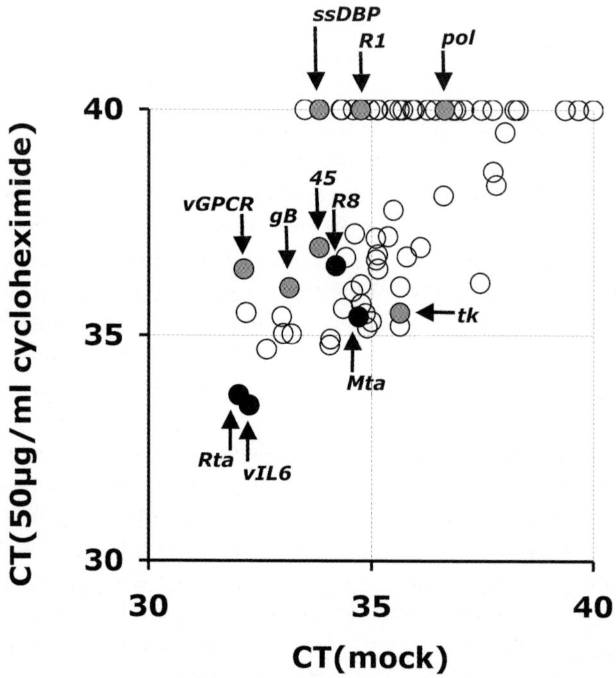

Fig. 1 Array analysis of RRV transcription at 6 h after de novo infection in the presence (*vertical*) or absence (*horizontal*) of cycloheximide in permissive fibroblasts (from Dittmer et al. 2005). Shown are relative mRNA levels on a log2 scale (*CT*). Known KSHV immediate-early genes are in *black*, known Rta targets in KSHV in *gray*, and all others in *open circles*

3 Lytic Reactivation

All gammaherpesviruses encode an Rta/Orf50 homologue, and each has been shown to play a pivotal role in the initiation of viral lytic gene expression and lytic reactivation from latency. Although the gene product of *Orf50*, named Rta (replication and transcription activator), is the only essential latent/lytic switch protein for the gamma-2-herpesviruses (rhadinoviruses), two proteins, Zta and Rta, independently can reactivate EBV, a gamma-1 or lymphocryptovirus, from latency. This difference in viral lytic switch proteins between the lymphocryptoviruses and rhadinoviruses indicates a marked difference in the precise molecular mechanisms of virus-mediated lytic reactivation of the two related subgroups of gammaherpesviruses.

3.1 Lymphocryptovirus—EBV

Many excellent reviews exist that describe EBV lytic reactivation in detail. Hence, we will only recount the basic tenets here to compare them vis-à-vis rhadinovirus lytic reactivation. The major viral lytic switch protein of EBV is considered to be Zta (also known as Zebra, *BZLF-1*, or Z protein) (Chen et al. 1999; Cox et al. 1990; Quinlivan et al. 1993; Ragoczy et al. 1998; Zalani et al. 1996). Zta is a bZIP-type transcriptional transactivator. The EBV Orf50 homologue, the *BRLF1* gene product Rta, is also a sequence-specific DNA-binding protein known to function as a transcriptional activator (Quinlivan et al. 1993; Ragoczy et al. 1998; Ragoczy and Miller 2001; Russo et al. 1996). Independently, both Zta and Rta initiate the expression of lytic genes and, in a somewhat cell type-specific manner, Rta can lead to the activation of DE promoters. It is believed that Zta and Rta proteins of EBV act synergistically (although Zta can suppress the transactivation function of Rta) to induce viral reactivation in latently infected B cells. However, EBV Rta alone can activate a subset of lytic promoters that do not require Zta, and EBV Zta alone can activate a subset of lytic promoters that do not require Rta (Ragoczy and Miller 1999). A deletion mutant of either Zta or Rta is defective for viral reactivation (Feederle et al. 2000). The KSHV Rta/Orf50 is the sequence homologue of the EBV Rta/BRLF1, whereas the KSHV KbZIP/K8 protein is the sequence homologue of EBV Zta/BZLF1. In contrast to EBV, the KSHV Rta/Orf50 is considered to be the only essential lytic switch protein to date. A KSHV Rta/Orf50 deletion mutant is incapable of reactivation from latency (Xu et al. 2005), although a deletion mutant of KSHV KbZIP remains to be evaluated. In this regard the two subclasses of gammaherpesviruses (lymphocryptovirus and rhadinovirus) differ in their molecular mechanisms of reactivation in that to date only one viral protein, Rta/Orf50, has been shown to mediate viral reactivation and induce lytic gene expression in the rhadinoviruses (KSHV, HVS, RRV, and MHV-68). In fact, expression of KSHV Rta/Orf50 precedes expression of K8 (K-bZip) and transactivates the K8 promoter (Lukac et al. 1998; Sun et al. 1998). Therefore, in an effort to minimize complexity and derive general principles for homologous molecular mechanisms, we will only discuss the Rta/Orf50 proteins of the rhadinoviruses.

4
Rhadinoviruses—KSHV, HVS, RRV, and MHV-68

4.1
Experimental Considerations

Before we delve into a detailed molecular description of the rhadinovirus Rta/Orf50 transactivator, it seems prudent to highlight some experimental constraints that affect the general conclusions we can draw from the many studies on Rta/Orf50 and rhadinovirus reactivation. KSHV, RRV, and MHV68 establish latency in B lymphocytes and are associated with B-cell hyperplasia, but unlike EBV and primate lymphocryptoviruses, KSHV, RRV, and MHV68 do not immortalize primary B cells with any efficiency in culture. Rta/Orf50 is sufficient to reactivate latent virus in each case, but the exact B-cell compartment and lineage-specific transcription factor makeup may be different for each virus. Only RRV and MHV68 establish a robust de novo infection in primary fibroblasts, which amplifies input virus. Hence, Rta/Orf50's role in primary infection under low MOI conditions can only be investigated in these lytic model systems. KSHV infects primary and immortalized endothelial cells, but to date only a single plaque has been published (Boshoff et al. 1995; Ciufo et al. 2001). Although KSHV can be serially propagated, input virus is not amplified and high MOI infection in the presence of polybrene is required to initiate the culture (Foreman et al. 1997; Lagunoff et al. 2002; Renne et al. 1998). Reactivation from latency in response to biological signals or chemical inducers such as phorbol ester is a low-frequency event for KSHV (Chang et al. 2000; Renne et al. 1996) and sets up an interesting paradox. In a latent culture every single cell is infected with KSHV, carries 10–50 copies of the viral episome, and expresses the latency-associated nuclear antigen (LANA/Orf73). LANA is necessary and sufficient for latent viral replication, which is analogous to Rta/Orf50's requirement for lytic viral replication (Ballestas et al. 1999; Godfrey et al. 2005). If a latent cell culture, for instance, BCBL-1 cells, is treated with phorbol ester, 100% of the cells receive the drug and are subject to drug action; however, not all cells will express Rta/Orf50, and of the cells that express Rta/Orf50 not all cells express delayed-early genes. Furthermore, within the subset of cells that express delayed-early genes even fewer yet express true late viral mRNAs, such as Orf29, a capsid component (Zoeteweij et al. 1999). Therefore, we speculate that additional constraints exist that regulate viral replication at each junction (IE, DE, E, L) in the regulatory cascade.

McAllister et al. showed that the cell cycle state influences how competent each host cell is to support viral replication (McAllister et al. 2005). At this point it is unresolved whether the host cell cycle state simply reflects responsiveness of the PKC downstream targets that activate the Rta/Orf50 promoter, the effectiveness of Rta/Orf50 to function as a transactivator, or steps dependent on viral DE genes. Alternatively, cells in S phase may remain viable for a longer time and can, therefore, accumulate more viral mRNA and proteins before dying because of the effects of viral capsid maturation and egress. Many cell lines do not tolerate stable Rta/Orf50 expression, but recently exciting new tools have become available to investigate Rta/Orf50 function and add to the existing tools of cDNA expression (Sun et al. 1998; Lukac et al. 1998) and Rta/Orf50 dominant-negative plasmids (Lukac et al. 1998): an *Orf50* k/o virus for MHV68 and KSHV, a KSHV-inducible system in BCBL-1 and 293 cells (Nakamura et al. 2003; Xu et al. 2005), a KSHV Rta/Orf50 recombinant adenovirus (Liang et al. 2002), and novel DNA-binding mutants of KSHV Rta/Orf50 (Chang et al. 2005). These tools will yield exciting new insights into the molecular function of Rta/Orf50 and the biology of the rhadinoviruses, although they may also lead to discrepancies with existing studies due to differences in the experimental approach.

4.2
Chemically Induced Viral Reactivation of Gammaherpesviruses

Reactivation of rhadinoviruses in latently infected cells can be achieved by treatment with chemicals that mimic BCR signaling such as *n* -butyrate, 12-*O*-tetradecanoylphorbol-13-acetate (TPA), or calcium ionophores. These types of treatment lead to expression of viral lytic genes, foremost among them *Orf50*. Although the entire cellular signaling cascades that lead to viral reactivation in response to these chemical treatments are unknown, we do know that induction of the KSHV *Orf50* promoter by TPA involves the cellular AP1 pathway and induction by sodium butyrate (NaB) involves cellular Sp1 (Wang et al. 2004a, 2003a, 2003b). Cannon et al. have recently shown that overexpression of the KSHV lytic protein vGPCR/Orf74 indirectly resulted in a decreased efficiency of chemical induction of the ORF50 transcript (Cannon et al. 2006). Because the KSHV vGPCR/Orf74 is known to modulate cellular signaling pathways, which, in extreme instances, can lead to transformation (Bais et al. 1998; Polson et al. 2002) this adds credence to a model of multilayered cross talk between KSHV and the host. In RRV, the histone deacetylase inhibitor *trichostatin A* (TSA) is also capable of reactivating RRV from latently infected cells, presumably by de-repressing the *Orf50* promoter (DeWire et al. 2002). Like TSA and butyrate, valproic acid (2-propylpentanoic acid) also has potent histone deacetylase activity. At the same time it is FDA approved as antiseizure medication. Treatment of KSHV-infected primary effusion lymphoma (PEL) cells with valproate induced lytic reactivation in culture (Shaw 2000; Klass et al. 2005) and is currently in clinical trials for the treatment of Kaposi's sarcoma.

4.3
Viral Induction of Lytic Reactivation of Rhadinoviruses

Although many viral proteins were assayed for their ability to reactivate KSHV in PEL cells, only ectopic expression of Rta/Orf50 was sufficient to disrupt latency and activate lytic replication, resulting in a complete productive viral life cycle (Gradoville et al. 2000; Lukac et al. 1998; Sun et al. 1998). The KSHV Zta and Mta homologues by themselves were not able to induce lytic reactivation, although, as in EBV, they may transactivate some DE promoters independently of Rta/Orf50. Expression of HVS Orf50a protein and MHV-68 Rta/Orf50 also induces lytic reactivation and production of infectious viral progeny in HVS and MHV-68 models of latency, respectively (Goodwin et al. 2001; Wu et al. 2000). The RRV Rta/Orf50 and MHV-68 Rta/Orf50 can reactivate KSHV from latency (Damania et al. 2004), and the KSHV Rta/Orf50 protein can reactivate the murine viral homologue, MHV-68, from latency (Rickabaugh et al. 2005). These studies demonstrate a strong conservation of function across evolution of the rhadinovirus Rta proteins. Although RRV Orf50 has been shown to be a potent transactivator of RRV DE promoters, it has not yet been demonstrated to reactivate RRV from latency (DeWire et al. 2002), because lytic replication rather than latency is the default pathway following primary RRV infection of cells in culture. Recently, a latently infected RRV system has been established, and we can expect formal demonstration of this in the near future (DeWire and Damania 2005).

5
Rta/Orf50 Transcription

The Orf50 transcripts from all gamma-2-herpesviruses share a similar architecture that is essentially comprised of two exons separated by one intron. A major IE transcript is observed after reactivation of KSHV that is a 3.6-kb tricistronic mRNA encoding the Orf50, K8, and K8.1 reading frames (Zhu et al. 1999). Splicing results in the major Orf50 transcript; however, other alternatively spliced mono-, bi-, and polycistronic Orf50-containing transcripts have been found (Saveliev et al. 2002; Tang and Zheng 2002; Wang et al. 2004). Presently, the significance of these isoforms is unclear. Splicing of the Orf50 transcript is a characteristic of gamma-2-herpesviruses and is in contrast to the Orf50 cDNA of the gamma-1 herpesvirus EBV, which is identical to its genomic open reading frame structure (Manet 1989). In the rhadinoviruses, Orf45 is located within the Rta/Orf50 intron in opposite orientation and the rhadinovirus Orf45 promoter is presumably located within the Rta/Orf50 open reading frame. Similarly in EBV, Orf45 (BRRF1/Na) is located 5′ of the

Rta/Orf50 open reading frame, also in the opposite orientation and also within the first intron (the first exon of EBV Rta is noncoding) (Hong et al. 2004).

The major 3.6-kb Orf50 transcript is induced within 4 h of *n*-butyrate treatment of latently infected PEL cells and is resistant to treatment with the protein synthesis inhibitor, cycloheximide (CHX), thus displaying IE kinetics (Sun et al. 1999; Zhu et al. 1999). Lukac at al. also observed expression of Orf50 mRNA within 1 h of inducing viral reactivation in latently infected BCBL-1 cells by treatment with the phorbol ester TPA (Lukac et al. 1998). MHV-68 *Orf50* is also an immediate-early gene, as is the RRV *Orf50*, further demonstrating conservation of *Orf50* transcription kinetics among the rhadinoviruses (Rochford et al. 2001; DeWire et al. 2002; Dittmer et al. 2005). Of the rhadinoviruses, lytic reactivation of HVS results in transcription of two distinct Orf50 mRNA species, called *Orf50a* and *Orf50b* (Whitehouse et al. 1997). The Orf50a transcript is identical to that described for the other rhadinoviruses and is detected at early times during viral replication. The HVS Orf50b transcript is expressed at later time points during replication and is produced from a promoter within the second exon. Its function remains to be elucidated. Because abolishing Rta/Orf50 function inhibits lytic replication at IE times, it has not been possible to determine experimentally whether Rta/Orf50 has additional functions at DE or late times in any of the rhadinoviridae.

5.1 Regulation of the KSHV Rta/Orf50 Promoter

Because Rta/Orf50 protein is the key regulator of KSHV lytic reactivation, much attention has focused on the promoter that regulates Rta/Orf50 expression. The KSHV Rta/Orf50 protein autoregulates its own promoter via an indirect mechanism, because no obvious Rta/Orf50-responsive element (RRE) or RBP-J-κ consensus binding sites are present within the *Orf50* promoter (these mechanisms are described in detail in later sections of this review) (Chang and Miller 2004; Deng et al. 2000; Gradoville et al. 2000). By comparison to other Rta/Orf50-responsive promoters, the *Orf50* promoter is only marginally activated by Rta/Orf50 protein expression. Within the *Orf50* promoter, a binding sequence for the cellular transcription factor octamer-binding protein (Oct-1) was shown to mediate autoregulation by Rta/Orf50 protein. Oct-1 bound to a specific region of DNA within the *Orf50* promoter, as demonstrated by electrophoretic mobility shift assay (EMSA) (Sakakibara et al. 2001). In addition, both Sp1 and Sp3 cellular transcription factors appear to be involved in Rta/Orf50 autoactivation (Chen et al. 2000; Zhang et al. 1998). This is similar to the EBV Rta protein, which also mediates autoregulation of its own promoter (Rp) via a non-DNA-binding mechanism involving cellular Sp1 and Sp3 proteins (Ragoczy and Miller 2001). KSHV Rta/Orf50 also interacts with the CCAAT/enhancer binding protein α(C/EBP-α) to upregulate Rta/Orf50 expression (Wang et al. 2003b).

The Orf50 promoter is heavily methylated in latently infected PEL cells, and treatment with TPA leads to demethylation of the promoter (Chen et al. 2001). In vivo, several biopsies from KSHV-related diseases, including Kaposi's sarcoma, multicentric Castleman disease (MCD), and PEL, showed decreased methylation of the *Orf50* promoter, although this promoter was still heavily methylated in samples obtained from a latently infected KSHV carrier. This evidences an additional layer of regulation of Rta/Orf50 expression and, as such, KSHV lytic reactivation. Progressive methylation of viral lytic promoters could skew KSHV infection of host cells toward latency. This is consistent with the decoration of the viral episome and Rta/Orf50 promoter with inhibitory histone complexes (Lu et al. 2003). Because much of the investigation of Rta/Orf50's function is based on transient transfection assays of unmethylated, histone-free promoter-reporter plasmids, we do not know how Rta/Orf50's *general* transactivation function (via TAFs and regulation of HDACs) may work together with Rta/Orf50's *specific* transactivation functions (via direct DNA binding or RBP-J-κ binding sites).

The propensity of KSHV to establish latency on primary infection (Ciufo et al. 2001; Grundhoff and Ganem 2004; Krishnan et al. 2004; Moses et al. 1999) is in contrast to the propensity of RRV and MHV-68 to establish lytic infection. The bias of KSHV toward the establishment of latency after primary infection may be due to Rta/Orf50's interaction with species-specific chromatin modules, because Rta/Orf50 proteins of any rhadinovirus are able to transactivate unmethylated viral promoters in transient transfection assays. However, it is interesting to note that both MHV-68 and RRV are impaired in their ability to reactivate KSHV in the context of viral gene expression (Damania and Jung 2001).

Micrococcal nucleosome mapping techniques by Chen et al. reported a nucleosome positioned on the Orf50 promoter that overlapped the transcription start site and a GC-rich region bound by both Sp1 and Sp3 (Lu et al. 2003). The Sp1/Sp3 region of the Orf50 promoter was also mapped as being highly responsive to two chemical compounds known to inhibit histone deacetylases, NaB and TSA (Lu 2003). In addition, NaB treatment led to the rapid recruitment of Ini1/Snf5, a component of the Swi/Snf family of chromatin remodeling proteins. These data describe complex, multitiered levels of transcriptional regulation of the Orf50 promoter within latently infected host cells.

6
Rta/Orf50 Protein

6.1
KSHV Rta/Orf50

The KSHV *Orf50* gene encodes a 691-amino acid protein that is highly phosphorylated and localizes to the nucleus of mammalian cells (Lukac et al. 2001, 1999; Seaman and Quinlivan 2003). All Rta homologues share a conserved C-terminal activation domain. Deletion of 160 amino acids in the C-terminal activation domain of the KSHV Rta/Orf50 results in production of a truncated, but stable, Rta/Orf50 protein that forms multimers with wild-type Rta/Orf50 in PEL cells and functions as a dominant-negative inhibitor of Rta/Orf50 transactivation (Lukac et al. 1998). Expression of this truncated Rta/Orf50 protein leads to suppression of both spontaneous and chemically induced viral reactivation. Expression of KSHV Rta/Orf50 and subsequent viral reactivation can also be efficiently knocked down by expression of human RNase P (Zhu et al. 2004). These data suggest that Rta/Orf50 is, indeed, the lytic switch protein of KSHV. Recently, Pari and colleagues reported genetic evidence for the Rta-lytic switch hypothesis by utilizing the KSHV genome cloned into a bacterial artificial chromosome (BAC36) (Gao et al. 2003; Zhou et al. 2002) and generating a deletion within the Orf50 open reading frame (Xu et al. 2005). After transfection of the Orf50-deficient BAC into HEK 293 cells, latent genes were expressed at wild-type levels; however, the virus was unable to reactivate on chemical treatment, unequivocally demonstrating that Rta/Orf50 is required for successful viral reactivation.

A provocative recent report showed that Rta/Orf50 was present in KSHV virions (Bechtel et al. 2005b), which would make it a virion transactivator much like the herpes simplex virus VP16 and thus ensure lytic replication on primary infection. However, herpesvirus virions are notorious for capturing a variety of proteins, and even mRNAs (Bechtel et al. 2005a), simply as scaffolds during assembly or because of sloppy egress. The composition of nonstructural proteins in the rhadinovirus virions is highly variable (Zhu and Yuan 2003; Zhu et al. 2005; O'Connor and Kedes 2006; Trus et al. 2001) and may not necessarily have a function in the next infection cycle.

The different functions of KSHV Rta/Orf50 are subject to posttranslational modifications. KSHV Rta/Orf50 is highly phosphorylated, a modification that is mediated at least in part by the ability of Rta/Orf50 to bind to and be phosphorylated by the cellular Ste20-like kinase hKFC (Gwack et al. 2003b; Lukac et al. 2001,, 1999). The phosphorylation of Rta/Orf50 by hKFC as well as Rta's poly(ADP)-ribosylation by cellular PARP-1 protein both result in decreased ability of KSHV Rta/Orf50 to transactivate viral promoters (Gwack et al. 2003b). Recently, KSHV Rta/Orf50 was shown to have E3 ubiquitin ligase activity and could direct polyubiquitination of cellular interferon regulatory factor 7 as well as polyubiquitination of itself (Yu et al. 2005). Point mutations in the Cys + His-rich N-terminal domain of Rta/Orf50 abolished the E3 ligase activity, which should be independent of Rta's transcriptional activity. Much more still needs to be learned, but the many posttranslational modifications of KSHV Rta/Orf50 most likely act to regulate the function, and possibly the stability, of this key viral transactivator protein.

KSHV Rta/Orf50 has been shown to modulate transcription of host genes as well. It has been reported to transactivate cellular interleukin 6 (IL-6) transcription (Deng et al. 2002a). KSHV Rta/Orf50 can also modulate the ability of cellular STAT3 to function as a transactivator, indirectly leading to modulation of host gene expression (Gwack et al. 2002). Furthermore, Rta/Orf50 interacts with RBP-J-κ (Chang et al. 2005) and may thereby regulate the transcription of RBP-J-κ-dependent host mRNAs.

6.2 MHV-68 Rta/Orf50

The Rta protein encoded by *Orf50* of MHV-68 is sufficient to induce lytic reactivation in latently infected cells (Wu et al. 2000). Investigations of the disruption of the MHV-68 Rta/Orf50 open reading frame demonstrated that Rta/Orf50 is also necessary for viral reactivation (Liu et al. 2000; Pavlova et al. 2005, 2003). The requirement for MHV-68 Rta/Orf50 in lytic reactivation was also demonstrated by other means, whereby a loss in viral reactivation was observed after efficient knockdown of MHV-68 Rta/Orf50 expression by RNAi (Jia et al. 2004). A mutant MHV-68 virus, called M50, was generated to constitutively express MHV-68 Rta/Orf50 by insertion of a new promoter element into the 5′-untranslated region (UTR) of the Orf50 promoter (May et al. 2004). Constitutive expression of Rta/Orf50 by the M50 mutant MHV-68 virus resulted in defective establishment of latency. The physiological relevance of this mutant phenotype was demonstrated by studies in which immunization of mice with the mutant M50 virus resulted in partial protection against challenge with wild-type virus, demonstrating the importance of proper transcriptional regulation of *Orf50* in the both the pathogenesis and establishment of latency by MHV-68 (Boname et al. 2004; May et al. 2004). Importantly, a Rta/ORF50-null mutant of MHV-68 established long-term latency in the lungs of infected mice but failed to vaccinate against a wild-type virus challenge, therefore implicating the necessity of lytic replication for generation of a protective immune response (Moser et al. 2006). In addition, gene array studies of a recombinant virus that overexpressed MHV-68 Orf50 found that nearly every MHV-68 gene assayed was upregulated by Rta/Orf50 overexpression. This phenotype is consistent with Rta/Orf50 being the first gene in the MHV-68 lytic transcriptional cascade and highlights the potent transactivating ability of the protein (Martinez-Guzman et al. 2003).

6.3 HVS Rta/Orf50

Much less is known about HVS Rta/Orf50, but what has been reported suggests that the properties of HVS Rta/Orf50 resemble those of the other rhadinoviruses. HVS Rta/Orf50 induces viral reactivation of latently infected cells (Goodwin et al. 2001), a function conserved throughout all rhadinoviruses. HVS Rta/Orf50 contains an AT-hook DNA binding domain that is required for transactivation of at least two delayed early viral promoters, Orf6 and Orf57 (Walters et al. 2004). Although this report demonstrates that HVS Rta/Orf50 can transactivate certain viral promoters via direct DNA binding, this viral transactivator most likely also works through indirect mechanisms as do the lytic switch proteins of the other gammaherpesviruses. HVS Rta was shown to bind TATA-binding protein (TBP) in vitro, which provides an alternative means to influence viral and cellular transcription (Hall et al. 1999). Of note, the amino acid sequence of both Rta/Orf50 isoforms, Orf50a and Orf50b of HVS A11 and HVS C488 (a low-passage transforming isolate), are among the most divergent open reading frames in these viral isolates (Ensser et al. 2003), with only ~70% amino acid identity compared to >90% amino acid identity for the other 60 of 75 (80%) open reading frames in the HVS genome. Whereas the Orf50b of A11 showed decreased transactivation capability compared to Orf50a of A11, the Orf50b of C488 demonstrated full transactivation capability. In contrast, both the A11 and C488 Orf50a proteins were capable of reactivating lytic replication in persistently infected cells, indicating a conservation of function for one Rta/Orf50 isoform and not the other. The same study reported that Rta/Orf50 allelic variation cosegregated with Stp and Tip in their ability to transform human T cells in culture. By inference based on this genetic data set alone, Rta/Orf50 could be considered an oncogene. On the other hand, there is no independent evidence that Rta/Orf50 can transform cells in culture or can cooperate with other oncogenes to do so. Rather, long-term stable expression of Rta/Orf50 seems incompatible with continued cell growth. No molecular mechanism has been described to support such classification of HVS Rta/Orf50 as an oncogene, but one can imagine that a promiscuous transactivator like Rta/Orf50 could reset the cellular transcription profile via interactions with RPB-J-κ or other via other mechanisms. This was recently explored by Nakamura et al. using a tetracycline-inducible

KSHV Rta/Orf50 cell line (Chang et al. 2005). Among the host genes that were induced by Rta/Orf50 via RPB-J-κ were CD21 and CD23a, which are involved in lymphocyte activation. KSHV Rta/Orf50 also inhibits the p53 tumor suppressor protein (Gwack et al. 2001b) and this capability is conserved among the primate rhadinoviruses, though not in MHV-68 (Damania et al. 2004). Overall, Rta/Orf50 is a key regulator of both viral and cellular transcription. An in-depth analysis of the molecular mechanisms of transactivation by Rta/Orf50 is described below.

7
Rta/Orf50 Function

7.1
Viral Promoters Transactivated by KSHV Rta/Orf50

KSHV Rta/Orf50 is a strong transactivator of many viral promoters in transient transfection assays. A summary of these promoters is reported in Table 1.

7.2
Viral Promoters Transactivated by MHV-68 Rta/Orf50

Ectopic expression of MHV-68 Rta/Orf50 in latently infected cells leads to expression of lytic proteins (as determined by Western blots of cell extracts probed with immune mouse serum) and is sufficient to drive viral reactivation of latently infected cells (Wu et al. 2001, 2000). The MHV-68 Orf57 promoter is transactivated by Rta/Orf50. MHV-68 RREs are found within the Orf57 promoter region of the viral genome, as they are in the KSHV Orf57 promoter (Pavlova et al. 2005). In addition, the MK3 promoter was reported to contain an MHV-68 Rta-responsive element (Coleman et al. 2003). Potential MHV-68 RREs bear significant homology to published KSHV RREs, but to date fewer MHV-68 promoters than KSHV promoters have been investigated in detail.

7.3
Viral Promoters Transactivated by RRV Orf50

RRV Orf50 has also been shown to transactivate promoter-reporter constructs in transient transfection assays. The R8, Orf57, and gB promoters of RRV were highly activated and the vIRF promoter was only slightly activated by Rta/Orf50 (DeWire et al. 2002; Lin et al. 2002). In contrast to KSHV, RRV Rta/Orf50 did not autoactivate its own promoter.

Table 1 Viral promoters transactivated by KSHV Rta/Orf50

KSHV viral promoter	Mechanism	Reference
Nut-1/PAN	C/EBP-α, DNA	Song et al. 2001; Wang et al. 2003b
Kaposin (K12)	DNA	Lukac et al. 1998; Song et al. 2003
K-bZip (K8)	Ap-1, DNA	Lukac et al. 1999; Wang et al. 2003a, 2004
MTA (ORF57)	RBP-J-κ, C/EBP-α, DNA	Duan et al. 2001; Liang et al. 2002; Lukac et al. 1999; Wang et al. 2003b
K6 (vMIP-1)	RBP-J-κ	Chang et al. 2005
K5	Unknown	Haque et al. 2000
K1	Unknown	Bowser et al. 2002
vIRF (K9)	Sp1	Chen et al. 2000; Ueda et al. 2002
ssDNA-binding (ORF6)	RBP-J-κ	Liang et al. 2002
DNA pol.-pross (ORF59)	Unknown	Nishimura et al. 2001
Thymidine kinase (ORF21)	Sp1	Zhang et al. 1998
vOX-2 (K14)	RBP-J-κ	Jeong et al. 2001; Liang and Ganem 2004
vGPCR (ORF74)	RBP-J-κ	Jeong et al. 2001; Liang and Ganem 2004
vIL-6 (K2)	DNA	Deng et al. 2002
LANA LT_i	RBP-J-κ	Matsumura et al. 2005; Lan et al. 2005; Staudt and Dittmer, 2006
Rta (ORF50)	Oct-1, Sp1, Sp3, C/EBP-α	Chen et al. 2000; Deng et al. 2000; Gradoville et al. 2000; Sakakibara et al. 2001; Wang et al. 2003a; Zhang et al. 1998

7.4
Viral Promoters Transactivated by HVS Orf50

The HVS Rta/Orf50a does transactivate its own promoter, as well as the DE Orf6, Orf57, and Orf9 viral promoters of HVS (Whitehouse et al. 1997; Thurau et al. 2000; (Walters et al. 2005, 2004; Byun et al. 2002). The HVS Rta/Orf50a contains an AT-hook DNA binding domain that is required for transactivation of the Orf6 and Orf57 promoters, and cellular C/EPB-α synergizes with HVS Rta/Orf50a to transactivate the Orf9 DNA polymerase promoter (Walters et al. 2004). HVS Rta/Orf50a was found to autoregulate its own promoter by use of a 36-bp RRE that has no significant homology to previously reported RREs of any of the gammaherpesviruses. The HVS RRE DNA sequence conferred Rta/Orf50a responsiveness to an enhancer-less SV40 minimal promoter (Walters et al. 2005).

8 Mechanisms of Rta/Orf50 Transactivation

8.1 KSHV Orf50-Responsive Elements and Direct DNA Binding

There have been many reports showing data that KSHV Rta/Orf50 protein transactivates viral promoters by direct DNA binding to sequences found within these promoters [termed Rta response elements (RREs) or Orf50 response element in *n* promoter ($50RE_n$)]. KSHV Rta/Orf50 can transactivate two delayed-early (DE) promoters, Orf57 and K8 (K-bZip), by direct DNA binding (Lukac et al. 2001; Duan et al. 2001; Song et al. 2002). The N-terminal 272 amino acids of Rta protein are sufficient to bind a 12-bp DNA sequence, 5′-AACAATAATGTT-3′, found within both DE promoters and termed the $50RE_{57}$. KSHV Rta/Orf50 also transactivates the PAN/nut-1 and K12 promoters (Chang et al. 2002) and was shown to directly bind DNA within these promoters. Intriguingly, the sequences of the 50RE found within the Pan/nut-1 and K12 promoters (5′-AAATGGGTGGCTAACCCCTACATAA-3′, PAN DNA sequence shown, K12 promoter sequence underlined) and the $50RE_{57}$ (5′-AACAATAATGTT-3′) share no significant homology. Yet another Orf50-responsive element was discovered in the vIL-6 promoter that contains a 26-bp sequence, 5′-AAACCCCGCCCCCTGGTGCTCACTTT-3′ (Deng et al. 2002b). Direct comparison of RRE-containing viral promoters revealed that the transcription initiation rate of these promoters, as opposed to transcript stability, is the major determinant of expression of these viral proteins (Song et al. 2003). Liao et al. reported that KSHV Rta/Orf50 forms oligomers and makes multiple contacts with a tandem array of phased A/T triplets in the configuration of $(A/T)_3$ $(G/C)_7$ repeats (Liao et al. 2003a). An RRE and TATA box was also found within the KSHV OriLyt, and this DNA region functioned as an Rta/Orf50-responsive promoter when cloned into a reporter vector (Wang et al. 2004b). This promoter regulated a late 1.4-kb polyadenylated mRNA that was sensitive to the viral DNA polymerase inhibitor *foscarnet* and has coding capacity for a 75 aa open reading frame whose gene product is of unknown function.

HMGB1 is a cellular protein belonging to the high-mobility group (HMG) box protein subfamily that affects transactivation function of both EBV Zta and EBV Rta (Ellwood et al. 2000; Mitsouras et al. 2002). HMG proteins are large chromosomal proteins thought to function to promote higher-order DNA-protein complexes by changing DNA conformation to be more easily accessible to transcriptional machinery. HMGB1 was recently shown to enhance direct DNA binding of KSHV Rta/Orf50 to RREs in vitro and to enhance transactivation functions of both KSHV and MHV-68 Rta/Orf50 proteins in transient transfection assays (Song et al. 2004).

KSHV Rta/Orf50 was reported to contain a protein domain that seemed to act in an autoregulatory fashion. This autoregulatory domain of KSHV Rta/Orf50 is contained within amino acids 521–534 and functions to control the direct DNA binding ability of the protein as well as protein stability (Chang and Miller 2004). Deletion of amino acids 521–534 or mutation of a basic motif (KKRK) at aa 527–530 dramatically enhanced DNA binding of Rta/Orf50. Although the DNA binding ability of the KKRK mutant was enhanced, its ability to transactivate the PAN promoter was impaired, suggesting that these two functions do not correlate synergistically on the PAN promoter. In addition, expression of autoregulation-domain mutants led to appearance of an alternative form of Rta, termed Orf50b, which showed decreased posttranslational modifications. At this time, investigations into the structure and posttranslational modifications of KSHV Rta/Orf50 are in the early stages, and we can expect more insights in the near future.

8.2
Interaction of Rta/Orf50 with RBP-J-κ

Although KSHV Rta/Orf50 binds to DNA containing the RREs described above, no obvious consensus sequence was found among other Rta/Orf50-responsive promoters: an observation that prompted many to look for an alternate mechanism of Rta transactivation. Ganem and colleagues used a yeast-two-hybrid approach to assay possible cellular binding proteins and found that the cellular protein RBP-J-κ (also called CSL or CBF-1) interacted with KSHV Rta/Orf50 (Liang et al. 2002). RBP-J-κ is a sequence-specific DNA binding protein and is the downstream effector of Notch signal transduction (Mumm and Kopan 2000). In uninfected cells RBP-J-κ functions as a transcriptional repressor until ligand-mediated Notch signaling occurs, which leads to the conversion of RBP-J-κ from a repressor to a transactivator of downstream cellular gene targets (such as HES, hairy and enhancer of split genes). KSHV can usurp the function of cellular RBP-J-κ without the requirement for Notch-ligand interaction, as the binding of KSHV Rta/Orf50 protein to RBP-J-κconverts RBP-J-κ from a transcriptional repressor to a transactivator. This mechanism has been demonstrated for the KSHV Mta/Orf57, SSB/Orf6, PAN/nut-1, vGPCR/K14, vMIP-1/K6, and LT_i promoters (Liang et al. 2002; Liang and Ganem 2003, 2004; Lan et al. 2005; Matsumura et al. 2005; Staudt and Dittmer 2006). Amino acids 170–400 of Rta/Orf50 mediate binding to RBP-J-κ, and there are two contiguous but distinct regions of RBP-J-κ to which Rta/Orf50 binds: one is within the central repressor domain and one is within the N-terminal domain of RBP-J-κ. It is striking that the central repressor domain of RBP-J-κ to which KSHV Rta/Orf50 binds is the same region to which Notch, the physiological effector protein of RBP-J-κ, binds as well. This suggests that Rta/Orf50 replaces Notch during lytic reactivation in B cells; however, this has yet to be demonstrated. Liang and Ganem propose that KSHV may employ the repressive function of RBP-J-κ bound to lytic promoters as a means of maintaining latency in the absence of appropriate reactivation stimuli (Liang and Ganem 2003). The interaction between KSHV Rta/Orf50 and cellular RBP-J-κ demonstrates an elegant mechanism the virus has developed to hijack an essential cellular signal transduction pathway as a means to obtain control over viral latency and lytic reactivation.

8.3
Interaction of Rta/Orf50 with Other Cellular Transcription Factors

The ubiquitously expressed cellular transcription factor Sp1 plays an important role in Rta/Orf50 transactivation of promoters, although to date there have been no reports of direct binding between KSHV Rta/Orf50 and cellular Sp1 protein. Sp1 binding sites within the *Orf50* promoter are essential for butyrate-induced Rta/Orf50 expression and lytic replication (Ye et al. 2005). Sp1 is also involved in Rta/Orf50 transactivation of other viral promoters, including vIRF/K9, thymidine kinase/Orf21, and Rta/Orf50 (Chen et al. 2000; Ye et al. 2005; Zhang et al. 1998).

Using a proteomics approach, Wang et al. identified a novel cellular protein, MGC2663, that stably bound to Rta/Orf50 after tandem immunoaffinity chromatography (Wang et al. 2001). MGC2663 was found to bind KSHV Rta/Orf50 and specifically synergized with Rta/Orf50 to activate viral transcription. The MGC2663 protein was previously uncharacterized, but Wang et al. found that it was expressed in every primate cell line tested and that it enhanced transactivation by Rta/Orf50, hence assigning MGC2663 the name *K-RBP* for KHSV Rta binding protein.

KSHV Rta/Orf50 also binds CBP [cyclic AMP (cAMP)-responsive element binding protein (CREB)-binding protein], which is a transcriptional coactivator that contains intrinsic histone acetyltransferase (HAT) activity (Gwack et al., 2003a, 2001a). Acetylation of histones is associated with relaxing nucleosomal structures, thus rendering regions of tightly packed DNA open and accessible to the transcriptional machinery. Binding of CBP to KSHV Rta/Orf50 increased the ability of Rta/Orf50 to transactivate viral promoters. Binding between these two proteins is mediated by the N-terminal basic domain of Rta/Orf50, which contains a conserved LxxLL CBP-binding motif, and the C/H3 domain and C-terminal transactivation domain of CBP. In addition, other cellular CBP-binding proteins, including CBP-BP and c-Jun, enhanced the ability of Rta/Orf50 to transactivate viral promoters. The transactivation function of EBV Rta is also enhanced by binding cellular CBP (Swenson et al. 2001). KSHV Rta/Orf50 also binds HDAC1, a cellular histone deacetylase, and this binding decreases the ability of Rta/Orf50 to transactivate viral promoters (Gwack et al. 2001a). In addition, Sp1 binding sites are involved in Rta/Orf50 transactivation of many viral promoters (see Table 1 for references). Sp1 itself binds to the CBP/p300 coactivator complex, and the activity of Sp1 is repressed by HDAC1 in the absence of viral infection (Doetzlhofer et al. 1999), suggesting a complex interplay among these important modifiers of basal transcription.

The development and characterization of KSHV Rta/Orf50 DNA-binding mutants, which display either enhanced or abolished DNA binding to RREs, has enabled classifications of Rta/Orf50-responsive promoters into either of two subgroups: those where Rta/Orf50 directly binds promoter DNA and those where Rta/Orf50 does not directly bind promoter DNA but rather transactivates by protein-protein interactions with cellular transcription factors, including RBP-J-κ (Chang et al. 2005). These mutants will no doubt facilitate further clarification of Rta/Orf50-responsive promoters as to which mechanism Rta/Orf50 transactivates and which cellular pathways are involved.

9 Repression of Rta/Orf50 Transactivation

Rta/Orf50 interacts with the viral early protein K-bZip, encoded by the K8 open reading frame. This protein-protein interaction leads to repression of Rta's ability to transactivate *in vitro* (Liao et al. 2003a, 2003b). K-bZip is a homologue of EBV Zta, and accumulating evidence suggests a role for K-bZip in DNA replication and transactivation. K-bZip repressed Rta/Orf50 transactivation of the Orf57/Mta and K8/Kb-Zip promoters but had no effect on Rta/Orf50's transactivation of the PAN/nut-1 promoter, demonstrating promoter-specific repression by K-bZip. The leucine zipper domain (aa 190–237) of K-bZip seems to be required for Rta/Orf50 binding (Liao et al. 2003b).

In addition to viral proteins, cellular interferon response factor 7 (IRF7) was reported to decrease transactivation of the Mta/Orf57 promoter by competing with Rta/Orf50 for binding to the RRE. Interferon-α was also shown to decrease transactivation of the Mta/Orf57 promoter by Rta/Orf50, with this process still involving IRF7 (Wang et al. 2005). This is consistent with the observation that interferon-α inhibits KSHV reactivation in PEL (Chang et al. 2000; Zoeteweij et al. 1999; Pozharskaya et al. 2004). In contrast, Hayward and colleagues have reported an E3 ubiquitin ligase activity of KSHV Rta/Orf50 and that Rta directs polyubiquitination of cellular IRF7 leading to proteosomal degradation of IRF7 and blockage of IRF7-mediated expression of type I interferon transcripts (Yu et al. 2005). Because Rta/Orf50 is the principal regulator of KSHV reactivation it seems logical that this protein serves as the nexus between virus and antivirus response.

The chemical compound methotrexate was shown to downregulate KSHV Rta/Orf50-mediated transactivation and inhibit lytic reactivation (Curreli et al. 2002). This is consistent with the previously reported antiviral activity of methotrexate for other herpesviruses (Lembo et al. 1999; Shanley and Debs 1989). Yet there are also reports that methotrexate induces EBV reactivation (Feng et al. 2004). Most likely the systemic antiviral activity of methotrexate is related to its action as an antimetabolite and dihydrofolate reductase inhibitor, while its effects on Rta/Orf50 transactivation may be a by-product of cellular stress signaling induced by imbalances in intracellular nucleotide pools.

Interestingly, the KSHV latency-associated nuclear antigen (LANA) protein is capable of repressing transcription of the Rta/Orf50 promoter (Lan et al. 2004). This repression is dependent on the presence of RBP-J-κ-responsive elements found within the Orf50 promoter (Lan et al. 2005). Lan et al. also showed that LANA bound Rta/Orf50 protein directly (Lan et al. 2005). RRV LANA (R-LANA), like KSHV LANA, represses the transactivation ability of RRV Rta/Orf50, and this repression is reversed by treatment with the histone deacetylase inhibitor TSA (DeWire and Damania 2005). The HVS C488 LANA also suppresses the HVS Rta/Orf50 (Schafer et al. 2003), which fits a model in which LANA, the latent transactivator, and Rta/Orf50, the lytic transactivator, counterbalance each other. This balance of power is evolutionarily conserved among the rhadinoviruses, although the governing molecular mechanisms may be different. The outcome of the Rta/Orf50–LANA/Orf73 power struggle eventually determines whether the virus persists latently or reactivates, which in turn results in a profound difference in overall viral persistence and pathogenesis within the infected host.

Acknowledgements Work in the authors' laboratory is supported by the NCI-designated UNC Lineberger Comprehensive Cancer Center, the NIH (Grants CA-109232, CA-110136, DE-017084), and a translational award from the Leukemia and Lymphoma Society. We apologize for any omissions due to space limitations and would like to thank Drs. Blossom Damania and Matthew Walters and all members of the Dittmer and Damania laboratories for helpful discussion.

References

Bais C, Santomasso B, Coso O, Arvanitakis L, Raaka EG, Gutkind JS, Asch AS, Cesarman E, Gershengorn MC, Mesri EA, Gerhengorn MC (1998). G-protein-coupled receptor of Kaposi's sarcoma-associated herpesvirus is a viral oncogene and angiogenesis activator [see comments] [published erratum appears in *Nature* 1998 Mar 12;392 (6672):210]. *Nature* 391 (6662):86–9.

Ballestas ME, Chatis PA, Kaye KM (1999). Efficient persistence of extrachromosomal KSHV DNA mediated by latency-associated nuclear antigen. *Science* 284 (5414):641–4.

Bechtel J, Grundhoff A, Ganem D (2005a). RNAs in the virion of Kaposi's sarcoma-associated herpesvirus. *J Virol* 79 (16):10138–46.

Bechtel JT, Winant RC, Ganem D (2005b). Host and viral proteins in the virion of Kaposi's sarcoma-associated herpesvirus. *J Virol* 79 (8):4952–64.

Boname JM, Coleman HM, May JS, Stevenson PG (2004). Protection against wild-type murine gammaherpesvirus-68 latency by a latency-deficient mutant. *J Gen Virol* 85 (Pt 1):131–5.

Boshoff C, Schulz TF, Kennedy MM, Graham AK, Fisher C, Thomas A, McGee JO, Weiss RA, O'Leary JJ (1995). Kaposi's sarcoma-associated herpesvirus infects endothelial and spindle cells. *Nat Med* 1 (12):1274–8.

Bowser BS, DeWire SM, Damania B (2002). Transcriptional regulation of the K1 gene product of Kaposi's sacroma-associated herpesvirus. J Virol 76 (24):1257–1283

Byun H, Gwack Y, Hwang S, Choe J (2002). Kaposi's sarcoma-associated herpesvirus open reading frame (ORF) 50 transactivates K8 and ORF57 promoters via heterogeneous response elements. *Mol Cells* 14 (2):185–91.

Cannon M, Cesarman E, Boshoff C (2006). KSHV G protein-coupled receptor inhibits lytic gene transcription in primary-effusion lymphoma cells via p21-mediated inhibition of Cdk2. Blood 107 (1):277–84.

Chang H, Gwack Y, Kingston D, Souvlis J, Liang X, Means RE, Cesarman E, Hutt-Fletcher L, Jung JU (2005). Activation of CD21 and CD23 gene expression by Kaposi's sarcoma-associated herpesvirus RTA *J Virol* 79 (8):4651–63.

Chang J, Renne R, Dittmer D, Ganem D (2000). Inflammatory cytokines and the reactivation of Kaposi's sarcoma- associated herpesvirus lytic replication. *Virology* 266 (1):17–25.

Chang PJ, Miller G (2004). Autoregulation of DNA binding and protein stability of Kaposi's sarcoma-associated herpesvirus ORF50 protein. *J Virol* 78 (19):10657–73.

Chang PJ, Shedd D, Gradoville L, Cho MS, Chen LW, Chang J, Miller G (2002). Open reading frame 50 protein of Kaposi's sarcoma-associated herpesvirus directly activates the viral PAN and K12 genes by binding to related response elements. *J Virol* 76 (7):3168–78.

Chen H, Lee JM, Wang Y, Huang DP, Ambinder RF, Hayward SD (1999). The Epstein-Barr virus latency BamHI-Q promoter is positively regulated by STATs and Zta interference with JAK/STAT activation leads to loss of BamHI-Q promoter activity. *Proc Natl Acad Sci USA* 96 (16):9339–44.

Chen J, Ueda K, Sakakibara S, Okuno T, Parravicini C, Corbellino M, Yamanishi K (2001). Activation of latent Kaposi's sarcoma-associated herpesvirus by demethylation of the promoter of the lytic transactivator. *Proc Natl Acad Sci USA* 98 (7):4119–24.

Chen J, Ueda K, Sakakibara S, Okuno T, Yamanishi K (2000). Transcriptional regulation of the Kaposi's sarcoma-associated herpesvirus viral interferon regulatory factor gene. *J Virol* 74 (18):8623–8634.

Ciufo DM, Cannon JS, Poole LJ, Wu FY, Murray P, Ambinder RF, Hayward GS (2001). Spindle cell conversion by Kaposi's sarcoma-associated herpesvirus: formation of colonies and plaques with mixed lytic and latent gene expression in infected primary dermal microvascular endothelial cell cultures. *J Virol* 75 (12):5614–26.

Coleman HM, Brierley I, Stevenson PG (2003). An internal ribosome entry site directs translation of the murine gammaherpesvirus 68 MK3 open reading frame. *J Virol* 77 (24):13093–105.

Cox MA, Leahy J, Hardwick JM (1990). An enhancer within the divergent promoter of Epstein-Barr virus responds synergistically to the R and Z transactivators. *J Virol* 64 (1):313–21.

Curreli F, Cerimele F, Muralidhar S, Rosenthal LJ, Cesarman E, Friedman-Kien AE, Flore O (2002). Transcriptional downregulation of ORF50/Rta by methotrexate inhibits the switch of Kaposi's sarcoma-associated herpesvirus/human herpesvirus 8 from latency to lytic replication. *J Virol* 76 (10):5208–19.

Damania B, Jeong JH, Bowser BS, DeWire SM, Staudt MR, Dittmer DP (2004). Comparison of the Rta/Orf50 transactivator proteins of gamma-2-herpesviruses. *J Virol* 78 (10):5491–9.

Damania B, Jung JU (2001). Comparative analysis of the transforming mechanisms of Epstein-Barr virus Kaposi's sarcoma-associated herpesvirus, herpesvirus saimiri. *Adv Cancer Res* **80,** 51–82.

Deng H, Chu JT, Rettig MB, Martinez-Maza O, Sun R (2002a). Rta of the human herpesvirus 8/Kaposi sarcoma-associated herpesvirus up-regulates human interleukin-6 gene expression. *Blood* 100 (5):1919–21.

Deng H, Song MJ, Chu JT, Sun R (2002b). Transcriptional regulation of the interleukin-6 gene of human herpesvirus 8 (Kaposi's sarcoma-associated herpesvirus). *J Virol* 76 (16):8252–64.

Deng H, Young A, Sun R (2000). Auto-activation of the rta gene of human herpesvirus-8/Kaposi's sarcoma-associated herpesvirus. *J Gen Virol* 81 (Pt 12):3043–8.

DeWire SM, Damania B (2005). The latency-associated nuclear antigen of rhesus monkey rhadinovirus inhibits viral replication through repression of Orf50/Rta transcriptional activation. *J Virol* 79 (5):3127–38.

DeWire SM, McVoy MA, Damania B (2002). Kinetics of expression of rhesus monkey rhadinovirus (RRV) and identification and characterization of a polycistronic transcript encoding the RRV Orf50/Rta RRV R8, and R8.1 genes. *J Virol* 76 (19):9819–31.

Dittmer DP, Gonzalez CM, Vahrson W, DeWire SM, Hines-Boykin R, Damania B (2005). Whole-genome transcription profiling of rhesus monkey rhadinovirus (RRV). *J Virol* 79 (13):8637–50.

Doetzlhofer A, Rotheneder H, Lagger G, Koranda M, Kurtev V, Brosch G, Wintersberger E, Seiser C (1999). Histone deacetylase 1 can repress transcription by binding to Sp1. *Mol Cell Biol* 19 (8):5504–11.

Duan W, Wang S, Liu S, Wood C (2001). Characterization of Kaposi's sarcoma-associated herpesvirus/human herpesvirus-8 ORF57 promoter. *Arch Virol* 146 (2):403–13.

Ellwood KB, Yen YM, Johnson RC, Carey M (2000). Mechanism for specificity by HMG-1 in enhanceosome assembly. *Mol Cell Biol* 20 (12):4359–70.

Ensser A, Thurau M, Wittmann S, Fickenscher H (2003). The genome of herpesvirus saimiri C488 which is capable of transforming human T cells. *Virology* 314 (2):471–87.

Fakhari FD, Dittmer DP (2002). Charting latency transcripts in Kaposi's sarcoma-associated herpesvirus by whole-genome real-time quantitative PCR *J Virol* 76 (12):6213–23.

Feederle R, Kost M, Baumann M, Janz A, Drouet E, Hammerschmidt W, Delecluse HJ (2000). The Epstein-Barr virus lytic program is controlled by the co-operative functions of two transactivators. *EMBO J* 19 (12):3080–9.

Feng WH, Cohen JI, Fischer S, Li L, Sneller M, Goldbach-Mansky R, Raab-Traub N, Delecluse HJ, Kenney SC (2004). Reactivation of latent Epstein-Barr virus by methotrexate: a potential contributor to methotrexate-associated lymphomas. *J Natl Cancer Inst* 96 (22):1691–702.

Foreman KE, Friborg J, Jr., Kong WP, Woffendin C, Polverini PJ, Nickoloff BJ, Nabel GJ (1997). Propagation of a human herpesvirus from AIDS-associated Kaposi's sarcoma [see comments]. *N Engl J Med* 336 (3):163–71.

Gao SJ, Deng JH, Zhou FC (2003). Productive lytic replication of a recombinant Kaposi's sarcoma-associated herpesvirus in efficient primary infection of primary human endothelial cells. *J Virol* 77 (18):9738–49.

Godfrey A, Anderson J, Papanastasiou A, Takeuchi Y, Boshoff C (2005). Inhibiting primary effusion lymphoma by lentiviral vectors encoding short hairpin RNA *Blood* 105 (6):2510–8.

Goodwin DJ, Walters MS, Smith PG, Thurau M, Fickenscher H, Whitehouse A (2001). Herpesvirus saimiri open reading frame 50 (Rta) protein reactivates the lytic replication cycle in a persistently infected A549 cell line. *J Virol* 75 (8):4008–4013.

Gradoville L, Gerlach J, Grogan E, Shedd D, Nikiforow S, Metroka C, Miller G (2000). Kaposi's sarcoma-associated herpesvirus open reading frame 50/Rta protein activates the entire viral lytic cycle in the HH-B2 primary effusion lymphoma cell line. *J Virol* 74 (13):6207–6212.

Grundhoff A, Ganem D (2004). Inefficient establishment of KSHV latency suggests an additional role for continued lytic replication in Kaposi sarcoma pathogenesis. *J Clin Invest* 113 (1):124–36.

Gwack Y, Baek HJ, Nakamura H, Lee SH, Meisterernst M, Roeder RG, Jung JU (2003a). Principal role of TRAP/mediator and SWI/SNF complexes in Kaposi's sarcoma-associated herpesvirus RTA-mediated lytic reactivation. *Mol Cell Biol* 23 (6):2055–67.

Gwack Y, Byun H, Hwang S, Lim C, Choe J (2001a). CREB-binding protein and histone deacetylase regulate the transcriptional activity of Kaposi's sarcoma-associated herpesvirus open reading frame 50. *J Virol* 75 (4):1909–17.

Gwack Y, Hwang S, Byun H, Lim C, Kim JW, Choi EJ, Choe J (2001b). Kaposi's sarcoma-associated herpesvirus open reading frame 50 represses p53-induced transcriptional activity and apoptosis. *J Virol* 75 (13):6245–8.

Gwack Y, Hwang S, Lim C, Won YS, Lee CH, Choe J (2002). Kaposi's Sarcoma-associated herpesvirus open reading frame 50 stimulates the transcriptional activity of STAT3. *J Biol Chem* 277 (8):6438–42.

Gwack Y, Nakamura H, Lee SH, Souvlis J, Yustein JT, Gygi S, Kung HJ, Jung JU (2003b). Poly(ADP-ribose) polymerase 1 and Ste20-like kinase hKFC act as transcriptional repressors for gamma-2 herpesvirus lytic replication. *Mol Cell Biol* 23 (22):8282–94.

Hall KT, Stevenson AJ, Goodwin DJ, Gibson PC, Markham AF, Whitehouse A (1999). The activation domain of herpesvirus saimiri R protein interacts with the TATA-binding protein. *J Virol* 73 (12):9756–7337.

Hong GK, Delecluse HJ, Gruffat H, Morrison TE, Feng WH, Sergeant A, Kenney SC (2004). The BRRF1 early gene of Epstein-Barr virus encodes a transcription factor that enhances induction of lytic infection by BRLF1. *J Virol* 78 (10):4983–92.

Jenner RG, Alba MM, Boshoff C, Kellam P (2001). Kaposi's sarcoma-associated herpesvirus latent and lytic gene expression as revealed by DNA arrays. *J Virol* 75 (2):891–902.

Jeong J, Papin J, Dittmer D (2001). Differential regulation of the overlapping Kaposi's sarcoma-associated herpesvirus vGCR (Orf74) and LANA (ORF73) promoters. *J Virol* 75 (4):1798–807

Jia Q, Wu TT, Liao HI, Chernishof V, Sun R (2004). Murine gammaherpesvirus 68 open reading frame 31 is required for viral replication. *J Virol* 78 (12):6610–20.

Klass CM, Krug LT, Pozharskaya VP, Offermann MK (2005). The targeting of primary effusion lymphoma cells for apoptosis by inducing lytic replication of human herpesvirus 8 while blocking virus production. *Blood* 105 (10):4028–34.

Krishnan HH, Naranatt PP, Smith MS, Zeng L, Bloomer C, Chandran B (2004). Concurrent expression of latent and a limited number of lytic genes with immune modulation and antiapoptotic function by Kaposi's sarcoma-associated herpesvirus early during infection of primary endothelial and fibroblast cells and subsequent decline of lytic gene expression. *J Virol* 78 (7):3601–20.

Lagunoff M, Bechtel J, Venetsanakos E, Roy A-M, Abbey N, Herndier B, McMahon M, Ganem D (2002). De novo infection and serial transmission of Kaposi's sarcoma-associated herpesvirus in cultured endothelial cells. *J Virol* 76 (5):2440–2448.

Lan K, Kuppers DA, Robertson ES (2005). Kaposi's sarcoma-associated herpesvirus reactivation is regulated by interaction of latency-associated nuclear antigen with recombination signal sequence-binding protein Jκ, the major downstream effector of the Notch signaling pathway. *J Virol* 79 (6):3468–78.

Lan K, Kuppers DA, Verma SC, Robertson ES (2004). Kaposi's sarcoma-associated herpesvirus-encoded latency-associated nuclear antigen inhibits lytic replication by targeting Rta: a potential mechanism for virus-mediated control of latency. *J Virol* 78 (12):6585–94.

Lan K, Kuppers DA, Verma SC, Sharma N, Murakami M, Robertson ES (2005). Induction of Kaposi's sarcoma-associated herpesvirus latency-associated nuclear antigen by the lytic transactivator RTA: a novel mechanism for establishment of latency. *J Virol* 79 (12):7453–65.

Lau LF, Nathans D (1985). Identification of a set of genes expressed during the G0/G1 transition of cultured mouse cells. *EMBO J* 4 (12):3145–51.

Lau LF, Nathans D (1987). Expression of a set of growth-related immediate early genes in BALB/c 3T3 cells: coordinate regulation with c-fos or c-myc. *Proc Natl Acad Sci USA* 84 (5):1182–6.

Lembo D, Cavallo R, Cornaglia M, Mondo A, Hertel L, Angeretti A, Landolfo S (1999). Overexpression of cellular dihydrofolate reductase abolishes the anticytomegaloviral activity of methotrexate. *Arch Virol* 144 (7):1397–403.

Liang Y, Chang J, Lynch SJ, Lukac DM, Ganem D (2002). The lytic switch protein of KSHV activates gene expression via functional interaction with RBP-Jκ (CSL), the target of the Notch signaling pathway. *Genes Dev* 16 (15):1977–89.

Liang Y, Ganem D (2003). Lytic but not latent infection by Kaposi's sarcoma-associated herpesvirus requires host CSL protein, the mediator of Notch signaling. *Proc Natl Acad Sci USA* 100 (14):8490–5.

Liang Y, Ganem D (2004). RBP-J (CSL) is essential for activation of the K14/vGPCR promoter of Kaposi's sarcoma-associated herpesvirus by the lytic switch protein RTA *J Virol* 78 (13):6818–26.

Liao W, Tang Y, Kuo YL, Liu BY, Xu CJ, Giam CZ (2003a). Kaposi's sarcoma-associated herpesvirus/human herpesvirus 8 transcriptional activator Rta is an oligomeric DNA-binding protein that interacts with tandem arrays of phased A/T-trinucleotide motifs. *J Virol* 77 (17):9399–411.

Liao W, Tang Y, Lin SF, Kung HJ, Giam CZ (2003b). K-bZIP of Kaposi's sarcoma-associated herpesvirus/human herpesvirus 8 (KSHV/HHV-8) binds KSHV/HHV-8 Rta and represses Rta-mediated transactivation. *J Virol* 77 (6):3809–15.

Lin SF, Robinson DR, Oh J, Jung JU, Luciw PA, Kung HJ (2002). Identification of the bZIP and Rta homologues in the genome of rhesus monkey rhadinovirus. *Virology* 298 (2):181–8.

Liu S, Pavlova IV, Virgin HWt., and Speck SH (2000). Characterization of gammaherpesvirus 68 gene 50 transcription. *J Virol* 74 (4):2029–37.

Lu F, Zhou J, Wiedmer A, Madden K, Yuan Y, Lieberman PM (2003). Chromatin remodeling of the Kaposi's sarcoma-associated herpesvirus ORF50 promoter correlates with reactivation from latency. *J Virol* 77 (21):11425–35.

Lukac DM, Garibyan L, Kirshner JR, Palmeri D, Ganem D (2001). DNA binding by Kaposi's sarcoma-associated herpesvirus lytic switch protein is necessary for transcriptional activation of two viral delayed early promoters. *J Virol* 75 (15):6786–99.

Lukac DM, Kirshner JR, Ganem D (1999). Transcriptional activation by the product of open reading frame 50 of Kaposi's sarcoma-associated herpesvirus is required for lytic viral reactivation in B cells. *J Virol* 73 (11):9348–61.

Lukac DM, Renne R, Kirshner JR, Ganem D (1998). Reactivation of Kaposi's sarcoma-associated herpesvirus infection from latency by expression of the ORF 50 transactivator, a homolog of the EBV R protein. *Virology* 252 (2):304–12.

Martinez-Guzman D, Rickabaugh T, Wu TT, Brown H, Cole S, Song MJ, Tong L, Sun R (2003). Transcription program of murine gammaherpesvirus 68. *J Virol* 77 (19):10488–503.

Matsumura S, Fujita Y, Gomez E, Tanese N, Wilson AC (2005). Activation of the Kaposi's sarcoma-associated herpesvirus major latency locus by the lytic switch protein RTA (ORF50). *J Virol* 79 (13):8493–505.

May JS, Coleman HM, Smillie B, Efstathiou S, Stevenson PG (2004). Forced lytic replication impairs host colonization by a latency-deficient mutant of murine gammaherpesvirus-68. *J Gen Virol* 85 (Pt 1):137–46.

McAllister SC, Hansen SG, Messaoudi I, Nikolich-Zugich J, Moses AV (2005). Increased efficiency of phorbol ester-induced lytic reactivation of Kaposi's sarcoma-associated herpesvirus during S phase. *J Virol* 79 (4):2626–30.

McKnight JL, Pellett PE, Jenkins FJ, Roizman B (1987). Characterization and nucleotide sequence of two herpes simplex virus 1 genes whose products modulate alpha-trans-inducing factor-dependent activation of alpha genes. *J Virol* 61 (4):992–1001.

Mitsouras K, Wong B, Arayata C, Johnson RC, Carey M (2002). The DNA architectural protein HMGB1 displays two distinct modes of action that promote enhanceosome assembly. *Mol Cell Biol* 22 (12):4390–401.

Moses AV, Fish KN, Ruhl R, Smith PP, Strussenberg JG, Zhu L, Chandran B, Nelson JA (1999). Long-term infection and transformation of dermal microvascular endothelial cells by human herpesvirus 8. *J Virol* 73 (8):6892–902.

Mumm JS, Kopan R (2000). Notch signaling: from the outside in. *Dev Biol* 228 (2):151–65.

Nakamura H, Lu M, Gwack Y, Souvlis J, Zeichner SL, Jung JU (2003). Global changes in Kaposi's sarcoma-associated virus gene expression patterns following expression of a tetracycline-inducible Rta transactivator. *J Virol* 77 (7):4205–20.

O'Connor C, M, Kedes DH (2006). Mass spectrometric analyses of purified rhesus monkey rhadinovirus reveal 33 virion-associated proteins. *J Virol* 80 (3):1574–83.

Paulose-Murphy M, Ha N-K, Xiang C, Chen Y, Gillim L, Yarchoan R, Meltzer P, Bittner M, Trent J, Zeichner S (2001). Transcription program of human herpesvirus 8 (Kaposi's sarcoma-associated herpesvirus). *J Virol* 75 (10):4843–4853.

Pavlova I, Lin CY, Speck SH (2005). Murine gammaherpesvirus 68 Rta-dependent activation of the gene 57 promoter. *Virology* 333 (1):169–79.

Pavlova IV, Virgin HWt., and Speck SH (2003). Disruption of gammaherpesvirus 68 gene 50 demonstrates that Rta is essential for virus replication. *J Virol* 77 (10):5731–9.

Polson AG, Wang D, DeRisi J, Ganem D (2002). Modulation of host gene expression by the constitutively active G protein-coupled receptor of Kaposi's sarcoma-associated herpesvirus. *Cancer Res* 62 (15):4525–30.

Pozharskaya VP, Weakland LL, Offermann MK (2004). Inhibition of infectious human herpesvirus 8 production by gamma interferon and alpha interferon in BCBL-1 cells. *J Gen Virol* 85 (Pt 10):2779–87.

Quinlivan EB, Holley-Guthrie EA, Norris M, Gutsch D, Bachenheimer SL, Kenney SC (1993). Direct BRLF1 binding is required for cooperative BZLF1/BRLF1 activation of the Epstein-Barr virus early promoter BMRF1. *Nucleic Acids Res* 21 (14):1999–2007.

Ragoczy T, Heston L, Miller G (1998). The Epstein-Barr virus Rta protein activates lytic cycle genes and can disrupt latency in B lymphocytes. *J Virol* 72 (10):7978–84.

Ragoczy T, Miller G (1999). Role of the epstein-barr virus RTA protein in activation of distinct classes of viral lytic cycle genes. *J Virol* 73 (12):9858–66.

Ragoczy T, Miller G (2001). Autostimulation of the Epstein-Barr virus BRLF1 promoter is mediated through consensus Sp1 and Sp3 binding sites. *J Virol* 75 (11):5240–51.

Renne R, Blackbourn D, Whitby D, Levy J, Ganem D (1998). Limited transmission of Kaposi's sarcoma-associated herpesvirus in cultured cells. *J Virol* 72 (6):5182–8.

Renne R, Zhong W, Herndier B, McGrath M, Abbey N, Kedes D, Ganem D (1996). Lytic growth of Kaposi's sarcoma-associated herpesvirus (human herpesvirus 8) in culture. *Nat Med* 2 (3):342–6.

Rickabaugh TM, Brown HJ, Wu TT, Song MJ, Hwang S, Deng H, Mitsouras K, Sun R (2005). Kaposi's sarcoma-associated herpesvirus/human herpesvirus 8 RTA reactivates murine gammaherpesvirus 68 from latency. *J Virol* 79 (5):3217–22.

Rochford R, Lutzke ML, Alfinito RS, Clavo A, Cardin RD (2001). Kinetics of murine gammaherpesvirus 68 gene expression following infection of murine cells in culture and in mice. *J Virol* 75 (11):4955–63.

Roizman B (1996). Herpesviridae. *In* "Virology" (BNFields DMKnipe, and PMHowley Eds.), Vol. 2, pp. 2221–2230. 2 vols. Lippincott-Raven Philadelphia.

Russo James J, Bohenzky Roy A, Chien M-C, Chen J, Yan M, Maddalena D, Parry JP, Peruzzi D, Edelman Isidore S, Chang Y, Moore Patrick S (1996). Nucleotide sequence of the Kaposi sarcoma-associated herpesvirus (HHV8). *Proc Natl Acad Sci USA* 93 (25):14862–14867.

Sakakibara S, Ueda K, Chen J, Okuno T, Yamanishi K (2001). Octamer-binding sequence is a key element for the autoregulation of Kaposi's sarcoma-associated herpesvirus ORF50/Lyta gene expression. *J Virol* 75 (15):6894–900.

Sarid R, Flore O, Bohenzky RA, Chang Y, Moore PS (1998). Transcription mapping of the Kaposi's sarcoma-associated herpesvirus (human herpesvirus 8) genome in a body cavity-based lymphoma cell line (BC-1). *J Virol* 72 (2):1005–12.

Saveliev A, Zhu F, Yuan Y (2002). Transcription mapping and expression patterns of genes in the major immediate-early region of Kaposi's sarcoma-associated herpesvirus. *Virology* 299 (2):301–14.

Schafer A, Lengenfelder D, Grillhosl C, Wieser C, Fleckenstein B, Ensser A (2003). The latency-associated nuclear antigen homolog of herpesvirus saimiri inhibits lytic virus replication. *J Virol* 77 (10):5911–25.

Seaman WT, Quinlivan EB (2003). Lytic switch protein (ORF50) response element in the Kaposi's sarcoma-associated herpesvirus K8 promoter is located within but does not require a palindromic structure. *Virology* 310 (1):72–84.

Shanley JD, Debs RJ (1989). The folate antagonist, methotrexate, is a potent inhibitor of murine and human cytomegalovirus in vitro. *Antiviral Res* 11 (2):99–106.

Shaw RN, Arbiser JL, Offermann MK (2000). Valproic acid induces human herpesvirus 8 lytic gene expression in BCBL-1 cells. *AIDS* 14 (7):899–902.

Song MJ, Deng H, Sun R (2003). Comparative study of regulation of RTA-responsive genes in Kaposi's sarcoma-associated herpesvirus/human herpesvirus 8. *J Virol* 77 (17):9451–62.

Song MJ, Hwang S, Wong W, Round J, Martinez-Guzman D, Turpaz Y, Liang J, Wong B, Johnson RC, Carey M, Sun R (2004). The DNA architectural protein HMGB1 facilitates RTA-mediated viral gene expression in gamma-2 herpesviruses. *J Virol* 78 (23):12940–50.

Song MJ, Li X, Brown HJ, Sun R (2002). Characterization of interactions between RTA and the promoter of polyadenylated nuclear RNA in Kaposi's sarcoma-associated herpesvirus/human herpesvirus 8. *J Virol* 76 (10):5000–13.

Sun R, Lin SF, Gradoville L, Yuan Y, Zhu F, Miller G (1998). A viral gene that activates lytic cycle expression of Kaposi's sarcoma-associated herpesvirus. *Proc Natl Acad Sci USA* 95 (18):10866–71.

Sun R, Lin SF, Staskus K, Gradoville L, Grogan E, Haase A, Miller G (1999). Kinetics of Kaposi's sarcoma-associated herpesvirus gene expression. *J Virol* 73 (3):2232–42.

Swenson JJ, Holley-Guthrie E, Kenney SC (2001). Epstein-Barr virus immediate-early protein BRLF1 interacts with CBP, promoting enhanced BRLF1 transactivation. *J Virol* 75 (13):6228–34.

Tang S, Zheng ZM (2002). Kaposi's sarcoma-associated herpesvirus K8 exon 3 contains three 5′-splice sites and harbors a K8.1 transcription start site. *J Biol Chem* 277 (17):14547–56.

Thurau M, Whitehouse A, Wittmann S, Meredith D, Fickenscher H (2000). Distinct transcriptional and functional properties of the R transactivator gene Orf50 of the transforming herpesvirus saimiri strain C488. *Virology* 268 (1):167–77.

Triezenberg SJ, LaMarco KL, McKnight SL (1988). Evidence of DNA: protein interactions that mediate HSV-1 immediate early gene activation by VP16. *Genes Dev* 2 (6):730–42.

Trus BL, Heymann JB, Nealon K, Cheng N, Newcomb WW, Brown JC, Kedes DH, Steven AC (2001). Capsid structure of Kaposi's sarcoma-associated herpesvirus, a gammaherpesvirus, compared to those of an alphaherpesvirus, herpes simplex virus type 1, and a betaherpesvirus, cytomegalovirus. *J Virol* 75 (6):2879–90.

Ueda K, Ishikawa K, Nishimura K, Sakakibara S, Do E, Yamanishi K (2002). Kaposi's sarcoma-associated herpesvirus (human herpesvirus 8) replication and transcription factor activates the K9 (vIRF) gen through two distinct cis elements by a non-DNA-binding mechanism. *J Virol* 76 (23):12044–54.

Walters MS, Hall KT, Whitehouse A (2004). The herpesvirus saimiri open reading frame (ORF) 50 (Rta) protein contains an AT hook required for binding to the ORF 50 response element in delayed-early promoters. *J Virol* 78 (9):4936–42.

Walters MS, Hall KT, Whitehouse A (2005). The herpesvirus saimiri Rta gene autostimulates via binding to a non-consensus response element. *J Gen Virol* 86 (Pt 3):581–7.

Wang J, Zhang J, Zhang L, Harrington W, Jr., West JT, Wood C (2005). Modulation of human herpesvirus 8/Kaposi's sarcoma-associated herpesvirus replication and transcription activator transactivation by interferon regulatory factor 7. *J Virol* 79 (4):2420–31.

Wang S, Liu S, Wu MH, Geng Y, Wood C (2001). Identification of a cellular protein that interacts and synergizes with the RTA (ORF50) protein of Kaposi's sarcoma-associated herpesvirus in transcriptional activation. *J Virol* 75 (24):11961–73.

Wang SE, Wu FY, Chen H, Shamay M, Zheng Q, Hayward GS (2004a). Early activation of the Kaposi's sarcoma-associated herpesvirus RTA, RAP, MTA promoters by the tetradecanoyl phorbol acetate-induced AP1 pathway. *J Virol* 78 (8):4248–67.

Wang SE, Wu FY, Fujimuro M, Zong J, Hayward SD, Hayward GS (2003a). Role of CCAAT/enhancer-binding protein α (C/EBPα) in activation of the Kaposi's sarcoma-associated herpesvirus (KSHV) lytic-cycle replication-associated protein (RAP) promoter in cooperation with the KSHV replication and transcription activator (RTA) and RAP *J Virol* 77 (1):600–23.

Wang SE, Wu FY, Yu Y, Hayward GS (2003b). CCAAT/enhancer-binding protein-α is induced during the early stages of Kaposi's sarcoma-associated herpesvirus (KSHV) lytic cycle reactivation and together with the KSHV replication and transcription activator (RTA) cooperatively stimulates the viral RTA, MTA, PAN promoters. *J Virol* 77 (17):9590–612.

Wang Y, Chong OT, Yuan Y (2004). Differential regulation of K8 gene expression in immediate-early and delayed-early stages of Kaposi's sarcoma-associated herpesvirus. *Virology* 325 (1):149–63.

Wang Y, Li H, Chan MY, Zhu FX, Lukac DM, Yuan Y (2004b). Kaposi's sarcoma-associated herpesvirus ori-Lyt-dependent DNA replication: *cis*-acting requirements for replication and ori-Lyt-associated RNA transcription. *J Virol* 78 (16):8615–29.

Whitehouse A, Carr IM, Griffiths JC, Meredith DM (1997). The herpesvirus saimiri ORF50 gene, encoding a transcriptional activator homologous to the Epstein-Barr virus R protein, is transcribed from two distinct promoters of different temporal phases. *J Virol* 71 (3):2550–4.

Wu TT, Tong L, Rickabaugh T, Speck S, Sun R (2001). Function of Rta is essential for lytic replication of murine gammaherpesvirus 68. *J Virol* 75 (19):9262–73.

Wu TT, Usherwood EJ, Stewart JP, Nash AA, Sun R (2000). Rta of murine gammaherpesvirus 68 reactivates the complete lytic cycle from latency. *J Virol* 74 (8):3659–67.

Xu Y, AuCoin DP, Huete AR, Cei SA, Hanson LJ, Pari GS (2005). A Kaposi's sarcoma-associated herpesvirus/human herpesvirus 8 ORF50 deletion mutant is defective for reactivation of latent virus and DNA replication. *J Virol* 79 (6):3479–87.

Ye J, Shedd D, Miller G (2005). An Sp1 response element in the Kaposi's sarcoma-associated herpesvirus open reading frame 50 promoter mediates lytic cycle induction by butyrate. *J Virol* 79 (3):1397–408.

Yu Y, Wang SE, Hayward GS (2005). The KSHV immediate-early transcription factor RTA encodes ubiquitin E3 ligase activity that targets IRF7 for proteosome-mediated degradation. *Immunity* 22 (1):59–70.

Zalani S, Holley-Guthrie E, Kenney S (1996). Epstein-Barr viral latency is disrupted by the immediate-early BRLF1 protein through a cell-specific mechanism. *Proc Natl Acad Sci USA* 93 (17):9194–9.

Zhang L, Chiu J, Lin JC (1998). Activation of human herpesvirus 8 (HHV-8) thymidine kinase (TK) TATAA-less promoter by HHV-8 ORF50 gene product is SP1 dependent. *DNA Cell Biol* 17 (9):735–42.

Zhong W, Wang H, Herndier B, Ganem D (1996). Restricted expression of Kaposi sarcoma-associated herpesvirus (human herpesvirus 8) genes in Kaposi sarcoma. *Proc Natl Acad Sci USA* 93 (13):6641–6.

Zhou FC, Zhang YJ, Deng JH, Wang XP, Pan HY, Hettler E, Gao SJ (2002). Efficient infection by a recombinant Kaposi's sarcoma-associated herpesvirus cloned in a bacterial artificial chromosome: application for genetic analysis. *J Virol* 76 (12):6185–96.

Zhu FX, Chong JM, Wu L, Yuan Y (2005). Virion proteins of Kaposi's sarcoma-associated herpesvirus. *J Virol* 79 (2):800–11.

Zhu FX, Cusano T, Yuan Y (1999). Identification of the immediate-early transcripts of Kaposi's sarcoma-associated herpesvirus. *J Virol* 73 (7):5556–67.

Zhu FX, Yuan Y (2003). The ORF45 protein of Kaposi's sarcoma-associated herpesvirus is associated with purified virions. *J Virol* 77 (7):4221–30.

Zhu J, Trang P, Kim K, Zhou T, Deng H, Liu F (2004). Effective inhibition of Rta expression and lytic replication of Kaposi's sarcoma-associated herpesvirus by human RNase P. *Proc Natl Acad Sci USA* 101 (24):9073–8.

Zoeteweij JP, Eyes ST, Orenstein JM, Kawamura T, Wu L, Chandran B, Forghani B, Blauvelt A (1999). Identification and rapid quantification of early- and late-lytic human herpesvirus 8 infection in single cells by flow cytometric analysis: characterization of antiherpesvirus agents. *J Virol* 73 (7):5894–902.

CTMI (2006) 312:101–136

Structure and Function of Latency-Associated Nuclear Antigen

S. C. Verma · K. Lan · E. Robertson (✉)

Department of Microbiology and Tumor Virology Program of the Abramson Comprehensive Cancer Center, University of Pennsylvania School of Medicine, 201E Johnson Pavilion, 3610 Hamilton Walk, Philadelphia, PA 19104, USA
erle@mail.med.upenn.edu

Abstract Latency-associated nuclear antigen (LANA) encoded by open reading frame 73 (ORF73) is the major latent protein expressed in all forms of KSHV-associated malignancies. LANA is a large (222–234 kDa) nuclear protein that interacts with various cellular as well as viral proteins. LANA has been classified as an oncogenic protein as it dysregulates various cellular pathways including tumor suppressor pathways associated with pRb and p53 and can transform primary rat embryo fibroblasts in cooperation with the cellular oncogene *Hras*. It associates with GSK-3β, an important modulator of Wnt signaling pathway leading to the accumulation of cytoplasmic β-catenin, which upregulates Tcf/Lef regulated genes after entering into the nucleus. LANA also blocks the expression of RTA, the reactivation transcriptional activator, which is critical for the latency to lytic switch, and thus helps in maintaining viral latency. LANA tethers the viral episomal DNA to the host chromosomes by directly binding to its cognate binding sequence within the TR region of the genome through its C terminus and to the nucleosomes through the N terminus of the molecule. Tethering to the host chromosomes helps in efficient partitioning of the viral episomes in the dividing cells. Disruptions of LANA expression led to reduction in the episomal copies of the viral DNA, supporting its role in persistence of the viral DNA. The functions known so far suggest that LANA is a key player in KSHV-mediated pathogenesis.

1 Introduction

The search for an etiological agent associated with Kaposi sarcoma (KS) lesions led to the identification of Kaposi sarcoma-associated herpesvirus (KSHV) as a likely candidate as biological cofactor in KS development (Chang et al. 1994). KS lesions comprise layers of spindle cells expressing lymphatic endothelial, smooth muscle, macrophage, and/or dendritic cell marker (Weninger et al. 1999). During the onset of KS, only 10% of the spindle cells are KSHV positive, but they progress to approximately 90% during the late stage of KS lesions (Boshoff et al. 1995; Staskus et al. 1997; Dupin et al. 1999; Sturzl et al. 1999). These infected cells express the latency-associated nuclear antigen (LANA) (Rainbow et al. 1997; Sarid et al. 1999).

The KSHV genome comprises an approximately 140-Kbp-long unique coding region that is flanked by multiple terminal repeats of 801 bp with high GC content (Russo et al. 1996). It is believed that the long unique region encodes for approximately 90 open reading frames (ORFs), although it is not known whether each of these encodes for a functional gene product (Russo et al. 1996). On infection, KSHV enters a quiescent period known as latent infection whereby the viral genome circularizes and exists as nuclear episomes through multiple host cell divisions (Renne et al. 1996). During this period, the virus limits gene expression to a specific subset of viral proteins, presumably to minimize the elicited host immune response (Zhong et al. 1996). This subset of protein includes LANA encoded by ORF73 (Kedes et al. 1997; Kellam et al. 1997; Rainbow et al. 1997), viral cyclin encoded by ORF72, and viral Fas-associated death domain (FAAD) interleukin-1β-converting enzyme (FLICE) inhibitory protein (v-FLIP) encoded by ORF71. These adjacent genes are cotranscribed on two polycistronic mRNAs, LT1 (LANA, v-cyclin, v-FLIP) or LT 2 (v-cyclin, v-FLIP) (Dittmer et al. 1998; Sarid et al. 1999; Talbot et al. 1999).

Analysis of the infected cells for latent viral infection by immunofluorescence and immunohistochemistry indicated that the LANA protein is present in almost all KSHV-infected cells of KS, primary effusion lymphoma (PEL), and multicentric Castleman disease (MCD) (Rainbow et al. 1997; Dupin et al. 1999; Katano et al. 2000; Parravicini et al. 2000). However, v-cyclin and v-FLIP have been shown to be predominantly expressed in PEL cell lines (Platt et al. 2000; Low et al. 2001). As the majority of the tumor cells are latently infected (Boshoff et al. 1995; Staskus et al. 1997; Dittmer et al. 1998; Chang et al. 2000), latent, not lytic, antigens are believed to be involved in tumorigenesis, although additional studies have shown that some lytic antigens including K1, GPCR, and ORF37, the DNA exonuclease involved in RNA degradation and initiation of host cell gene expression (Lee et al. 1998; Couty et al. 2001; Jeong et al. 2001; Nador et al. 2001; Glaunsinger and Ganem 2004), are important mediators of KSHV-induced pathogenesis. Thus the viral components expressed during this phase of the viral life cycle are most likely the players for disease manifestation (Schulz 1999). This places attention on LANA as a major effector of KSHV pathogenesis.

2
Expression of LANA

2.1
Tissue-Specific Expression of LANA

KSHV becomes latent after infection in the majority of the cells it infects. During this quiescent stage of the viral life cycle, the virus exists in the nucleus as an episomal DNA (Renne et al. 1996). The viral DNA is typically copied by the host DNA replication machinery during cell division. During mitosis, KSHV DNA is tethered through specific sequences on the TR and to histone H1 on host chromatin via the KSHV-encoded LANA. Therefore, LANA is thought to play a central role in maintaining viral latency (Ballestas et al. 1999b; Cotter and Robertson 1999; Ballestas and Kaye 2001).

In more than 90% of infected tumor cells, KSHV is locked into latent infection, which is likely to be stringently controlled. Protein expression is restricted to only a few viral genes. Most prominent among these is LANA, which is involved in episome maintenance and transcriptional regulation associated with viral oncogenesis (Zhong et al. 1996; Kedes et al. 1997; Kellam et al. 1997; Rainbow et al. 1997). LANA mRNA transcription is shown to be under the control of LANA promoter (LANAp) (Sarid et al. 1998; Jeong et al. 2001; Renne et al. 2001). LANAp is active in the absence of other viral gene products in transient assays and in latent PEL and KS tumor cells (Jeong et al. 2001, 2004; Renne et al. 2001). ORF73 (encoding LANA), ORF72 (v-cyclinD), and ORF71 (vFLICE) are all expressed from the same locus in polycistronic, differentially spliced mRNAs whose transcription is coordinately regulated by a common promoter; these three ORFs form a major latent cassette (Cesarman et al. 1996; Dittmer et al. 1998; Sarid et al. 1999; Talbot et al. 1999; Grundhoff and Ganem 2001). Moreover, the LANAp is bidirectional, controlling the constitutive expression of the latent genes to the left but lytic, TPA-inducible expression of the K14 and ORF74/vGPCR to the right (Fig. 1). Interestingly, LANA can autoactivate its own promoter during latent infection, so ensuring that the levels of LANA are adequate for maintenance of latency (Jeong et al. 2001, 2004). Recently, Pearce et al. and Samols et al. demonstrated the encoding of microRNAs from the latency transcriptional units, ORF71, -72, and -73 (Pearce et al. 2005; Samols et al. 2005). microRNAs are noncoding regulatory RNA molecules that bind to 3′-untranslated regions (UTRs) of mRNAs to either prevent their translation or induce their degradation, thereby regulating gene expression (Ambros 2004; Bartel 2004).

Generally, KSHV persists in KS and PEL tumor cells, which are of endothelial and B-cell origin, respectively. In asymptomatic carriers, KSHV is found in CD19-positive peripheral blood mononuclear cells (Mesri et al. 1996;

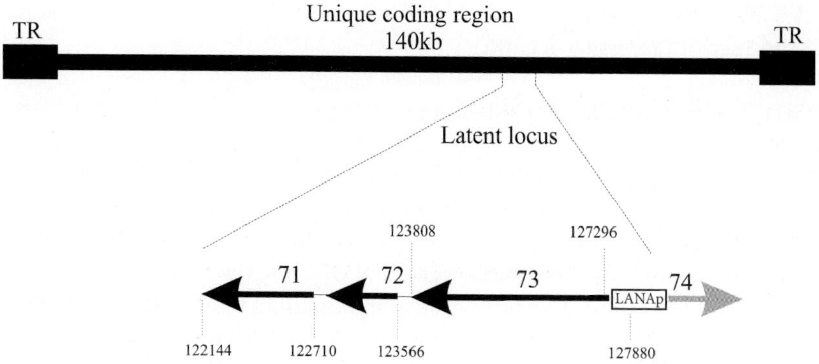

Fig. 1 Schematic diagram of the major KSHV latency cassette. Location of the major latency transcripts cluster in KSHV genome is shown. *Numbers* indicate nucleotide positions, based on the database sequence NC_003409 (Russo et al. 1996). *Black boldface arrows* represent ORF71, 72, and 73; *gray boldface arrow* represents ORF74

Jeong et al. 2002). However, KSHV can infect various human cell lines such as primary fibroblasts, 293 cells, and human primary vascular endothelial cell preparations with low efficiency as well as animal cells such as baby hamster kidney (BHK) cells in tissue culture conditions (Renne et al. 1998; Lagunoff et al. 2002; Bechtel et al. 2003). The KSHV LANA protein is predominantly expressed in the context of primary infection, whereas lytic genes are infrequently transcribed (Sun et al. 1999; Paulose-Murphy et al. 2001). Similarly, LANAp is active in a wide range of cell lines after transient transfection in culture (Jeong et al. 2001, 2004). However, in a SCID-hu Thy/Liv mouse model of de novo KSHV infection, LANA mRNA was transcribed in CD19-positive B cells but not in the more abundant T-cell subsets (Dittmer et al. 1999). Dittmer's group also found that a region extending to −1299 bp upstream of the LANA transcription start site was able to drive *lac* Z-reporter gene expression in several lines of transgenic mice. In agreement with KSHV's natural tropism, reporter gene expression was detected in CD19-positive B cells but not in CD3-positive T cells. Surprisingly, reporter gene expression was also detected in the kidney and in the liver at a lower level. In contrast to KS tumors, transgene expression was localized to kidney tubular epithelium rather than vascular endothelial cells (Jeong et al. 2002). This suggests that this particular promoter fragment contains all *ci* s-regulatory elements sufficient for B-cell specificity but is likely to lack those required for endothelial specificity. These in vivo studies reflect a conundrum in KSHV biology: In culture, KSHV is able to infect a variety of cell types indiscriminantly, whereas in healthy latent carriers KSHV is found in B lymphocytes (Lagunoff et al. 2002). The transgenic mouse experiments suggest that there are tissue-restricted mechanisms regulating LANA gene expression, providing one possible explanation for B cell-specific viral persistence (Jeong et al. 2002).

2.2
Levels of LANA mRNA in Infected Cells

Like other herpesvirus, KSHV also displays a tightly controlled program of gene expression. KSHV can enter one of two modes of infection: (a) latent infection, in which the viral genome persists in its host cell with restricted gene expression without cell destruction (Zhong et al. 1996), and (b) lytic infection, which produces virion progeny and destroys the host cell (Sarid et al. 1998). Lytic infection, in turn, can be divided into immediate-early, early, and late stages. With KSHV DNA fragments as a probe, three classes of differentially transcribed messages were defined in a KSHV-infected PEL cell line (Sarid et al. 1998). Class I mRNAs can be detected in untreated cultures and are not increased after 12-*O*-tetradecanoylphorbol-13-acetate (TPA) treatment, which reactivates the KSHV lytic replication cycle. Class II mRNAs can be detected in untreated cultures but are greatly increased by TPA treatment. Class III mRNAs can only be detected in TPA-treated cultures (Sarid et al. 1998). Translated into the customary herpesvirus classification, class III mRNAs correspond to lytic transcripts and class I mRNAs correspond to latent transcripts. Class II mRNAs appear to be a unique set of transcripts, because it is not clear whether the mRNAs detected in untreated cultures stem from the few percent of cells that undergo spontaneous lytic reactivation in culture. KSHV-specific cDNA arrays have also confirmed and expanded these data (Jenner et al. 2001; Paulose-Murphy et al. 2001). Although KSHV lytic transcription appears to be an all-or-nothing response, which proceeds to completion once orf50 (RTA) is expressed, KSHV can enter very distinct latency programs in terms of temporal regulation (latency type one or two) and tissue specificity (KS endothelial vs. B cell lineage). The type one latency mRNAs are resistant to the effects of the known chemical inducers such as ionomycin or butyrate, as well as TPA (Fakhari and Dittmer 2002). This suggests that LANAp, the promoter for LANA, v-FLIP, and v-cyclin, and the LANA-2/v-IRF-3 promoter share common regulatory features. The fact that latency type one mRNAs do not increase on viral reactivation in BCBL-1 cells suggests that their expression might be latency specific and that latency type one mRNAs might be expressed in latently infected cells to the exclusion of lytic mRNAs. Alternatively, latency type one mRNAs might be constitutively transcribed, but their promoters are unaffected or protected from the orf50-mediated upregulation of neighboring lytic transcription units (Fakhari and Dittmer 2002).

A number of studies showed that the mRNA level for LANA is unchanged after viral reactivation induced by TPA (Dittmer et al. 1998; Dittmer 2003). However, Jung and colleagues demonstrated opposite results by using a RTA tetracycline-inducible BCBL1 cell line (Nakamura et al. 2003). Under induced conditions, RTA expression is turned on, and in this system they were able to monitor changes in all viral transcripts. Interestingly, LANA mRNA levels are dramatically upregulated (Nakamura et al. 2003). RTA was shown to play a critical role in KSHV lytic reactivation, and forced expression of RTA was shown to initiate the full program of lytic replication (Lukac et al. 1998, 1999). These results suggest that LANA expression level may be changed during the natural reactivation process. In support of these studies, Chandran and colleagues showed that LANA transcripts were turned on by 48 h after infection (Krishnan et al. 2004).

3
Structure and Characterization of LANA

Sera obtained from HHV-8-infected patients were shown to recognize a specific nuclear antigen in latently infected PEL cell lines or KS tissue. This was characterized by a punctate nuclear immunofluorescence pattern (Lennette et al. 1996) (Fig. 2). Screening of cDNA libraries with serum from a HHV8-positive patient identified this nuclear antigen as the product of the viral gene open reading frame 73 (ORF73), and the encoded protein was designated LANA (Kedes et al. 1997; Kellam et al. 1997; Rainbow et al. 1997). LANA is expressed from a latently controlled 5.32-kb transcript that also encodes the viral cyclin (v-Cyc) and v-FLIP (Dittmer et al. 1998). The 5.32-kb latent transcript is spliced to form a 1.7-kb transcript that encodes only v-Cyc and v-FLIP (Kellam et al. 1997; Dittmer et al. 1998). The predicted LANA protein is 1,162 amino acids (aa) and has a theoretical molecular mass of 135 kDa. In contrast to LANA's predicted molecular size, this protein has a much higher molecular size of 220–230 kDa when analyzed by Western blotting (Gao et al. 1996; Kellam et al. 1997; Rainbow et al. 1997). LANA is divided into three distinct domains: an N-terminal 337-amino acid domain; an extremely hydrophilic central domain of 585 aa consisting of multiple repeat elements predominantly containing the charged polar amino acids glutamine, glutamic acid, and asparagine; and a C-terminal 240-amino acid domain (Russo et al. 1996). In contrast to the central domain, the N- and C-terminal domains are rich in basic amino acids (Fig. 3). The differently charged domains of the protein may in part explain the aberrant running of LANA on sodium dodecyl sulfate (SDS)-polyacrylamide gels.

To date, a number of studies have investigated the structural and functional aspects of LANA. In strain BC-1, LANA contains 1,162 aa (Russo et al. 1996) and has an apparent molecular mass of 224–234 kDa on SDS polyacrylamide gels (Rainbow et al. 1997). However, its size varies greatly among HHV-8

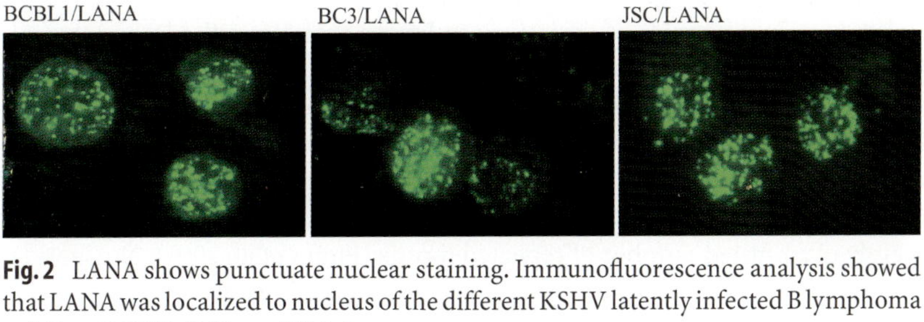

Fig. 2 LANA shows punctuate nuclear staining. Immunofluorescence analysis showed that LANA was localized to nucleus of the different KSHV latently infected B lymphoma cell lines BCBL1, BC3, and JSC-1

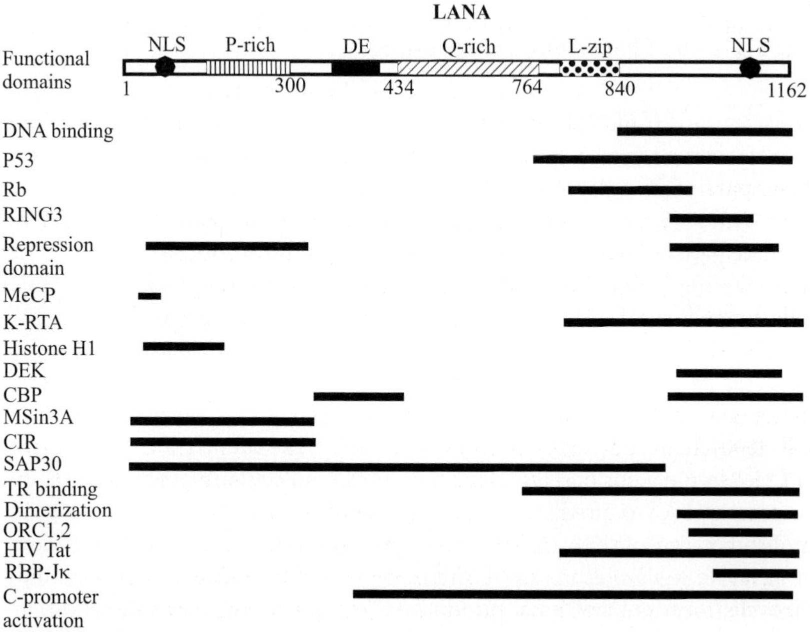

Fig. 3 Schematic showing the structure and functional domains of LANA. LANA is a 1162-amino acid protein nuclear protein (BC-1 strain). Numbers indicate the amino acids (*aa*). The putative domains include an N-terminal proline-rich domain (*P-rich*); aspartic acid, glutamic acid repeat region (*DE*); glutamine-rich domain (*Q-rich*); leucine zipper (*LZ*); LANA also has two nuclear localization signals (*NLS*) in both the N- and C-terminal regions. The approximate position of each functional domain is shown by a *black bar*, corresponding to the function shown in the column on the left (Cotter and Robertson 1999; Friborg et al. 1999; Platt et al. 1999; Krithivas et al. 2000; Schwam et al. 2000; Cotter et al. 2001; Groves et al. 2001; Hyun et al. 2001; Lim et al. 2001; Krithivas et al. 2002; Mattsson et al. 2002; Shinohara et al. 2002; Borah et al. 2004; Lan et al. 2004, 2005a)

isolates, mainly because of variations in length that affect the large internal acidic domain (Neipel et al. 1997; Rainbow et al. 1997; Gao et al. 1999; Glenn et al. 1999; Zhang et al. 2000). Sequence alignments of LANA from BC1, BBG1, KS1, and KS2 cell lines showed that LANA can be divided into two conserved regions, spanning aa 1 to 337 (the unique left region) and 798 to 1036 (the unique right region), that are separated by an internal region composed mainly of a limited number of repeated motifs (Piolot et al. 2001). Nucleotide alignments revealed that only four nucleotide positions are variable in the unique left region, whereas the unique right region was completely conserved

(Piolot et al. 2001). In contrast, high variability was observed in the central acidic domain, at both the nucleotide and amino acid levels. The region encoding aa 338–797 is composed of a series of short conserved tandem repeats whose number is characteristic of each HHV-8 isolate. This region is interrupted by a conserved nonrepeated sequence (the unique central region) encompassing aa 439–452 (Piolot et al. 2001).

Studies have shown that LANA is localized in the nuclei of KSHV latently infected cells and that it interacts with the border of heterochromatin during interphase and with chromosomes during mitosis (Szekely et al. 1998, 1999; Ballestas et al. 1999b; Cotter and Robertson 1999). A unique nuclear localization signal (NLS) sequence was mapped within aa 24–30 of LANA (Szekely et al. 1999; Schwam et al. 2000; Piolot et al. 2001). The LANA NLS is highly homologous to the NLSs that have been identified in EBV EBNA1 (Schwam et al. 2000). In particular, these sequences have a phosphorylation site for protein kinase A and a cdc2-type kinase in common. Because phosphorylation of residues close to or within the NLS has been shown to modify the nuclear import of numerous cellular proteins, this suggests that LANA nuclear transport may be regulated by a similar kinase(s), notably during the course of the cell cycle (Boulikas 1993). Except for the N terminus, which was shown to be localized to nucleus, Wilson and colleagues demonstrated that both the N and C termini of LANA can localize to the nucleus. They showed that the N terminus gives a diffuse pattern in interphase nuclei typical of many transcription factors; however the C terminus accumulates as discrete nuclear speckles reminiscent of the punctate pattern shown by full-length LANA (Schalling et al. 1995; Friborg et al. 1999; Glenn et al. 1999).

It has also been shown that LANA self-associates to form dimers in the absence of viral DNA or other viral gene products (Schwam et al. 2000). Self-association is mediated by the C terminus of LANA, and, interestingly, the sequences required for self-association of LANA are conserved. The amino acid sequence of the C terminus of KSHV LANA with the similar regions of LANA homologs encoded by four other rhadinoviruses, ateline herpesvirus-3 (Ensser et al. 1997), herpesvirus saimiri (Albrecht et al. 1992), macaque rhadinovirus 17577 (Searles et al. 1999), and murine herpesvirus 68 (Virgin et al. 1997), when compared showed that the greatest conservation of the C-terminal sequence corresponds to residues 934–1072 of KSHV LANA, with a more limited conservation between residues 885 and 922 (Russo et al. 1996). Clearly, the C terminus of LANA (residues 884–1089) is required for self-association and includes the majority of the residues shown to be conserved between members of the Rhadinovirus family (Schwam et al. 2000).

Like EBNA1 of EBV, LANA tethers the viral episomes to host chromosomes, ensuring that the viral genome is segregated to daughter cells on mitosis (Cotter and Robertson 1999; Schwam et al. 2000). LANA interacts with the mitotic chromosomes in KSHV-infected cells. The major chromosome-binding site of LANA was mapped to the N terminus between aa 1 and 32 (Schwam et al. 2000). This region also contains the LANA NLS; therefore it is possible that nuclear targeting and binding to mitotic chromosomes were functionally related (Schwam et al. 2000).

4
The Multifunctional Role of LANA

4.1
LANA as a Transcriptional Modulator

Numerous studies from a number of laboratories have demonstrated that LANA can activate as well as repress transcription (Groves et al. 2001; Hyun et al. 2001; Jeong et al. 2001; Knight et al. 2001; Lim et al. 2001; Renne et al. 2001; An et al. 2002). LANA has also been shown to autoactivate transcription of the major latent promoter (Renne et al. 2001; Jeong et al. 2004), suggesting a role in maintenance of expression of these latent transcripts while the majority of the viral genome remains silent. Transcriptional activation by LANA is probably independent of direct DNA binding as no specific LANA binding sequences have been identified in the major promoter elements (Russo et al. 1996). LANA activation of transcription is directed by a wide array of simple, synthetic promoters containing binding sites for a range of cellular proteins including ATF, AP-1, CAAT, or Sp1, linked to a TATA box (Renne et al. 2001). Although the TATA box alone was shown to be activated by LANA in these studies (Renne et al. 2001), activation of the cellular IL-6 promoter by LANA required both a TATA box and an upstream AP-1 element (An et al. 2002). Additionally, activation of the HIV-1 LTR in BJAB cells requires both the TATA box and the core enhancer elements (NF–IL-6, Ets, NF-κB, and Sp1 sites) (Hyun et al. 2001). However, LANA can activate the HIV-1 LTR independently of the core enhancer if the Tat protein is coexpressed; this effect is mediated by the C-terminal 400 aa of LANA, which includes the third repeat region (Hyun et al. 2001). Truncated LANA containing the internal repeat domain (IRD) plus the C terminus activates the transcription of an EBV latency promoter containing PU.1, ATF, and Sp1 elements (Groves et al. 2001). Mechanistically, LANA can modify the DNA binding activity of Sp1 to activate the transcription of the telomerase reverse transcriptase promoter (Knight et al. 2001). Furthermore, LANA was shown to interact with Sp1 directly as well as associated in KSHV-infected cells, suggesting an upregulation of hTERT by directly interacting with Sp1 at the transcription domains (Verma et al. 2004).

LANA can also contribute to broad repressive effects on transcription. In Cos cells, LANA represses the transcriptional activity of NF-κB. Additionally, repression of the HIV-1 LTR suggested that the interaction of LANA with the cellular transcription apparatus is cell specific (Renne et al. 2001). LANA also represses transcription controlled by the ATF4/CREB2 protein, independently of ATF4 DNA binding. This effect is mediated by LANA aa 751–1162, which includes the third repeat region (Lim et al. 2001). Similarly, the first acidic repeat (aa 340–431) of LANA is one of two domains required to competitively bind the cellular cyclic AMP-responsive element binding protein (CREB) binding protein (CBP) and block its histone acetyltransferase activity, thus inhibiting the ability of CBP to coactivate transcription with c-fos (Lim et al. 2001). This mechanism reflects the general ability of LANA to repress transcription driven by other cellular and viral CBP-dependent transactivators, including the KSHV lytic switch protein ORF50/Rta (Gwack et al. 2001). To determine the role of LANA on cellular pathways, an overexpression system followed by a microarray analysis revealed the induction of a large number of genes in LANA-expressing cells including pRb and p53 signaling molecules (An et al. 2005).

4.1.1
LANA Contributes to Maintenance of Viral Latency by Regulating RTA Function

4.1.1.1
LANA Represses RTA Expression

Recently, studies in our laboratory showed that LANA is capable of maintaining viral latency through repression of the transcriptional activity of RTA (replication and transcription activator) promoter (Lan et al. 2004). RTA is encoded by ORF50 of KSHV and is an immediate-early protein. Exogenous expression of RTA from a heterologous promoter or induction of endogenous RTA expression with chemical inducers can initiate lytic reactivation of KSHV (Lukac et al. 1998, 1999). HHV-8 RTA is the key transcriptional regulator that controls the switch from viral latency to lytic replication and is sufficient to drive the viral lytic cascade to completion, resulting in the production of viral progeny (Lukac et al. 1998, 1999). The RTA protein is made as an IE gene product; it autoregulates its own expression, and activates the transcription of various viral and cellular genes (Deng et al. 2000). It was shown that LANA downmodulates Rta's promoter activity in transient reporter assays, thus repressing the Rta-mediated transactivation (Lan et al. 2004). This eventually resulted in a decrease in the production of KSHV progeny (Lan et al. 2004). LANA physically interacts with Rta in vitro and associates with RTA in KSHV-infected cells (Lan et al. 2004). These results demonstrated that LANA is capable of inhibiting viral lytic replication by repressing transcription and antagonizing the function of RTA (Lan et al. 2004). Therefore LANA provides a critical role in controlling the switch between viral latency and lytic replication (Lan et al. 2004). Similarly, Schafer and colleagues in a recent report also indicated that the LANA homologue in HVS can repress the expression of ORF50 encoded by HVS (Schafer et al. 2003). Most recently, DeWire and Damania also showed that the LANA homologue of rhesus monkey rhadinovirus (RRV) can dramatically decrease the transactivation function of RRV Orf50 (RTA) and inhibit RRV lytic replication in vitro (DeWire and Damania 2005). Interestingly, a recent report showed that KSHV could undergo spontaneous lytic replication, and RTA transcription level was increased when LANA expression was knocked down with specific siRNA (Godfrey et al. 2005). These findings provide one explanation as to why KSHV is predominantly maintained in the latent state in infected cells as well as why after the primary lytic infection, KSHV rapidly establishes latent infection. Moreover, it may be a common and conserved mechanism whereby LANA inhibits lytic reactivation through downmodulation of the immediate-early transactivator, RTA expression. Further, it has been shown that LANA downregulates RTA's promoter activity through the RBP-Jκ binding sites within the promoter (Lan et al. 2005a). RBP-Jκ is a sequence-specific DNA-binding protein in the CSL/CBF-1 family, normally thought to function as a major repressor of transcription in response to activation of the Notch pathway signaling. Our results once again showed that viral proteins can regulate gene expression through direct targeting of important cellular signaling pathways. Interestingly, Liang and colleagues also demonstrated that RTA can functionally interact with RBP-Jκ and mimic cellular Notch function to activate lytic gene expression (Liang et al. 2002; Liang and Ganem 2003).

4.1.1.2
RTA Upregulates LANA After De Novo Infection

Most recently, it has been shown that RTA is detected in association with KSHV viral particles (Bechtel et al. 2005; Lan et al. 2005a, 2005b). This raises the possibility that RTA may contribute to latency establishment by inducing LANA expression upon entry. Interestingly, it was shown that RTA activates the LANA promoter and induces LANA expression in transient transfection experiments (Lan et al. 2005b). Moreover, the transcription of RTA correlates with that of LANA during the early stages of de novo infection of KSHV and LANA transcription is responsive to induction of RTA expression (Lan et al. 2005b). Additionally, the RBP-Jκ binding site within the LANA promoter is also shown to be critical for regulation by RTA known to be associated

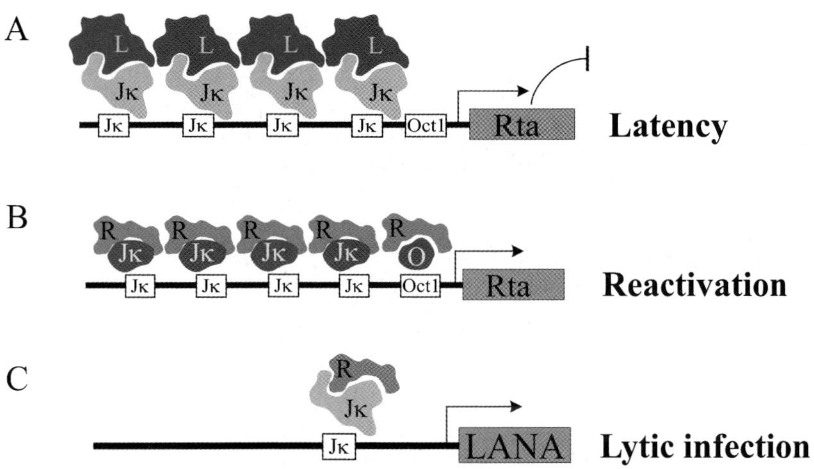

Fig. 4A–C Hypothetical model to show interplay between LANA and RTA. **A** During the latent infection of KSHV, LANA represses RTA expression through physical interaction with RBP-Jκ. **B** On reactivation of the virus, RTA can autoactivate its own expression then initiate the cascade of lytic gene expression. **C** In early stage of primary infection, RTA can activate LANA expression through functional association with RBP-Jκ

with a number of cellular regulatory proteins including RBP-Jκ. In RBP-Jκ-knockout cells KSHV was unable to establish latent infection efficiently compared to wild-type cells (Lan et al. 2005b). These findings therefore suggest that RTA can contribute to the establishment of latency through activation of LANA expression in the early stages of infection. This represents a feedback loop whereby LANA and Rta regulate each other and is likely to be, at least in part, a key event in establishment of KSHV latent infection. A potential model can be proposed that suggests a feedback regulatory loop between RTA and LANA and implies that the interplay between RTA and LANA could be the master control for the switch between lytic replication and latency of KSHV utilizing the major cellular Notch signaling pathway through targeting of the RBP-Jκ effector protein (Fig. 4).

4.2 Deregulation of Cellular Factors by LANA

4.2.1 Role of LANA on Action of GSK-3β and β-Catenin in Cell Cycle Deregulation

LANA, which is expressed in all forms of KS tumors, modulates distribution of glycogen synthase kinase (GSK)-3β, a negative regulator of β-catenin and Wnt signaling (Fujimuro et al. 2003). In normal cells cytosolic amounts of β-catenin are normally limiting because it leads to targeted degradation after phosphorylation with GSK-3β (Polakis 2000). The nuclear effector protein β-catenin binds to the cytoplasmic domain of E-cadherin, which is the transmembrane protein present at adhesion junctions, and also to the cytoplasmic complex comprised of the serine-threonine kinase GSK-3β, the tumor suppressor adenomatous polyposis coli (APC) and the scaffold protein Axin (Polakis 2000). Association of β-catenin with the above complex results in phosphorylation of β-catenin by GSK-3β through association with the F-box protein β-TrCP leading to ubiquitination and thus proteosomal degradation by the specific E5 ligases (Kitagawa et al. 1999). Activation of Wnt signaling by binding of Wnt ligand to a frizzled (Fz)/low-density lipoprotein receptor-related protein (LRP) complex activates the cytoplasmic protein Disheveled (Dsh in *Drosophila* and Dvl in vertebrates), possibly through phosphorylation by casein kinase 1 (CK1) and casein kinase 2 (CK2) (Willert et al. 1997; Sakanaka et al. 1999). Dsh/Dvl then inhibits the activity of the multiprotein complex β-catenin-Axin-APC-GSK-3β, which targets β-catenin for proteosomal degradation by phosphorylation, resulting in accumulation of cytosolic forms of β-catenin (Polakis 2000). Stabilized β-catenin translocates into the nucleus and binds to members of the T-cell factor (Tcf)/lymphoid enhancing factor (Lef) family of DNA binding proteins, leading to transcription of Wnt target genes (Boshoff 2003).

Fujimuro and Hayward reported increased levels of β-catenin in PEL cells (BC2, BC3, and BCBL1) (Fujimuro et al. 2003; Fujimuro and Hayward 2004). Immunohistochemical detection of β-catenin showed elevated expression of β-catenin in KS tumors compared to the surrounding normal tissues (Fujimuro et al. 2003). LANA, which is primarily expressed in all KSHV-infected cells, induced the accumulation of β-catenin (Fujimuro et al. 2003). The accumulation of β-catenin is possibly through interaction of LANA with GSK-3β, which is important for phosphorylation of β-catenin and its degradation. LANA changes the distribution of GSK-3β from predominantly cytoplasmic to nuclear, thus preventing phosphorylation of cytoplasmic β-catenin, which in turn translocates to the nucleus and forms a complex with the Tcf and Lef family of transcription factors to activate genes containing Tcf and Lef binding sites (He et al. 1998; Barker et al. 2000; Fujimuro and Hayward 2003, 2004; Fujimuro et al. 2003). These target genes include *MYC*, *JUN*, and *CCND1*, known to be critical in regulating cell proliferation (cyclin D1) (He et al. 1998; Boshoff 2003; Fujimuro et al. 2003) (Fig. 5).

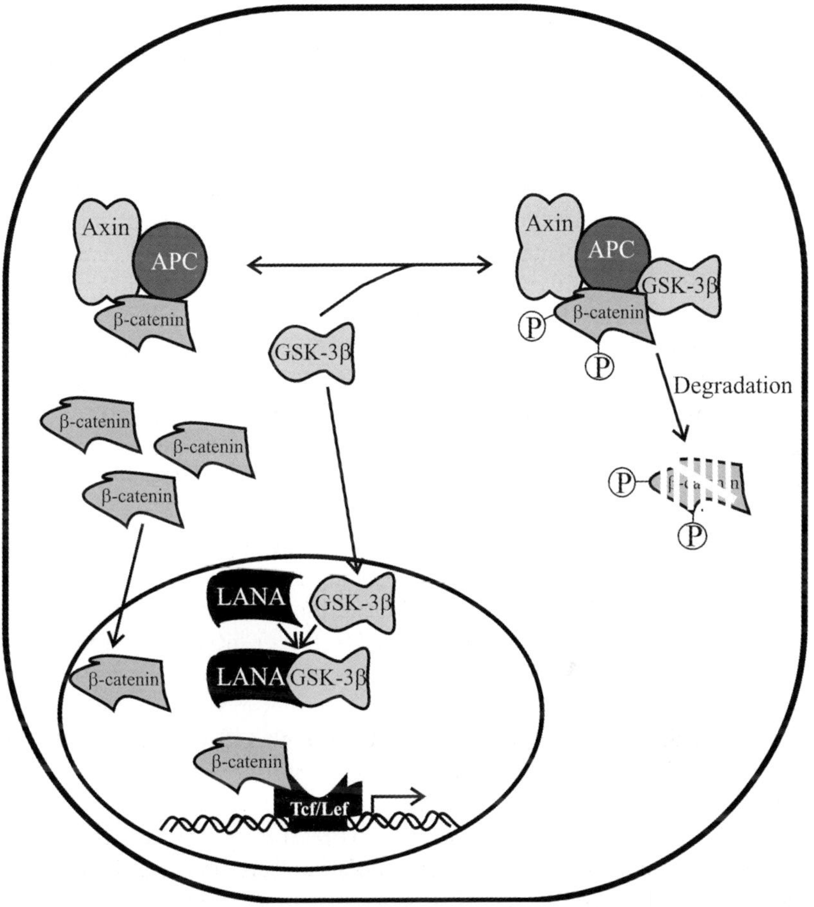

Fig. 5 Model showing modulation of Wnt signaling pathways through interaction with LANA. LANA binds to GSK-3β and translocates this to the nucleus, leading to the depletion of cytoplasmic GSK-3β. This results in accumulation of unphosphorylated β-catenin, which in turn translocates to the nucleus and associates with the LEF and TCF family of transcription factors, thus modulating gene expression controlled by these transcription factors

4.2.2
LANA Regulates p53 Activity

p53 was the first identified tumor suppressor gene, and the induction of this pathway eliminates or inhibits the proliferation of abnormal cells, thereby preventing neoplastic development (Hollstein et al. 1991; Friend 1994; Levine

1997). Mutations identified in the p53 gene are estimated to occur in 50% of all cancers, and loss of p53 function leads to cell transformation and oncogenesis (Hollstein et al. 1991; Levine 1997). Besides mutation in the gene, several viral oncoproteins have been shown to interact with p53 and modulate its transcriptional activity (Chen et al. 1996; Ko and Prives 1996). HPV-16-encoded E6 binds with ubiquitin ligase to mediate p53 ubiquitination and degradation (Werness et al. 1990). SV40 large T antigen modulates p53-mediated apoptosis by physically interacting with and inactivating p53 (Jay et al. 1981). More recently, Nabel and colleagues have shown that LANA interacts with p53 and downregulates its transcriptional activity as well as blocking its ability to induce apoptosis (Friborg et al. 1999). Modulation of p53 transcriptional activity by LANA is neither through degradation nor inhibition of its ability to bind DNA. This raises the possibility that additional factors are required for its function (Friborg et al. 1999). The region of LANA that binds to p53 was mapped to aa 441–1162, which is further narrowed down to aa 751–1162 (Borah et al. 2004) (Fig. 1). The herpesvirus positional homologue of LANA also binds to p53 and most probably performs a similar function (Borah et al. 2004). The domain of p53 that maps to bind KSHV-encoded LANA and its HVS ORF73 homologue is the tetramerization domain essential for enzymatic activity, but the presence of LANA did not affect the tetramerization of p53, suggesting that inactivation of p53 function by LANA may involve another potential mechanism besides that which involves tetramer disruption (Borah et al. 2004). Recently, Si and Robertson reported that LANA-expressing cells had increased chromosomal instability, detected by the presence of increased multinucleation, micronuclei, and aberrant centrosomes (Si and Robertson 2006). They proposed that repression of p53 functional activity contributes to the chromosomal instability, thereby facilitating KSHV-mediated pathogenesis (Si and Robertson 2006).

4.2.3
LANA Inhibits Rb-Induced Growth Arrest

Retinoblastoma (pRB) acts as a negative regulator of cell cycle progression by blocking the cells from entering into S phase from G_1 (Sherr and McCormick 2002). In early G_1, pRB is hypophosphorylated and associates with specific members of the E2F transcription factor, finally resulting in transcriptional repression (Kiess et al. 1995). Cell cycle progression from G_1 to S phase occurs when pRB is hyperphosphorylated by complexes of D-type cyclins/CDK4/CDK6, thereby releasing the E2F transcription factor. This results in activation of the DNA replication machinery and nucleotide synthesis to prepare the cell for division (Dyson 1998). LANA has been shown to bind and

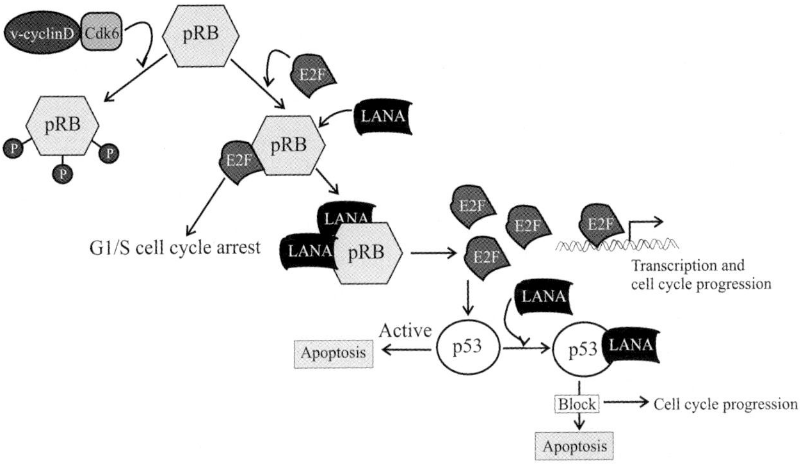

Fig. 6 Modulation of pRb and p53 pathways in KSHV-infected cells. Latency-associated nuclear antigen (LANA) competes with E2F transcription factor for binding to pRb and releases the E2F transcription, which leads to the upregulation of E2F-modulated genes. LANA also interacts with p53 and blocks the p53-mediated apoptosis, thus helping in cell cycle progression and tumorigenesis

inactivate pRb, thereby upregulating transcription of the promoter containing the cell cycle transcription factor E2F DNA-binding sequences (Radkov et al. 2000). Exogenous expression of LANA in SAOS-2 cells overcomes pRb-induced growth arrest and drives the cells into S phase (Radkov et al. 2000). LANA can transform primary rat embryo fibroblasts and make them tumorigenic in cooperation with the cellular oncogene Harvey rat sarcoma viral oncogene homologue (*Hras*) (Radkov et al. 2000). Additionally, the LANA homologue of herpesvirus saimiri ORF73 can also bind to pRB, resulting in upregulation of E2F responsive promoter activity (Borah et al. 2004) (Fig. 6).

4.2.4
LANA Associates with RING3/Brd2

RING3 is a ubiquitously expressed group of proteins in mammalian cells (Beck et al. 1992; Apone et al. 1996). They are bromodomain-containing proteins that include the SWI/SNF-type chromatin remodeling proteins that have elevated activity in human leukemia (Denis and Green 1996). LANA binds to RING3 through the extraterminal (ET) domain, characteristic of fsh-related proteins, suggesting a highly conserved function in terms of protein-protein interactions (Beck et al. 1992; Nomura et al. 1994; Jones et al. 1997). Proteins belonging to this group have been shown to be involved in cell cycle control.

Interaction of RING3/Brd2 shown with nuclear extract protein complex has been hypothesized to promote G_1-S transition (Denis et al. 2000; Guo et al. 2000). The interaction between LANA and RING3 leads to the phosphorylation of serine and threonine residues in the C terminus of LANA, between residues 951 and 1107 (Platt et al. 1999). This group of proteins also includes the yeast BDF proteins that regulate transcription through interaction with general transcription factors (Guo et al. 2000). RING3 localizes to euchromatin regions in the interphase nucleus even in the absence of KSHV viral genome and is released to the cytoplasm during mitosis. In KSHV-infected cells most of the RING3 proteins colocalized with LANA, suggesting that RING3 may contribute to KSHV genome persistence by local euchromatin microenvironment around the viral episomes tethered to the heterochromatin modulating transcription or promoter replication-segregation of the KSHV genome (Mattsson et al. 2002). Very recently RING3/Brd2 binding domain of LANA was mapped lying between aa 1007 and 1055 (Viejo-Borbolla et al. 2005). Mutants of LANA capable of supporting replication and dimerization were able to interact with RING3/Brd2, suggesting that RING3 may be important for interaction of C-terminal domain of LANA with heterochromatin and its functional activities (Viejo-Borbolla et al. 2005).

4.2.5
The Interferon-Inducible Protein MNDA Binds to LANA

MDNA is a nuclear protein expressed selectively in hematopoietic cells and has been shown to be induced by interferon (IFN) (Johnstone and Trapani 1999). This protein, like other proteins of this family, human IFI16, AIM2 and murine p202, p203, 204, and D3 protein (hematopoietic interferon-inducible nuclear protein), contains a conserved 200-aa region (Johnstone and Trapani 1999). LANA binds to MNDA through its amino-terminal region (aa 22–274) (Fukushi et al. 2003). This region of LANA has also been shown to be involved in transcriptional repression (Lim et al. 2000; Schwam et al. 2000). LANA colocalizes with MNDA in PEL cells and also is overexpressed in HEK cells (Fukushi et al. 2003). The primary function of IFN in infected cells is the inhibition of the growth of virus-infected cells, but exogenous expression of IFN did not affect growth of KSHV-infected cells (Fukushi et al. 2003). These studies suggest that MDNA alone is not enough for IFN-induced growth inhibition of KSHV-infected cells (Fukushi et al. 2003).

4.2.6
LANA Associates with the Acetyl Transferase pCBP and p300

CREB binding protein (CBP) and p300 were both identified through protein-protein interaction assays. CBP was identified in association with the transcription factor CREB (Chrivia et al. 1993) and p300 through its interaction with the adenoviral transforming protein E1A (Qiu et al. 1998). These proteins share high sequence homology and have been shown to be involved in a variety of cellular functions including modulation of tumor suppressor pathways (Wu et al. 2002; Schiller et al. 2003; Iyer et al. 2004). CBP and p300 may also serve as tumor suppressors through more indirect mechanisms; for example, CBP has been shown to be a negative regulator of the Wnt pathway in *Drosophila* (Waltzer and Bienz 1998), and activation of the Wnt pathway induces c- *myc* and cyclin D1 (at least in mammalian cells) (He et al. 1998). Therefore the loss of CBP would be expected to activate Wnt targets, with a resultant induction of tumorigenesis (Fig. 7).

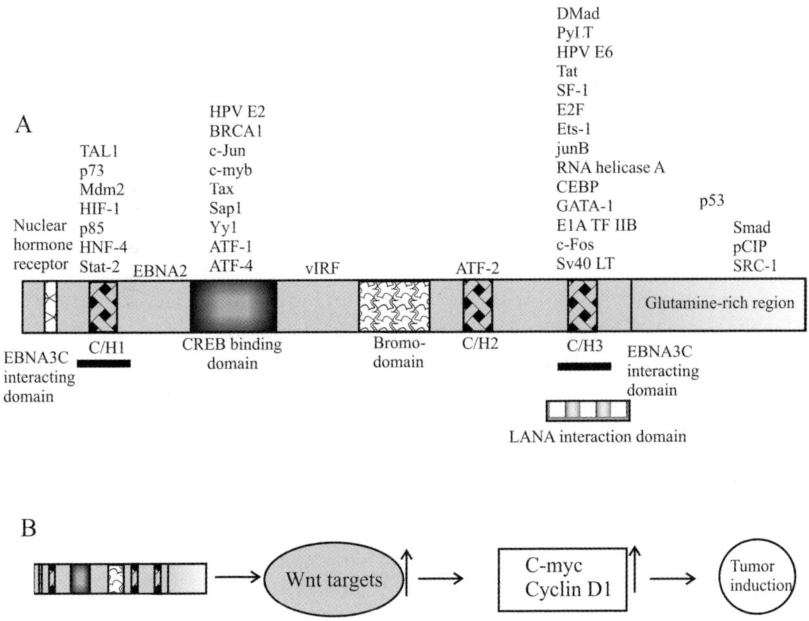

Fig. 7A, B pCBP/p300. **A** Schematic representation of domains involved in binding to various cellular and viral proteins. LANA interacts with C/H3 domain of the protein, which is also the binding region of other viral encoded proteins including HPV E6, E1A SV40 LT, EBNA3C. **B** Downregulation of CBP leads to the upregulation of c-myc and cyclin D1 and thus induction of tumorigenesis

LANA has been shown to bind with CBP and interferes with the interaction of CBP with c-Fos, a representative C/H3 region binding cellular transcription factor, in vivo and in vitro (Lim et al. 2001). Additionally, LANA inhibits transcriptional activity and also the in vitro histone acetyltransferase activity of CBP, suggesting that LANA modulates the global transcriptional activities of infected cells at least in part through interaction with CBP (Lim et al. 2001).

4.2.7 LANA Induces Expression of the Id Proteins Involved in Regulating Cell Differentiation

Id proteins belong to a family of four helix-loop-helix (HLH) proteins (Id-1 to Id-4) that were initially recognized as growth factor-inducible genes that inhibited cell differentiation (Benezra et al. 1990). These proteins are also referred to as inhibitors of DNA binding (Id) because they lack the basic amino acid sequence characteristic of basic bHLH transcription factors necessary for DNA binding. They also form heterodimers with transcription factors, inhibiting their ability to bind DNA (Benezra et al. 1990; Norton 2000). Recently various biological roles have been assigned to the Id family of proteins, including cell cycle regulation, embryonic development, cell death, and tumorigenesis (Nickoloff et al. 2000).These recently described functions may be through interaction with non-HLH proteins including the pocket domain proteins, pRb, p107 and p130, MIDA1, and ETS transcription factors (Hollnagel et al. 1999; Yates et al. 1999). Id proteins also promote cell progression through dysregulation of expression of p21, inhibitor of the cell cycle-dependent kinase (CDKI). Moreover, Id-1 regulates cell proliferation, possibly through direct interactions with the p16 promoter and/or inactivation of ETS2 transcription factors (Alani et al. 2001). Expression profile of Id-1 in human tumors, including squamous cell carcinoma and endometrial, breast, and cervical cancers, showed a significant level of expression compared to normal cells (Langlands et al. 2000; Schindl et al. 2001; Tang et al. 2003). Higher levels of expression have also been found in invasive or high-grade tumors, indicating a possible correlation with the more aggressive form of the disease (Lin et al. 2000). Overexpression of Id-1 in keratinocytes and endothelial cells (ECs) has been shown to delay onset of cellular senescence (Tang et al. 2003). KS tumors showed elevated levels of Id-1 expression, and infection with KSHV drastically increased Id-1 expression (Tang et al. 2003). Ectopic expression of LANA in endothelial cells increased Id-1 expression, suggesting that LANA may drive cell proliferation by inducing Id-1 expression (Tang et al. 2003).

4.3
Role of LANA in Episome Maintenance

Persistence of KSHV through successive rounds of cell division in new daughter cells is enhanced by the partitioning of the viral DNA to host daughter cells during mitosis to prevent loss of the viral DNA after the cell divides. Studies have demonstrated that LANA associates with the KSHV genome in infected cells and colocalizes with the viral genome both in interphase nuclei and on mitotic chromatin in a punctate pattern (Ballestas et al. 1999b; Cotter and Robertson 1999). The ability of LANA to support persistence of the KSHV genome is crucial for the establishment of latency. LANA has been shown to target heterochromatin-associated nuclear bodies and selectively associates with human chromatin in human mouse hybrids containing a single fused nucleus (Szekely et al. 1999). The presence of LANA was shown to be essential for the long-term maintenance of the KSHV-derived Z6 cosmids (Ballestas et al. 1999b; Cotter and Robertson 1999). The above reports showed colocalization of LANA with the viral genome in the nucleus, demonstrated its role in episomal persistence, and additionally suggested its role in partitioning of the viral genome during mitosis. Recombinant KSHV virus cloned in a bacterial artificial chromosome (BAC36ΔLANA) disrupted for LANA expression did not maintain viral episome, and the cells became virus free after 2 weeks of selection, confirming LANA's role in viral episome maintenance (Ye et al. 2004). Recently, Godfrey and colleagues used a short hairpin RNA (shRNA) approach to knock down the expression of the oncogenic latent gene cluster including LANA (Godfrey et al. 2005). Independent silencing of LANA did not completely knock down expression of LANA or v-cyclin but reduced it to 20% compared to control treatments. The reduction in the expression of LANA gradually reduced KSHV copy number until it achieved a lower but steady copy number (Godfrey et al. 2005). Indirect binding studies with KSHV DNA fragments indicated that LANA displays preferential binding to different regions of the KSHV genome, with at least three regions showing relatively stronger binding compared to the entire genome (Cotter and Robertson 1999). A small 2-Kbp region at the left end was shown to have strongest binding to LANA, and the binding sequence was narrowed down to a 13-bp cognate sequence in the TR region of the KSHV genome (Ballestas and Kaye 2001; Cotter et al. 2001; Renne et al. 2001). The binding region of LANA was initially mapped to the last 200 aa of the carboxy terminus and is now further mapped to residues 996–1139 (Komatsu et al. 2004). Scanning deletions within this region ablated both LANA oligomerization and DNA binding, consistent with a requirement for oligomerization to bind DNA (Komatsu et al. 2004; Srinivasan et al. 2004). A region of the KSHV genome at the left end as well as a single and three copies of the TR elements persisted in cells stably expressing LANA in *trans*, suggesting that this region in fact contained the *cis*-acting DNA element (Cotter and Robertson 1999; Ballestas and Kaye 2001; Renne et al. 2001). However, this does not exclude other regions of the genome as alternate elements capable of functioning as origins of replication during persistent infection. Interestingly, LANA showed affinity to two other regions of the KSHV genome, albeit to a lesser degree than the left end (Cotter and Robertson 1999). LANA was also shown to bind to histone H1 but not core histones and to tether the viral episomes to the host chromatin (Cotter and Robertson 1999). A deletion in the chromosome binding domain

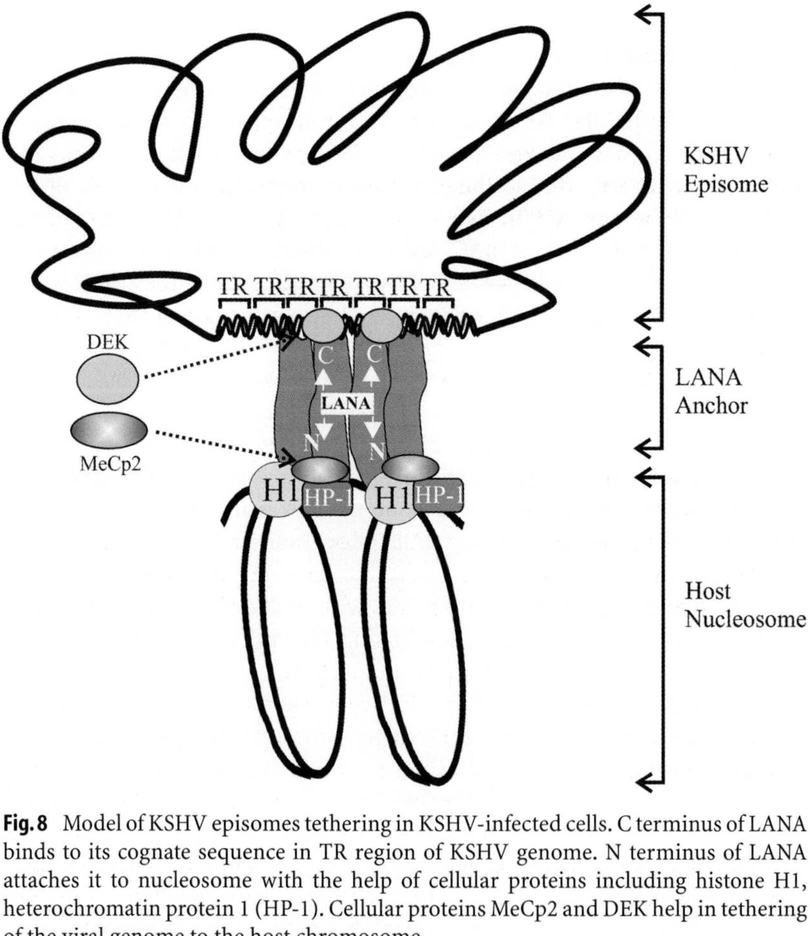

Fig. 8 Model of KSHV episomes tethering in KSHV-infected cells. C terminus of LANA binds to its cognate sequence in TR region of KSHV genome. N terminus of LANA attaches it to nucleosome with the help of cellular proteins including histone H1, heterochromatin protein 1 (HP-1). Cellular proteins MeCp2 and DEK help in tethering of the viral genome to the host chromosome

(CBD), aa 5–22 of the N-terminal region of LANA, abolished episomal maintenance but was restored by replacing the mutation with histone H1 protein (Shinohara et al. 2002). LANA therefore simultaneously associates with mitotic chromosomes via the CBD of its N terminus and through binding of its C terminus to the KSHV TR (Ballestas et al. 1999a; Piolot et al. 2001). These domains mediate chromosomal tethering, which may involve interactions with cellular proteins including the methyl CpG binding protein (MeCP2), heterochromatin proteins (HP-1) at N terminus, and DEK at C terminus of LANA protein (Krithivas et al. 2002) (Fig. 8). These studies demonstrate that LANA can function similar to EBNA1 in maintenance of the EBV episome in the infected cell (Cotter and Robertson 1999).

4.4 Viral DNA Replication

Replication of viral DNA is required for efficient segregation of viral episomes. LANA has been shown to be essential for the replication of the viral cosmid clone along with tethering to the host chromosomes (Ballestas et al. 1999b; Cotter and Robertson 1999). The TR elements of the KSHV genome have been proposed to be the potential site of initiation of replication by various groups using short-term replication assays (Hu et al. 2002; Lim et al. 2002; Grundhoff and Ganem 2003; Barbera et al. 2004; Komatsu et al. 2004; Hu and Renne 2005), although the replicative potential of other regions has not been thoroughly evaluated. Plasmids containing TR elements have been shown to replicate in LANA-expressing cells (Hu et al. 2002; Lim et al. 2002; Grundhoff and Ganem 2003; Hu and Renne 2005). Renne and colleagues performed deletion mapping of the TR region to identify minimal replicator sequences, and it was shown that a 71-bp-long minimal replicator region contained two distinctive sequence elements, LANA binding sites (LBS1/2). Additionally, adjacent 29- to 32-bp-long GC-rich sequences were essential for replication of the TR elements and thus could be termed a replication element (Hu and Renne 2005). In terms of proteins required for the replication of the viral episome, a number of cellular proteins have been proposed to be involved in this process. Human origin recognition complexes (hsORCs) have been shown to interact with LANA in in vitro experiments, most probably similar to EBV encoded EBV, which is believed to recruit cellular replication machinery at the Ori site (Dhar et al. 2001; Schneider 2001; Ritzi et al. 2003). The detailed mechanism of replication in KSHV is not well understood. However, recently it was shown that KSHV forms chromatin structures and the latent replication origin within the TR is bound by the LANA-associated proteins CBP, double-bromodomain-containing protein 2 (BRD2), and origin recognition complex 2 protein (ORC2) (Stedman et al. 2004). This region is enriched in hyperacetylated histones H3 and H4 relative to other regions of the latent genome (Stedman et al. 2004). In slight contrast to previous studies, chromatin immunoprecipitation assays performed in our lab have shown that the association of cellular replication machinery proteins ORC2 and MCM3 occurs at various regions along the KSHV genome, suggesting that additional regions of the KSHV genome may be capable of functioning as origins of replication (Verma 2006).

5 Overview of LANA's Role in KSHV Pathogenesis

Numerous studies over the last 10 years have shown that LANA is a multifunctional protein ubiquitously expressed in KSHV-infected cells (Ballestas et al. 1999b; Cotter and Robertson 1999; Friborg et al. 1999; Radkov et al. 2000; An et al. 2002; Boshoff 2003; Cannon et al. 2003; Fujimuro et al. 2003; Fukushi et al. 2003; Verma and Robertson 2003; Borah et al. 2004; Cloutier et al. 2004; Lan et al. 2004; An et al. 2005). LANA modulates various cellular pathways both positively and negatively to drive cell proliferation. LANA also downregulates tumor suppressor pathways regulated by p53 and pRB as discussed earlier in this chapter. Interestingly, one study showed that KSHV can transform human umbilical vein endothelial cells (HUVEC) and that the levels of telomerase in these transformed cells were elevated (Flore et al. 1998). Telomerase is a holoenzyme that prevents chromosome degradation, end-to-end chromosome fusions, and chromosome loss and maintains telomere length by the addition of new repeat sequences at the ends of the chromosomes, thus deriving cell immortalization (Greider 1991a, b; Harrington and Greider 1991). In support of these studies LANA has also been shown to upregulate the human telomerase promoter through interaction with Sp1 transcription factor, supporting the role of LANA as an oncoprotein (Knight et al. 2001; Verma et al. 2004). Additional evidence suggests that LANA cooperates with HIV Tat and upregulates the HIV LTR promoter, thus supporting the hypothesis that KSHV coinfection promotes HIV propagation (Hyun et al. 2001). Furthermore, LANA can also activate the major latent EBV promoter elements (LMP1 and Cp1) in transient studies (Groves et al. 2001).

KSHV prevents the cell surface expression of MHC chains and protects itself from CTL cell lysis during the lytic cell cycle. However, as only a subset of the infected cells undergo lytic replication (1%–5%), KSHV is likely to have a means to protect itself from the immune response during latent infection (Staskus et al. 1997; Zhong and Ganem 1997; Dittmer et al. 1998; Sarid et al.

1998; Katano et al. 1999; Talbot et al. 1999; Parravicini et al. 2000). Infection of primary dermal microvascular endothelial cells with KSHV resulted in downregulation of cell surface MHC class I, CD31 (PE-CAM), and CD54 (ICAM-I) immune response modulators. Moreover, this downregulation was correlated with increased levels of LANA, suggesting that during latency LANA may have a role in the downregulation of MHC class I molecules (Tomescu et al. 2003).

More recent, interesting studies have demonstrated a role for LANA in regulating KSHV reactivation, and this involves modulation of the major Notch cellular signaling pathway (Lan et al. 2005b). In our most recent studies, we demonstrated that intracellular activated domain of Notch1 (ICN) is accumulated in KSHV latently infected cells through stabilization mediated by LANA (Lan et al., unpublished data). This increase in levels of ICN is crucial for oncogenesis induced by KSHV as shown by the significant increase in proliferation of KSHV-infected cells. The ability of ICN to potentially reactivate the virus through upregulation of RTA can be effectively antagonized by KSHV LANA. This provides a finely tuned model of virus-host interaction in which KSHV selectively usurps the function of a conserved cellular signaling pathway to achieve oncogenesis while dispensing additional function to strictly maintain latency. Taken together, these studies now clearly show a feedback mechanism whereby KSHV can establish latency by repressing the expression of its major transactivator. Further work is needed to explore this new functional role of LANA. However, so far we can say with confidence that LANA plays a critical role in KSHV-mediated tumorigenesis by regulating transcription, chromatin remodeling, episomal maintenance, replication of genome, and control of reactivation and thus maintenance of KSHV latency.

References

Alani RM, Young AZ, Shifflett CB (2001) Id1 regulation of cellular senescence through transcriptional repression of p16/Ink4a. Proc Natl Acad Sci USA 98:7812–7816

Albrecht JC, Nicholas J, Biller D, Cameron KR, Biesinger B, Newman C, Wittmann S, Craxton MA, Coleman H, Fleckenstein B, et al. (1992) Primary structure of the herpesvirus saimiri genome. J Virol 66:5047–5058

Ambros V (2004) The functions of animal microRNAs. Nature 431:350–355

An FQ, Compitello N, Horwitz E, Sramkoski M, Knudsen ES, Renne R (2005) The latency-associated nuclear antigen of Kaposi's sarcoma-associated herpesvirus modulates cellular gene expression and protects lymphoid cells from p16 INK4A-induced cell cycle arrest. J Biol Chem 280:3862–3874

An J, Lichtenstein AK, Brent G, Rettig MB (2002) The Kaposi sarcoma-associated herpesvirus (KSHV) induces cellular interleukin 6 expression: role of the KSHV latency-associated nuclear antigen and the AP1 response element. Blood 99:649–654

Apone LM, Virbasius CM, Reese JC, Green MR (1996) Yeast TAF(II)90 is required for cell-cycle progression through G2/M but not for general transcription activation. Genes Dev 10:2368–2380

Ballestas ME, Chatis PA, Kaye KM (1999a) Efficient persistence of extrachromosomal KSHV DNA mediated by latency-associated nuclear antigen. Science 284:641–644

Ballestas ME, Chatis PA, Kaye KM (1999b) Efficient persistence of extrachromosomal KSHV DNA mediated by latency-associated nuclear antigen. Science 284:641–644

Ballestas ME, Kaye KM (2001) Kaposi's sarcoma-associated herpesvirus latency-associated nuclear antigen 1 mediates episome persistence through *cis*-acting terminal repeat (TR) sequence and specifically binds TR DNA. J Virol 75:3250–3258

Barbera AJ, Ballestas ME, Kaye KM (2004) The Kaposi's sarcoma-associated herpesvirus latency-associated nuclear antigen 1 N terminus is essential for chromosome association, DNA replication, and episome persistence. J Virol 78:294–301

Barker N, Morin PJ, Clevers H (2000) The Yin-Yang of TCF/β-catenin signaling. Adv Cancer Res 77:1-24

Bartel DP (2004) MicroRNAs: genomics, biogenesis, mechanism, and function. Cell 116:281–297

Bechtel JT, Liang Y, Hvidding J, Ganem D (2003) Host range of Kaposi's sarcoma-associated herpesvirus in cultured cells. J Virol 77:6474–6481

Bechtel JT, Winant RC, Ganem D (2005) Host and viral proteins in the virion of Kaposi's sarcoma-associated herpesvirus. J Virol 79:4952–4964

Beck S, Hanson I, Kelly A, Pappin DJ, Trowsdale J (1992) A homologue of the *Drosophila* female sterile homeotic (fsh) gene in the class II region of the human MHC. DNA Seq 2:203–210

Benezra R, Davis RL, Lockshon D, Turner DL, Weintraub H (1990) The protein Id: a negative regulator of helix-loop-helix DNA binding proteins. Cell 61:49–59

Borah S, Verma SC, Robertson ES (2004) ORF73 of herpesvirus saimiri, a viral homolog of Kaposi's sarcoma-associated herpesvirus, modulates the two cellular tumor suppressor proteins p53 and pRb. J Virol 78:10336–10347

Boshoff C (2003) Kaposi virus scores cancer coup. Nat Med 9:261–262

Boshoff C, Schulz TF, Kennedy MM, Graham AK, Fisher C, Thomas A, McGee JO, Weiss RA, O'Leary JJ (1995) Kaposi's sarcoma-associated herpesvirus infects endothelial and spindle cells. Nat Med 1:1274–1278

Boulikas T (1993) Nuclear localization signals (NLS). Crit Rev Eukaryot Gene Expr 3:193–227

Cannon M, Philpott NJ, Cesarman E (2003) The Kaposi's sarcoma-associated herpesvirus G protein-coupled receptor has broad signaling effects in primary effusion lymphoma cells. J Virol 77:57–67

Cesarman E, Nador RG, Bai F, Bohenzky RA, Russo JJ, Moore PS, Chang Y, Knowles DM (1996) Kaposi's sarcoma-associated herpesvirus contains G protein-coupled receptor and cyclin D homologs which are expressed in Kaposi's sarcoma and malignant lymphoma. J Virol 70:8218–8223

Chang J, Renne R, Dittmer D, Ganem D (2000) Inflammatory cytokines and the reactivation of Kaposi's sarcoma-associated herpesvirus lytic replication. Virology 266:17–25

Chang Y, Cesarman E, Pessin MS, Lee F, Culpepper J, Knowles DM, Moore PS (1994) Identification of herpesvirus-like DNA sequences in AIDS-associated Kaposi's sarcoma. Science 266:1865–1869
Chen X, Ko LJ, Jayaraman L, Prives C (1996) p53 levels, functional domains, and DNA damage determine the extent of the apoptotic response of tumor cells. Genes Dev 10:2438–2451
Chrivia JC, Kwok RP, Lamb N, Hagiwara M, Montminy MR, Goodman RH (1993) Phosphorylated CREB binds specifically to the nuclear protein CBP. Nature 365:855–859
Cloutier N, Gravel A, Flamand L (2004) Multiplex detection and quantitation of latent and lytic transcripts of human herpesvirus-8 using RNase protection assay. J Virol Methods 122:1-7
Cotter MA 2nd, Robertson ES (1999) The latency-associated nuclear antigen tethers the Kaposi's sarcoma-associated herpesvirus genome to host chromosomes in body cavity-based lymphoma cells. Virology 264:254–264
Cotter MA 2nd, Subramanian C, Robertson ES (2001) The Kaposi's sarcoma-associated herpesvirus latency-associated nuclear antigen binds to specific sequences at the left end of the viral genome through its carboxy-terminus. Virology 291:241–259
Couty JP, Geras-Raaka E, Weksler BB, Gershengorn MC (2001) Kaposi's sarcoma-associated herpesvirus G protein-coupled receptor signals through multiple pathways in endothelial cells. J Biol Chem 276:33805–33811
Deng H, Young A, Sun R (2000) Auto-activation of the rta gene of human herpesvirus-8/Kaposi's sarcoma-associated herpesvirus. J Gen Virol 81:3043–3048
Denis GV, Green MR (1996) A novel, mitogen-activated nuclear kinase is related to a *Drosophila* developmental regulator. Genes Dev 10:261–271
Denis GV, Vaziri C, Guo N, Faller DV (2000) RING3 kinase transactivates promoters of cell cycle regulatory genes through E2F. Cell Growth Differ 11:417–424
DeWire SM, Damania B (2005) The latency-associated nuclear antigen of rhesus monkey rhadinovirus inhibits viral replication through repression of Orf50/Rta transcriptional activation. J Virol 79:3127–3138
Dhar SK, Yoshida K, Machida Y, Khaira P, Chaudhuri B, Wohlschlegel JA, Leffak M, Yates J, Dutta A (2001) Replication from oriP of Epstein-Barr virus requires human ORC and is inhibited by geminin. Cell 106:287–296
Dittmer D, Lagunoff M, Renne R, Staskus K, Haase A, Ganem D (1998) A cluster of latently expressed genes in Kaposi's sarcoma-associated herpesvirus. J Virol 72:8309–8315
Dittmer D, Stoddart C, Renne R, Linquist-Stepps V, Moreno ME, Bare C, McCune JM, Ganem D (1999) Experimental transmission of Kaposi's sarcoma-associated herpesvirus (KSHV/HHV-8) to SCID-hu Thy/Liv mice. J Exp Med 190:1857–1868
Dittmer DP (2003) Transcription profile of Kaposi's sarcoma-associated herpesvirus in primary Kaposi's sarcoma lesions as determined by real-time PCR arrays. Cancer Res 63:2010–2015
Dupin N, Fisher C, Kellam P, Ariad S, Tulliez M, Franck N, van Marck E, Salmon D, Gorin I, Escande JP, Weiss RA, Alitalo K, Boshoff C (1999) Distribution of human herpesvirus-8 latently infected cells in Kaposi's sarcoma, multicentric Castleman's disease, and primary effusion lymphoma. Proc Natl Acad Sci USA 96:4546–4551
Dyson N (1998) The regulation of E2F by pRB-family proteins. Genes Dev 12:2245–2262

Ensser A, Pflanz R, Fleckenstein B (1997) Primary structure of the alcelaphine herpesvirus 1 genome. J Virol 71:6517–6525

Fakhari FD, Dittmer DP (2002) Charting latency transcripts in Kaposi's sarcoma-associated herpesvirus by whole-genome real-time quantitative PCR. J Virol 76:6213–6223

Flore O, Rafii S, Ely S, O'Leary JJ, Hyjek EM, Cesarman E (1998) Transformation of primary human endothelial cells by Kaposi's sarcoma-associated herpesvirus. Nature 394:588–592

Friborg J, Jr., Kong W, Hottiger MO, Nabel GJ (1999) p53 inhibition by the LANA protein of KSHV protects against cell death. Nature 402:889–894

Friend S (1994) p53: a glimpse at the puppet behind the shadow play. Science 265:334–335

Fujimuro M, Hayward SD (2003) The latency-associated nuclear antigen of Kaposi's sarcoma-associated herpesvirus manipulates the activity of glycogen synthase kinase-3β. J Virol 77:8019–8030

Fujimuro M, Hayward SD (2004) Manipulation of glycogen-synthase kinase-3 activity in KSHV-associated cancers. J Mol Med 82:223–231

Fujimuro M, Wu FY, ApRhys C, Kajumbula H, Young DB, Hayward GS, Hayward SD (2003) A novel viral mechanism for dysregulation of β-catenin in Kaposi's sarcoma-associated herpesvirus latency. Nat Med 9:300–306

Fukushi M, Higuchi M, Oie M, Tetsuka T, Kasolo F, Ichiyama K, Yamamoto N, Katano H, Sata T, Fujii M (2003) Latency-associated nuclear antigen of Kaposi's sarcoma-associated herpesvirus interacts with human myeloid cell nuclear differentiation antigen induced by interferon α. Virus Genes 27: 237–247

Gao SJ, Kingsley L, Li M, Zheng W, Parravicini C, Ziegler J, Newton R, Rinaldo CR, Saah A, Phair J, Detels R, Chang Y, Moore PS (1996) KSHV antibodies among Americans, Italians and Ugandans with and without Kaposi's sarcoma. Nat Med 2:925–928

Gao SJ, Zhang YJ, Deng JH, Rabkin CS, Flore O, Jenson HB (1999) Molecular polymorphism of Kaposi's sarcoma-associated herpesvirus (Human herpesvirus 8) latent nuclear antigen: evidence for a large repertoire of viral genotypes and dual infection with different viral genotypes. J Infect Dis 180:1466–1476

Glaunsinger B, Ganem D (2004) Lytic KSHV infection inhibits host gene expression by accelerating global mRNA turnover. Mol Cell 13:713–723

Glenn M, Rainbow L, Aurade F, Davison A, Schulz TF (1999) Identification of a spliced gene from Kaposi's sarcoma-associated herpesvirus encoding a protein with similarities to latent membrane proteins 1 and 2A of Epstein-Barr virus. J Virol 73:6953–6963

Godfrey A, Anderson J, Papanastasiou A, Takeuchi Y, Boshoff C (2005) Inhibiting primary effusion lymphoma by lentiviral vectors encoding short hairpin RNA. Blood 105:2510–2518

Greider CW (1991a) Chromosome first aid. Cell 67:645–647

Greider CW (1991b) Telomerase is processive. Mol Cell Biol 11:4572–4580

Groves AK, Cotter MA, Subramanian C, Robertson ES (2001) The latency-associated nuclear antigen encoded by Kaposi's sarcoma-associated herpesvirus activates two major essential Epstein-Barr virus latent promoters. J Virol 75:9446–9457

Grundhoff A, Ganem D (2001) Mechanisms governing expression of the v-FLIP gene of Kaposi's sarcoma-associated herpesvirus. J Virol 75:1857–1863
Grundhoff A, Ganem D (2003) The latency-associated nuclear antigen of Kaposi's sarcoma-associated herpesvirus permits replication of terminal repeat-containing plasmids. J Virol 77:2779–2783
Guo N, Faller DV, Denis GV (2000) Activation-induced nuclear translocation of RING3. J Cell Sci 113 (Pt 17): 3085–3091
Gwack Y, Byun H, Hwang S, Lim C, Choe J (2001) CREB-binding protein and histone deacetylase regulate the transcriptional activity of Kaposi's sarcoma-associated herpesvirus open reading frame 50. J Virol 75:1909–1917
Harrington LA, Greider CW (1991) Telomerase primer specificity and chromosome healing. Nature 353:451–454
He TC, Sparks AB, Rago C, Hermeking H, Zawel L, da Costa LT, Morin PJ, Vogelstein B, Kinzler KW (1998) Identification of c-MYC as a target of the APC pathway. Science 281:1509–1512
Hollnagel A, Oehlmann V, Heymer J, Ruther U, Nordheim A (1999) Id genes are direct targets of bone morphogenetic protein induction in embryonic stem cells. J Biol Chem 274:19838–19845
Hollstein M, Sidransky D, Vogelstein B, Harris CC (1991) p53 mutations in human cancers. Science 253:49–53
Hu J, Garber AC, Renne R (2002) The latency-associated nuclear antigen of Kaposi's sarcoma-associated herpesvirus supports latent DNA replication in dividing cells. J Virol 76:11677–11687
Hu J, Renne R (2005) Characterization of the minimal replicator of Kaposi's sarcoma-associated herpesvirus latent origin. J Virol 79:2637–2642
Hyun TS, Subramanian C, Cotter MA, 2nd, Thomas RA, Robertson ES (2001) Latency-associated nuclear antigen encoded by Kaposi's sarcoma-associated herpesvirus interacts with Tat and activates the long terminal repeat of human immunodeficiency virus type 1 in human cells. J Virol 75:8761–8771
Iyer NG, Ozdag H, Caldas C (2004) p300/CBP and cancer. Oncogene 23:4225–4231
Jay G, Khoury G, DeLeo AB, Dippold WG, Old LJ (1981) p53 transformation-related protein: detection of an associated phosphotransferase activity. Proc Natl Acad Sci USA 78:2932–2936
Jenner RG, Alba MM, Boshoff C, Kellam P (2001) Kaposi's sarcoma-associated herpesvirus latent and lytic gene expression as revealed by DNA arrays. J Virol 75:891–902
Jeong J, Papin J, Dittmer D (2001) Differential regulation of the overlapping Kaposi's sarcoma-associated herpesvirus vGCR (orf74) and LANA (orf73) promoters. J Virol 75:1798–1807
Jeong JH, Hines-Boykin R, Ash JD, Dittmer DP (2002) Tissue specificity of the Kaposi's sarcoma-associated herpesvirus latent nuclear antigen (LANA/orf73) promoter in transgenic mice. J Virol 76:11024–11032
Jeong JH, Orvis J, Kim JW, McMurtrey CP, Renne R, Dittmer DP (2004) Regulation and autoregulation of the promoter for the latency-associated nuclear antigen of Kaposi's sarcoma-associated herpesvirus. J Biol Chem 279:16822–16831
Johnstone RW, Trapani JA (1999) Transcription and growth regulatory functions of the HIN-200 family of proteins. Mol Cell Biol 19:5833–5838

Jones MH, Numata M, Shimane M (1997) Identification and characterization of BRDT: a testis-specific gene related to the bromodomain genes RING3 and *Drosophila* fsh. Genomics 45:529–534

Katano H, Sato Y, Kurata T, Mori S, Sata T (1999) High expression of HHV-8-encoded ORF73 protein in spindle-shaped cells of Kaposi's sarcoma. Am J Pathol 155:47–52

Katano H, Sato Y, Kurata T, Mori S, Sata T (2000) Expression and localization of human herpesvirus 8-encoded proteins in primary effusion lymphoma, Kaposi's sarcoma, and multicentric Castleman's disease. Virology 269:335–344

Kedes DH, Lagunoff M, Renne R, Ganem D (1997) Identification of the gene encoding the major latency-associated nuclear antigen of the Kaposi's sarcoma-associated herpesvirus. J Clin Invest 100:2606–2610

Kellam P, Boshoff C, Whitby D, Matthews S, Weiss RA, Talbot SJ (1997) Identification of a major latent nuclear antigen, LNA-1, in the human herpesvirus 8 genome. J Hum Virol 1:19–29

Kiess M, Gill RM, Hamel PA (1995) Expression and activity of the retinoblastoma protein (pRB)-family proteins, p107 and p130, during L6 myoblast differentiation. Cell Growth Differ 6:1287–1298

Kitagawa M, Hatakeyama S, Shirane M, Matsumoto M, Ishida N, Hattori K, Nakamichi I, Kikuchi A, Nakayama K, Nakayama K (1999) An F-box protein, FWD1, mediates ubiquitin-dependent proteolysis of β-catenin. EMBO J 18:2401–2410

Knight JS, Cotter MA, 2nd, Robertson ES (2001) The latency-associated nuclear antigen of Kaposi's sarcoma-associated herpesvirus transactivates the telomerase reverse transcriptase promoter. J Biol Chem 276:22971–22978

Ko LJ, Prives C (1996) p53: puzzle and paradigm. Genes Dev 10:1054–1072

Komatsu T, Ballestas ME, Barbera AJ, Kelley-Clarke B, Kaye KM (2004) KSHV LANA1 binds DNA as an oligomer and residues N-terminal to the oligomerization domain are essential for DNA binding, replication, and episome persistence. Virology 319:225–236

Krishnan HH, Naranatt PP, Smith MS, Zeng L, Bloomer C, Chandran B (2004) Concurrent expression of latent and a limited number of lytic genes with immune modulation and antiapoptotic function by Kaposi's sarcoma-associated herpesvirus early during infection of primary endothelial and fibroblast cells and subsequent decline of lytic gene expression. J Virol 78:3601–3620

Krithivas A, Fujimuro M, Weidner M, Young DB, Hayward SD (2002) Protein interactions targeting the latency-associated nuclear antigen of Kaposi's sarcoma-associated herpesvirus to cell chromosomes. J Virol 76:11596–11604

Krithivas A, Young DB, Liao G, Greene D, Hayward SD (2000) Human herpesvirus 8 LANA interacts with proteins of the mSin3 corepressor complex and negatively regulates Epstein-Barr virus gene expression in dually infected PEL cells. J Virol 74:9637–9645

Lagunoff M, Bechtel J, Venetsanakos E, Roy AM, Abbey N, Herndier B, McMahon M, Ganem D (2002) De novo infection and serial transmission of Kaposi's sarcoma-associated herpesvirus in cultured endothelial cells. J Virol 76:2440–2448

Lan K, Kuppers DA, Robertson ES (2005a) Kaposi's sarcoma-associated herpesvirus reactivation is regulated by interaction of latency-associated nuclear antigen with recombination signal sequence-binding protein Jκ, the major downstream effector of the Notch signaling pathway. J Virol 79:3468–3478
Lan K, Kuppers DA, Verma SC, Robertson ES (2004) Kaposi's sarcoma-associated herpesvirus-encoded latency-associated nuclear antigen inhibits lytic replication by targeting Rta: a potential mechanism for virus-mediated control of latency. J Virol 78:6585–6594
Lan K, Kuppers DA, Verma SC, Sharma N, Murakami M, Robertson ES (2005b) Induction of Kaposi's sarcoma-associated herpesvirus latency-associated nuclear antigen by the lytic transactivator RTA: a novel mechanism for establishment of latency. J Virol 79:7453–7465
Langlands K, Down GA, Kealey T (2000) Id proteins are dynamically expressed in normal epidermis and dysregulated in squamous cell carcinoma. Cancer Res 60:5929–5933
Lee H, Veazey R, Williams K, Li M, Guo J, Neipel F, Fleckenstein B, Lackner A, Desrosiers RC, Jung JU (1998) Deregulation of cell growth by the K1 gene of Kaposi's sarcoma-associated herpesvirus. Nat Med 4:435–440
Lennette ET, Blackbourn DJ, Levy JA (1996) Antibodies to human herpesvirus type 8 in the general population and in Kaposi's sarcoma patients. Lancet 348:858–861
Levine AJ (1997) p53, the cellular gatekeeper for growth and division. Cell 88:323–331
Liang Y, Chang J, Lynch SJ, Lukac DM, Ganem D (2002) The lytic switch protein of KSHV activates gene expression via functional interaction with RBP-Jκ (CSL), the target of the Notch signaling pathway. Genes Dev 16:1977–1989
Liang Y, Ganem D (2003) Lytic but not latent infection by Kaposi's sarcoma-associated herpesvirus requires host CSL protein, the mediator of Notch signaling. Proc Natl Acad Sci USA 100:8490–8495
Lim C, Gwack Y, Hwang S, Kim S, Choe J (2001) The transcriptional activity of cAMP response element-binding protein-binding protein is modulated by the latency associated nuclear antigen of Kaposi's sarcoma-associated herpesvirus. J Biol Chem 276:31016–31022
Lim C, Sohn H, Gwack Y, Choe J (2000) Latency-associated nuclear antigen of Kaposi's sarcoma-associated herpesvirus (human herpesvirus-8) binds ATF4/CREB2 and inhibits its transcriptional activation activity. J Gen Virol 81:2645–2652
Lim C, Sohn H, Lee D, Gwack Y, Choe J (2002) Functional dissection of latency-associated nuclear antigen 1 of Kaposi's sarcoma-associated herpesvirus involved in latent DNA replication and transcription of terminal repeats of the viral genome. J Virol 76:10320–10331
Lin CQ, Singh J, Murata K, Itahana Y, Parrinello S, Liang SH, Gillett CE, Campisi J, Desprez PY (2000) A role for Id-1 in the aggressive phenotype and steroid hormone response of human breast cancer cells. Cancer Res 60:1332–1340
Low W, Harries M, Ye H, Du MQ, Boshoff C, Collins M (2001) Internal ribosome entry site regulates translation of Kaposi's sarcoma-associated herpesvirus FLICE inhibitory protein. J Virol 75:2938–2945
Lukac DM, Kirshner JR, Ganem D (1999) Transcriptional activation by the product of open reading frame 50 of Kaposi's sarcoma-associated herpesvirus is required for lytic viral reactivation in B cells. J Virol 73:9348–9361

Lukac DM, Renne R, Kirshner JR, Ganem D (1998) Reactivation of Kaposi's sarcoma-associated herpesvirus infection from latency by expression of the ORF 50 transactivator, a homolog of the EBV R protein. Virology 252:304–312

Mattsson K, Kiss C, Platt GM, Simpson GR, Kashuba E, Klein G, Schulz TF, Szekely L (2002) Latent nuclear antigen of Kaposi's sarcoma herpesvirus/human herpesvirus-8 induces and relocates RING3 to nuclear heterochromatin regions. J Gen Virol 83:179–188

Mesri EA, Cesarman E, Arvanitakis L, Rafii S, Moore MA, Posnett DN, Knowles DM, Asch AS (1996) Human herpesvirus-8/Kaposi's sarcoma-associated herpesvirus is a new transmissible virus that infects B cells. J Exp Med 183:2385–2390

Nador RG, Milligan LL, Flore O, Wang X, Arvanitakis L, Knowles DM, Cesarman E (2001) Expression of Kaposi's sarcoma-associated herpesvirus G protein-coupled receptor monocistronic and bicistronic transcripts in primary effusion lymphomas. Virology 287:62–70

Nakamura H, Lu M, Gwack Y, Souvlis J, Zeichner SL, Jung JU (2003) Global changes in Kaposi's sarcoma-associated virus gene expression patterns following expression of a tetracycline-inducible Rta transactivator. J Virol 77:4205–4220

Neipel F, Albrecht JC, Fleckenstein B (1997) Cell-homologous genes in the Kaposi's sarcoma-associated rhadinovirus human herpesvirus 8: determinants of its pathogenicity? J Virol 71:4187–4192

Nickoloff BJ, Chaturvedi V, Bacon P, Qin JZ, Denning MF, Diaz MO (2000) Id-1 delays senescence but does not immortalize keratinocytes. J Biol Chem 275:27501–27504

Nomura N, Nagase T, Miyajima N, Sazuka T, Tanaka A, Sato S, Seki N, Kawarabayasi Y, Ishikawa K, Tabata S (1994) Prediction of the coding sequences of unidentified human genes. II. The coding sequences of 40 new genes (KIAA0041–KIAA0080) deduced by analysis of cDNA clones from human cell line KG-1. DNA Res 1:223–229

Norton JD (2000) ID helix-loop-helix proteins in cell growth, differentiation and tumorigenesis. J Cell Sci 113 (Pt 22): 3897–3905

Parravicini C, Chandran B, Corbellino M, Berti E, Paulli M, Moore PS, Chang Y (2000) Differential viral protein expression in Kaposi's sarcoma-associated herpesvirus-infected diseases: Kaposi's sarcoma, primary effusion lymphoma, and multicentric Castleman's disease. Am J Pathol 156:743–749

Paulose-Murphy M, Ha NK, Xiang C, Chen Y, Gillim L, Yarchoan R, Meltzer P, Bittner M, Trent J, Zeichner S (2001) Transcription program of human herpesvirus 8 (Kaposi's sarcoma-associated herpesvirus). J Virol 75:4843–4853

Pearce M, Matsumura S, Wilson AC (2005) Transcripts encoding K12, v-FLIP, v-cyclin, and the microRNA cluster of Kaposi's sarcoma-associated herpesvirus originate from a common promoter. J Virol 79:14457–14464

Piolot T, Tramier M, Coppey M, Nicolas JC, Marechal V (2001) Close but distinct regions of human herpesvirus 8 latency-associated nuclear antigen 1 are responsible for nuclear targeting and binding to human mitotic chromosomes. J Virol 75:3948–3959.

Platt GM, Cannell E, Cuomo ME, Singh S, Mittnacht S (2000) Detection of the human herpesvirus 8-encoded cyclin protein in primary effusion lymphoma-derived cell lines. Virology 272:257–266

Platt GM, Simpson GR, Mittnacht S, Schulz TF (1999) Latent nuclear antigen of Kaposi's sarcoma-associated herpesvirus interacts with RING3, a homolog of the *Drosophila* female sterile homeotic (fsh) gene. J Virol 73:9789–9795

Polakis P (2000) Wnt signaling and cancer. Genes Dev 14:1837–1851

Qiu Y, Sharma A, Stein R (1998) p300 mediates transcriptional stimulation by the basic helix-loop-helix activators of the insulin gene. Mol Cell Biol 18:2957–2964

Radkov SA, Kellam P, Boshoff C (2000) The latent nuclear antigen of Kaposi sarcoma-associated herpesvirus targets the retinoblastoma-E2F pathway and with the oncogene Hras transforms primary rat cells. Nat Med 6:1121–1127

Rainbow L, Platt GM, Simpson GR, Sarid R, Gao SJ, Stoiber H, Herrington CS, Moore PS, Schulz TF (1997) The 222- to 234-kilodalton latent nuclear protein (LNA) of Kaposi's sarcoma-associated herpesvirus (human herpesvirus 8) is encoded by orf73 and is a component of the latency-associated nuclear antigen. J Virol 71:5915–5921

Renne R, Barry C, Dittmer D, Compitello N, Brown PO, Ganem D (2001) Modulation of cellular and viral gene expression by the latency-associated nuclear antigen of Kaposi's sarcoma-associated herpesvirus. J Virol 75:458–468

Renne R, Blackbourn D, Whitby D, Levy J, Ganem D (1998) Limited transmission of Kaposi's sarcoma-associated herpesvirus in cultured cells. J Virol 72:5182–5188

Renne R, Lagunoff M, Zhong W, Ganem D (1996) The size and conformation of Kaposi's sarcoma-associated herpesvirus (human herpesvirus 8) DNA in infected cells and virions. J Virol 70:8151–8154

Ritzi M, Tillack K, Gerhardt J, Ott E, Humme S, Kremmer E, Hammerschmidt W, Schepers A (2003) Complex protein-DNA dynamics at the latent origin of DNA replication of Epstein-Barr virus. J Cell Sci 116:3971–3984

Russo JJ, Bohenzky RA, Chien MC, Chen J, Yan M, Maddalena D, Parry JP, Peruzzi D, Edelman IS, Chang Y, Moore PS (1996) Nucleotide sequence of the Kaposi sarcoma-associated herpesvirus (HHV8). Proc Natl Acad Sci USA 93:14862–14867

Sakanaka C, Leong P, Xu L, Harrison SD, Williams LT (1999) Casein kinase iepsilon in the wnt pathway: regulation of β-catenin function. Proc Natl Acad Sci USA 96:12548–12552

Samols MA, Hu J, Skalsky RL, Renne R (2005) Cloning and identification of a microRNA cluster within the latency-associated region of Kaposi's sarcoma-associated herpesvirus. J Virol 79:9301–9305

Sarid R, Flore O, Bohenzky RA, Chang Y, Moore PS (1998) Transcription mapping of the Kaposi's sarcoma-associated herpesvirus (human herpesvirus 8) genome in a body cavity-based lymphoma cell line (BC-1). J Virol 72:1005–1012

Sarid R, Wiezorek JS, Moore PS, Chang Y (1999) Characterization and cell cycle regulation of the major Kaposi's sarcoma-associated herpesvirus (human herpesvirus 8) latent genes and their promoter. J Virol 73:1438–1446

Schafer A, Lengenfelder D, Grillhosl C, Wieser C, Fleckenstein B, Ensser A (2003) The latency-associated nuclear antigen homolog of herpesvirus saimiri inhibits lytic virus replication. J Virol 77:5911–5925

Schalling M, Ekman M, Kaaya EE, Linde A, Biberfeld P (1995) A role for a new herpes virus (KSHV) in different forms of Kaposi's sarcoma. Nat Med 1:707–708

Schiller M, Verrecchia F, Mauviel A (2003) Cyclic adenosine 3′,5′-monophosphate-elevating agents inhibit transforming growth factor-β-induced SMAD3/4-dependent transcription via a protein kinase A-dependent mechanism. Oncogene 22:8881–8890

Schindl M, Oberhuber G, Obermair A, Schoppmann SF, Karner B, Birner P (2001) Overexpression of Id-1 protein is a marker for unfavorable prognosis in early-stage cervical cancer. Cancer Res 61:5703–5706

Schneider TD (2001) Strong minor groove base conservation in sequence logos implies DNA distortion or base flipping during replication and transcription initiation. Nucleic Acids Res 29:4881–4891

Schulz TF (1999) Epidemiology of Kaposi's sarcoma-associated herpesvirus/human herpesvirus 8. Adv Cancer Res 76:121–160

Schwam DR, Luciano RL, Mahajan SS, Wong L, Wilson AC (2000) Carboxy terminus of human herpesvirus 8 latency-associated nuclear antigen mediates dimerization, transcriptional repression, and targeting to nuclear bodies. J Virol 74:8532–8540

Searles RP, Bergquam EP, Axthelm MK, Wong SW (1999) Sequence and genomic analysis of a Rhesus macaque rhadinovirus with similarity to Kaposi's sarcoma-associated herpesvirus/human herpesvirus 8. J Virol 73:3040–3053

Sherr CJ, McCormick F (2002) The RB and p53 pathways in cancer. Cancer Cell 2:103–112

Shinohara H, Fukushi M, Higuchi M, Oie M, Hoshi O, Ushiki T, Hayashi J, Fujii M (2002) Chromosome binding site of latency-associated nuclear antigen of Kaposi's sarcoma-associated herpesvirus is essential for persistent episome maintenance and is functionally replaced by histone H1. J Virol 76:12917–12924

Si H, Robertson ES (2006) Kaposi's sarcoma-associated herpesvirus-encoded latency-associated nuclear antigen induces chromosomal instability through inhibition of p53 function. J Virol 80:697–709

Srinivasan V, Komatsu T, Ballestas ME, Kaye KM (2004) Definition of sequence requirements for latency-associated nuclear antigen 1 binding to Kaposi's sarcoma-associated herpesvirus DNA. J Virol 78:14033–14038

Staskus KA, Zhong W, Gebhard K, Herndier B, Wang H, Renne R, Beneke J, Pudney J, Anderson DJ, Ganem D, Haase AT (1997) Kaposi's sarcoma-associated herpesvirus gene expression in endothelial (spindle) tumor cells. J Virol 71:715–719

Stedman W, Deng Z, Lu F, Lieberman PM (2004) ORC, MCM, and histone hyperacetylation at the Kaposi's sarcoma-associated herpesvirus latent replication origin. J Virol 78:12566–12575

Sturzl M, Wunderlich A, Ascherl G, Hohenadl C, Monini P, Zietz C, Browning PJ, Neipel F, Biberfeld P, Ensoli B (1999) Human herpesvirus-8 (HHV-8) gene expression in Kaposi's sarcoma (KS) primary lesions: an in situ hybridization study. Leukemia 13 Suppl 1: S110–112

Sun R, Lin SF, Staskus K, Gradoville L, Grogan E, Haase A, Miller G (1999) Kinetics of Kaposi's sarcoma-associated herpesvirus gene expression. J Virol 73:2232–2242

Szekely L, Chen F, Teramoto N, Ehlin-Henriksson B, Pokrovskaja K, Szeles A, Manneborg-Sandlund A, Lowbeer M, Lennette ET, Klein G (1998) Restricted expression of Epstein-Barr virus (EBV)-encoded, growth transformation-associated antigens in an EBV- and human herpesvirus type 8-carrying body cavity lymphoma line. J Gen Virol 79 (Pt 6): 1445–1452

Szekely L, Kiss C, Mattsson K, Kashuba E, Pokrovskaja K, Juhasz A, Holmvall P, Klein G (1999) Human herpesvirus-8-encoded LNA-1 accumulates in heterochromatin-associated nuclear bodies. J Gen Virol 80 (Pt 11): 2889–2900

Talbot SJ, Weiss RA, Kellam P, Boshoff C (1999) Transcriptional analysis of human herpesvirus-8 open reading frames 71, 72, 73, K14, and 74 in a primary effusion lymphoma cell line. Virology 257:84–94

Tang J, Gordon GM, Muller MG, Dahiya M, Foreman KE (2003) Kaposi's sarcoma-associated herpesvirus latency-associated nuclear antigen induces expression of the helix-loop-helix protein Id-1 in human endothelial cells. J Virol 77:5975–5984

Tomescu C, Law WK, Kedes DH (2003) Surface downregulation of major histocompatibility complex class I, PE-CAM, and ICAM-1 following de novo infection of endothelial cells with Kaposi's sarcoma-associated herpesvirus. J Virol 77:9669–9684

Verma SC, Borah S, Robertson ES (2004) Latency-associated nuclear antigen of Kaposi's sarcoma-associated herpesvirus up-regulates transcription of human telomerase reverse transcriptase promoter through interaction with transcription factor Sp1. J Virol 78:10348–10359

Verma SC, Choudhuri, T., Kaul, R., and Robertson, ES (2006) Latency associated nuclear antigen of Kaposi's sarcoma associated herpesvirus interacts with origin recognition complexes at the LANA binding sequence within the terminal repeats. J Virology 80: In Press

Verma SC, Robertson ES (2003) Molecular biology and pathogenesis of Kaposi sarcoma-associated herpesvirus. FEMS Microbiol Lett 222:155–163

Viejo-Borbolla A, Ottinger M, Bruning E, Burger A, Konig R, Kati E, Sheldon JA, Schulz TF (2005) Brd2/RING3 interacts with a chromatin-binding domain in the Kaposi's sarcoma-associated herpesvirus latency-associated nuclear antigen 1 (LANA-1) that is required for multiple functions of LANA-1. J Virol 79:13618–13629

Virgin HW, Latreille P, Wamsley P, Hallsworth K, Weck KE, Dal Canto AJ, Speck SH (1997) Complete sequence and genomic analysis of murine gammaherpesvirus 68. J Virol 71:5894–5904

Waltzer L, Bienz M (1998) *Drosophila* CBP represses the transcription factor TCF to antagonize Wingless signalling. Nature 395:521–525

Weninger W, Partanen TA, Breiteneder-Geleff S, Mayer C, Kowalski H, Mildner M, Pammer J, Sturzl M, Kerjaschki D, Alitalo K, Tschachler E (1999) Expression of vascular endothelial growth factor receptor-3 and podoplanin suggests a lymphatic endothelial cell origin of Kaposi's sarcoma tumor cells. Lab Invest 79:243–251

Werness BA, Levine AJ, Howley PM (1990) Association of human papillomavirus types 16 and 18 E6 proteins with p53. Science 248:76–79

Willert K, Brink M, Wodarz A, Varmus H, Nusse R (1997) Casein kinase 2 associates with and phosphorylates dishevelled. EMBO J 16:3089–3096

Wu WS, Xu ZX, Ran R, Meng F, Chang KS (2002) Promyelocytic leukemia protein PML inhibits Nur77-mediated transcription through specific functional interactions. Oncogene 21:3925–3933

Yates PR, Atherton GT, Deed RW, Norton JD, Sharrocks AD (1999) Id helix-loop-helix proteins inhibit nucleoprotein complex formation by the TCF ETS-domain transcription factors. Embo J 18:968–976

Ye FC, Zhou FC, Yoo SM, Xie JP, Browning PJ, Gao SJ (2004) Disruption of Kaposi's sarcoma-associated herpesvirus latent nuclear antigen leads to abortive episome persistence. J Virol 78:11121–11129

Zhang YJ, Deng JH, Rabkin C, Gao SJ (2000) Hot-spot variations of Kaposi's sarcoma-associated herpesvirus latent nuclear antigen and application in genotyping by PCR-RFLP. J Gen Virol 81:2049–2058

Zhong W, Ganem D (1997) Characterization of ribonucleoprotein complexes containing an abundant polyadenylated nuclear RNA encoded by Kaposi's sarcoma-associated herpesvirus (human herpesvirus 8). J Virol 71:1207–1212

Zhong W, Wang H, Herndier B, Ganem D (1996) Restricted expression of Kaposi sarcoma-associated herpesvirus (human herpesvirus 8) genes in Kaposi sarcoma. Proc Natl Acad Sci USA 93:6641–6646

CTMI (2006) 312:137–156

The KSHV and Other Human Herpesviral G Protein-Coupled Receptors

M. Cannon (✉)

Cancer Research UK Viral Oncology Group, Wolfson Institute for Biomedical Research University College London, The Cruciform Building, London WC1E 6BT, UK
m.cannon@ucl.ac.uk

Abstract Kaposi sarcoma-associated herpesvirus (KSHV) is a γ2-herpesvirus discovered in 1994 and is the agent responsible for Kaposi sarcoma (KS), an endothelial cell malignancy responsible for significant morbidity and mortality worldwide. Over time, KSHV has pirated many human genes whose products regulate angiogenesis, inflammation, and the cell cycle. One of these encodes for a mutated G protein-coupled receptor (GPCR) that is a homologue of the human IL-8 receptor. GPCRs are the largest family of signaling molecules and respond to a wide array of ligands. Unlike its normal counterpart, the mutations present in KSHV vGPCR result in constitutive, ligand-independent signaling activity. Signaling by the KSHV vGPCR results in the elaboration of many mitogenic and angiogenic cytokines that are vital to the biology of KS and other KSHV-driven malignancies. Several other herpesviruses also encode GPCRs, the functions of which are under ongoing investigation. In addition, several human diseases are associated with mutated mammalian GPCRs in germline or somatic cells.

1 Introduction

Kaposi sarcoma (KS) is a multicellular tumor consisting of spindle cells of lymphatic endothelial origin and infiltrating hematopoietic cells (Kaaya et al. 1995; Wang et al. 2004). It is a cytokine-driven process that begins as a polyclonal proliferation with a low number of spindle cells and a robust inflammatory component. Although they induce an inflammatory and angiogenic reaction, KS spindle cells are not fully transformed in that they are dependent on exogenous growth factors in culture and do not form tumors in animal models (Salahuddin et al. 1988). Over time, KS lesions may evolve into a true clonal malignancy as evidenced by X chromosome inactivation studies and comparisons of multiple lesions from a single patient (Rabkin et al. 1997). Kaposi sarcoma-associated herpesvirus (KSHV), or human herpesvirus-8 (HHV-8), is a γ2-human herpesvirus that was discovered in 1994 and is the etiologic agent in all epidemiologic forms of KS (Chang et al. 1994, 1996b; Boshoff et al. 1995; Dupin et al. 1995; Whitby et al. 1995). It is also present in all forms of primary effusion lymphoma (PEL) and is associated with a more aggressive phenotype of multicentric Castleman disease (MCD) (Soulier et al. 1995; Cesarman et al. 1996a; Dupin et al. 2000). Like Epstein-Barr virus (EBV), KSHV is lymphotropic; its ability to infect B cells and monocytes is vital to new virion production and to the pathobiology of all KSHV-mediated disease.

One of the first potential KSHV oncogenes described was the viral G protein-coupled receptor (vGPCR) encoded by ORF 74 (Arvanitakis et al. 1997; Guo et al. 1997). GPCRs are also known as seven-transmembrane domain receptors or serpentine receptors and are the largest family of signal-transducing molecules, estimated to have a thousand members. By some accounts, GPCRs are the target of up to 30% of all clinically available medications (Gershengorn and Osman 2001). The amino terminus of GPCRs is situated extracellularly, whereas the carboxy terminus sits intracellularly. The seven α-helices span the cell membrane and make up the transmembrane bundle. Potential ligands include neurotransmitters, growth factors, hormones, calcium ions, and photons. Binding of ligand induces a relative shift in the positions of the transmembrane α-helices. Small ligands may bind within the transmembrane bundle, whereas larger ligands bind the amino termini and extracellular loops. Ligand-induced movement of the α-helices causes conformational changes of the intracellular loops that activate the heterotrimeric GTP-binding (G) protein. Once activated, the G protein exchanges GDP for GTP and splits into two separate signaling effectors: the α-subunit and the βγ-subunit. Normal G proteins contain an intrinsic

GTPase activity that converts GTP back to GDP, resulting in reassembly of the α and βγ subunits. Depending on which type(s) of heterotrimeric G proteins it binds, a given GPCR can signal via phospholipase C (PLC), ion channels or adenylyl cyclase (for review, see Estep et al. 2003).

The KSHV vGPCR is different from most GPCRs in that it signals constitutively. Although its activity can be modified up or down by various ligands, it does not require an agonist to signal. Furthermore, vGPCR is a promiscuous receptor in that it binds chemokines of various families and activates a broad range of signaling cascades depending on the cellular context. The result is a complicated network of transcription factor activation affecting both cellular and viral transcription patterns. Some models suggest that vGPCR is directly transforming and can be considered a viral oncogene. Other studies argue that vGPCR influences KSHV-mediated disease in a paracrine fashion via its ability to elicit the production of various cytokines, thereby altering the tumor microenvironment in KS, PEL, and MCD. As discussed below, there is good evidence that vGPCR may play multiple roles depending on the requirements of the KSHV viral life cycle and whether it is signaling in the context of endothelial or hematopoietic cells.

2
Overview of the KSHV vGPCR

During its coevolution with its human host, KSHV has acquired many homologues of human genes involved in cell cycle regulation, angiogenesis, and inflammation (Chang et al. 1996a; Russo et al. 1996; Cheng et al. 1997; Neipel et al. 1997; Swanton et al. 1997; Wong 1998). KSHV vGPCR has the potential to affect all these processes. Like ORF 74 of murine gammaherpesvirus 68, KSHV vGPCR is a constitutively active variant of the human IL-8 receptors CXCR1 and CXCR2, as well as herpesvirus saimiri ECRF3 (Ahuja and Murphy 1993; Cesarman et al. 1996b; Arvanitakis et al. 1997; Guo et al. 1997; Wakeling et al. 2001). vGPCR was first shown to have oncogenic potential in fibroblasts, but later work also proved that vGPCR could immortalize human umbilical vein endothelial cells (HUVEC) and protect them from serum deprivation (Bais et al. 1998; Couty et al. 2001; Montaner et al. 2001). In vivo, KSHV infects cells of endothelial and hematopoietic origin. Signaling molecules can have very different effects depending on the cellular context, so these initial studies in endothelial cells were important in establishing vGPCR as a vital component of KSHV-mediated cellular proliferation.

KSHV vGPCR can influence the expression of various cytokines involved in the biology of KS and PEL. For example, KS depends on vascular endothelial growth factor (VEGF) for its highly vascular morphology, and KS spindle cells secrete and respond in an autocrine manner to VEGF (Masood et al. 1997; Samaniego et al. 1998). Sodhi et al. show that vGPCR mediates VEGF secretion by stimulating the transcription factor hypoxia-inducible factor (HIF)-1α in fibroblasts (Sodhi et al. 2000). Conditioned medium from vGPCR-expressing fibroblasts in turn stimulates endothelial cell growth and the switch to an angiogenic phenotype (Bais et al. 1998). A fascinating study by Bais et al. in HUVEC shows that vGPCR induces immortalization with constitutive VEGF receptor-2/ KDR expression and activation. This was associated with antisenescence mediated by alternative lengthening of telomeres and an antiapoptotic response (Bais et al. 2003). vGPCR expression also increases VEGF production from hematopoietic cells (Cannon et al. 2003); this is significant in that VEGF is essential for PEL tumor growth and ascites production in mice and furthermore contributes to a more aggressive phenotype of MCD (Aoki and Tosato 1999, 2001; Nishi and Maruyama 2000).

In addition to VEGF, vGPCR induces the expression of many proinflammatory cytokines important in KSHV-mediated disease. The monocytic line THP-1 expresses IL-1β, TNFα, and IL-6, and Jurkat cells elaborate IL-2 and IL-4 when transfected with vGPCR (Schwarz and Murphy 2001). In PEL cell lines, vGPCR upregulates KSHV vIL-6, an important growth factor in PEL and MCD (Aoki et al. 1999; Jones et al. 1999; Liu et al. 2001). In endothelial cells, vGPCR upregulates the NFκB-dependent inflammatory cytokines RANTES, IL-6, IL-8, and granulocyte-macrophage colony-stimulating factor (GM-CSF). It also induces expression of the adhesion molecules VCAM-1, ICAM-1, and E-selectin. In the same study, supernatants from transfected KS cells activated NF-κB signaling in untransfected cells and caused the chemotaxis of monocytoid and T-lymphoid cells (Pati et al. 2001). This suggests that vGPCR may help to recruit hematopoietic cells that make up the inflammatory component of KS lesions.

Although it is directly transforming in fibroblasts, and perhaps responsible for an immortalizing autocrine loop in endothelial cells (Arvanitakis et al. 1997; Bais et al. 1998; Bais et al. 2003), the ability of vGPCR to cause such robust and broad cytokine activity argues for an important paracrine role in KSHV-mediated disease. This is supported by the first vGPCR mouse model in which vGPCR was transcribed under the CD2 promoter (primarily T cell) and resulted in multicentric, angioproliferative lesions histologically similar to KS (Yang et al. 2000). The lesions ranged from erythematous maculae to vascular tumors; they contained spindle and inflammatory cells and expressed CD34 and VEGF. Despite causing endothelial cell tumors, vGPCR was expressed from relatively few infiltrating T cells. Other mouse models and evidence for a paracrine role for vGPCR are discussed in Sect. 3.2.

KSHV vGPCR signaling may help to explain why AIDS-associated KS is a more aggressive tumor than classic or iatrogenic KS. It has been known for several years that the HIV-1 transactivator protein Tat activates KS cell growth and contributes directly to KS pathogenesis (Barillari et al. 1993; Ensoli et al. 1993; Cantaluppi et al. 2001). Pati et al. have recently shown that HIV-1 Tat and vGPCR synergistically activate the transcription factor nuclear factor of activated T cells (NFAT). In turn, NFAT activation is responsible for production of IL-2, IL-4, GM-CSF, and TNF-αas well as expression of ICAM-1, CD25, CD29, and Fas ligand by the T cell line HUT 78 (Pati et al. 2003). The expression of these surface molecules and inflammatory cytokines results in increased endothelial cell-T cell adherence and has important implications for vGPCR as a key molecule in the interaction between KSHV and HIV-1. Indeed, later work by the same group confirmed that HIV-1 Tat increases activation of NFκB and NFAT and accelerates tumor formation in mice that are implanted with a cell line derived from a vGPCR transgenic mouse (Guo et al. 2004).

3 KSHV vGPCR Signaling and Function

The signaling cascades that vGPCR utilizes to effect changes in viral and host cell transcription are varied and in some cases cell type specific. Despite increased understanding of vGPCR's signaling potential, correlation of these findings with biological significance remains more problematic. In this section I shall discuss what is known about the structure of vGPCR vis-à-vis its constitutive activity, the diversity of vGPCR signaling, as well as the potential roles in KSHV-mediated disease attributable to this unique receptor.

3.1 Multiple Ligands and Many Signaling Cascades

KSHV vGPCR is in the rhodopsin/β_2-adrenergic subfamily of GPCRs. The other two smaller subfamilies include the secretin-like receptors and the calcium-sensing/metabotropic glutamate receptors. Subfamilies are based on structural and genetic characteristics but share little intrafamilial sequence homology. Unlike normal mammalian GPCRs, vGPCR signals in the absence of ligand. This constitutive activity is largely due to abnormal transmembrane helices two and three, as well as its cytoplasmic tail (Ho et al. 2001; Schwarz and Murphy 2001). Additional mutational studies show that although the amino terminus is required for ligand binding, it is not necessary for constitutive signaling (Ho et al. 1999; Rosenkilde and Schwartz 2000). Interestingly, despite its constitutive activity, vGPCR retains its ability to respond to various CXC and CC-type chemokines; these include agonists, inverse agonists, and neutral ligands. Agonists that increase signaling above constitutive levels include IL-8 and growth-related oncogene-α (GROα) (Arvanitakis et al. 1997; Guo et al. 1997; Geras-Raaka et al. 1998b, 1998c; Gershengorn et al. 1998; Rosenkilde and Schwartz 2000). Neutral ligands have no effect on constitutive signaling but compete for binding, thereby inhibiting the effects of other ligands. These include neutrophil-activating peptide-2 (NAP-2) and epithelial cell-derived neutrophil-activating 78 (ENA-78) (Rosenkilde and Schwartz 2000). Inverse agonists inhibit constitutive signaling and include interferon-γ-inducible protein-10 (IP-10), stromal cell-derived factor-1 (SDF-1) ,and KSHV vMIP-II, a viral CC chemokine (Geras-Raaka et al. 1998b, 1998c). Of note, it is the angiogenic ELR-positive (Glu-Leu-Arg prior to the first cysteine) CXC chemokines that are vGPCR agonists, while the non-ELR chemokines are inverse agonists (Strieter et al. 1995). This has implications for vGPCR function and is discussed below. Another mechanism by which vGPCR may be modulated in vivo is via the GPCR-specific kinases (GRK), particularly the family containing GRK-4, -5, and −6 (Premont et al. 1995; Geras-Raaka et al. 1998a). GPCRs in an active state can be bound by GRKs that lead to interaction with arrestin and inhibition of G protein interaction. The biologic importance of GRK-mediated downregulation of vGPCR remains to be studied.

KSHV vGPCR not only binds a huge array of ligands but utilizes a surprising number of downstream effectors. These are cell type specific, so I shall concentrate on the data in endothelial and hematopoietic cells because they are the natural targets for KSHV infection. vGPCR can signal via Gαi and Gαq in both cell types (Couty et al. 2001; Cannon and Cesarman 2004). Studies in HeLa cells suggest that coupling to Gα13 and RhoA may also be possible, but this remains to be confirmed in a biologically relevant cell type (Shepard et al. 2001). The βγ-subunits of both pertussis-sensitive (Gαi) and pertussis-insensitive α-subunits also mediate vGPCR signaling (Couty et al. 2001; Montaner et al. 2001; Smit et al. 2002; Cannon and Cesarman 2004). The result is the activation of several kinases involved in endothelial and B cell proliferation. For example, vGPCR-mediated activation of a $G\alpha_i$-PI3K/Akt axis promotes endothelial cell survival; this is likely via activation of NFκB (Couty et al. 2001; Montaner et al. 2001; Pati et al. 2001). Although NFκB is also activated in B cells, we have found that it is neither $G\alpha_i$- nor PI3K/Akt dependent but rather mediated by a non-pertussis-sensitive G protein such as $G\alpha_q$ (Cannon and Cesarman 2004). vGPCR also activates the mitogen- and stress-activated protein kinases (MAPK, SAPK) involved in cellular proliferation, angiogenesis, and inflammation via both $G\alpha_q$ and $G\alpha_i$ G proteins (Bais et al. 1998; Smit et al. 2002; Cannon et al. 2003). This results in additional transcription factor activation including AP-1, CREB, and HIF1-α; the latter has been shown to result in VEGF expression from COS-7 cells and may well apply to endothelial and hematopoietic cells (Sodhi et al. 2000).

NFAT is a transcription factor that acts in conjunction with AP-1 to enhance the expression of many proteins in the productive immune response; furthermore, it is required for the calcium-mediated latent-to-lytic switch in KSHV transcription pattern (Macian et al. 2001; Zoeteweij et al. 2001). We and others have shown that vGPCR induces NFAT activity via combined interaction with $G\alpha_q$ and $G\alpha_i$ and that this is augmented by HIV-1 Tat (Cannon et al. 2003; Pati et al. 2003). Chiou et al. discovered that vGPCR upregulates transcription via the T1.1/PAN, K1, and LLP latency promoters of KSHV, while we show a vGPCR-induced increase in the lytic products ORF 50 and ORF 57 in latently infected PEL cells (Chiou et al. 2002; Cannon et al. 2003). So, in addition to its effects on host cell transcription, it is clear that vGPCR has important effects on KSHV transcription patterns; as discussed below, this has important implications for the role of vGPCR in the KSHV viral life cycle.

3.2
KSHV vGPCR and Disease Pathogenesis

KSHV vGPCR is a promiscuous receptor that activates cell survival signaling cascades, results in the elaboration of various cytokines, and can affect KSHV transcription patterns. Any or all of these functions may be vital to the pathogenesis of KSHV-mediated endothelial and B cell hyperproliferative syndromes. Some of the first studies showed that vGPCR transforms fibroblasts (Arvanitakis et al. 1997). However, arguing for a directly transforming role in KSHV-mediated disease is difficult: Unlike the latent transforming viral genes of EBV, vGPCR is expressed in the immediate-early lytic phase (Kirshner et al. 1999; Nador et al. 2001). A lytically activated host cell is destined to die in the process of new virion production, so to express a viral oncogene under such circumstances appears at odds with promoting an autocrine survival advantage. Some have argued that early vGPCR expression in endothelial cells may set up a self-perpetuating growth-promoting paracrine loop via upregulation of VEGF and VEGF receptors (Bais et al. 2003); this is a very enticing scenario but requires further investigation.

The bulk of evidence indicates that vGPCR has its most potent effects via a paracrine mechanism. KS and PEL tumors show that a relatively small subset of cells are KSHV infected and even fewer express lytic genes such as vGPCR (Staskus et al. 1997; Teruya-Feldstein et al. 1998; Dupin et al. 1999; Linderoth et al. 1999). Likewise, vGPCR-driven KS-like tumors in mice result from vGPCR expression by a subset of tumor cells (Yang et al. 2000; Guo

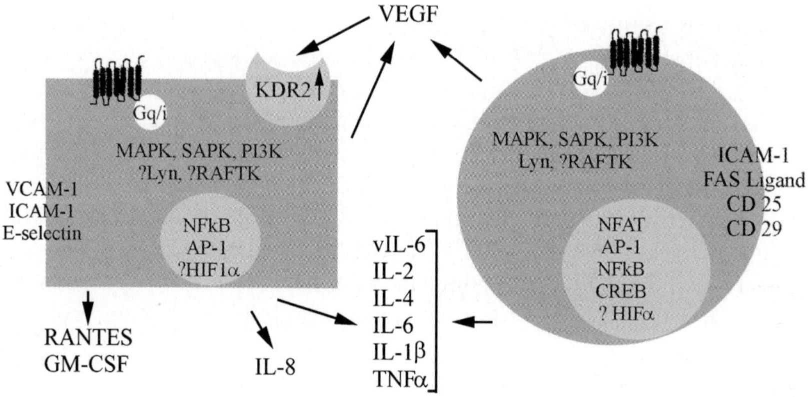

Fig. 1 Signaling intermediates, transcription factors, cytokines, and cell surface molecules upregulated by KSHV vGPCR in endothelial cells (*left*) and hematopoietic cells (*right*). In both cell types, vGPCR signals via $G\alpha_q$ and $G\alpha_i$ G proteins and activates a broad array of effector kinases that result in enhanced transcription factor activity. In addition to autocrine cell survival signals, various differentiation, angiogenic, and immunomodulatory factors are elaborated. Elements with *question marks* indicate that studies were not done in these specific cell types

et al. 2003). All KSHV-driven tumors display some level of lytic replication. Given the instability of KSHV infection in most cell types, this is probably required to maintain ongoing infection and new virion production (Lagunoff et al. 2002). When expressed from these lytically activated host cells, vGPCR could provide growth-promoting cytokine expression; and as a chemokine receptor involved in cell migration, it could also recruit new infectable cells to the tumor microenvironment. Figure 1 summarizes some of the cytokines and cell adhesion molecules upregulated by vGPCR that could explain its paracrine-driven proliferative effect on both hematopoietic and endothelial cells (also see Sect. 2).

Despite its obvious potential to affect the tumor microenvironment, if vGPCR is expressed only briefly during the early lytic phase, the question remains as to how relevant its effects are in vivo. However, some evidence suggests that vGPCR can be expressed in a dysregulated way, outside the normal KSHV lytic program (Sodhi et al. 2004). For example, HIV Tat can increase the expression of vGPCR, and vGPCR transcription is upregulated by RBP-J, a transcription factor and target of the Notch pathway (Yen-Moore et al. 2000; Liang and Ganem 2004). If sustained upregulation of vGPCR were to occur in an abnormal or abortive lytic phase, it would become easier to reconcile its in vitro potential with tumorigenesis and effects on the viral life cycle. Interestingly, abortive lytic cycle progression in which a subset of lytic genes are expressed has been found in other herpesviruses. Dysregulated vGPCR expression would have implications for our recent work. We have shown that vGPCR signaling in PEL cells leads to a p21-mediated inhibition of cyclin-dependent kinase-2 (Cdk2) and consequent cell cycle arrest (Cannon et al. 2006). Furthermore, inhibition of Cdk2 drastically suppresses the efficiency of chemically induced reactivation of KSHV as evidenced by transcription of the lytic products ORF 50 and ORF 26. This opens up the possibility that dysregulated vGPCR expression leads to inhibition of full productive lytic replication. In this state of limbo between normal latent and lytic phases, vGPCR-mediated elaboration of angiogenic and mitogenic chemokines would have a more prolonged and biologically important effect.

Reconciling all the data on KSHV vGPCR into one well-circumscribed function is not yet possible. In fact, current concepts in virology and tumorigenesis may require us to interpret vGPCR as having multiple seemingly discrete functions. It is clear, however, that KSHV has evolved toward tight regulation of vGPCR: Its expression is restricted in that it is transcribed within the 3′-end of a bicistronic message; furthermore, KSHV encodes its own inverse agonist of vGPCR, vMIP II. Such fine-tuning of vGPCR signaling suggests that vGPCR may play different roles at different points in the KSHV life cycle. Work by Dezube et al. shows that during early de novo KSHV infection of endothelial cells, vGPCR transcription fluctuates in a cyclic pattern every 48–72 h, consistent with viral replication (Dezube et al. 2002). So, in addition to maintaining a suitable tumor microenvironment as discussed above, it may be that vGPCR has yet another role in early infection, perhaps to establish successful latency or encourage the initial rounds of KSHV replication.

4 Other Pathogenic GPCRs

Virus-encoded GPCRs have been described in several DNA viruses including poxviruses and the β- and γ-herpesvirus (Table 1). As exemplified by KSHV vGPCR, these viral GPCRs are more promiscuous in their use of ligands and downstream signaling cascades than their mammalian homologues. Such broad signaling has resulted in many postulated roles for viral GPCRs: immune evasion, viral entry and replication, cell migration, and even direct cell transformation. In addition to virally encoded constitutive GPCRs, several mutated mammalian GPCRs are also responsible for human disease.

Table 1 Human herpesvirus G protein-coupled receptors

Virus	Family	Gene	Constitutive	References
CMV	β	US27	–	Fraile-Ramos et al. 2002
		US28	+	Kledal et al. 1998
		UL33	+	Marguiles et al. 1996
		UL78	–	Bankier, DNA Seq., 1991 2;1
HHV-6	β	U12	–	Gompels et al. 1995
				Isegawa et al. 1998
		U51	–	Milne et al. 2000
HHV-7	β	U12	–	Nicholas et al. 1996
				Tadagaki et al. 2005
	β	U51	–	Nicholas et al. 1996
				Tadagaki et al. 2005
EBV	γ1	BILF1	+	Beisser et al. 2005
				Paulsen et al. 2005
KSHV	γ2	ORF 74	+	Bais et al. 1998
				Rosenkilde et al. 1999

4.1 Human Herpesviral GPCRs

EBV is a ubiquitous lymphotropic γ1-herpesvirus that infects and remains latent in over 90% of people by the time they reach middle age. Manifestations of acute infection often go unnoticed in children but can cause infectious mononucleosis in adolescents and adults. EBV infects the oropharyngeal epithelium and surface B cells of the tonsils; via the latter it disseminates throughout the reticuloendothelial system. Although infection persists throughout life, the immune-competent host generally suffers no long-term ill effects after the EBV-driven polyclonal B cell proliferation subsides, thanks in part to a vigorous cytotoxic T cell response. EBV can, however, contribute to more sinister sequelae: nasopharyngeal carcinoma is most prevalent in southeastern China; Burkitt lymphoma is one of the most common childhood malignancies in sub-Saharan Africa; posttransplant lymphoproliferative disorder is EBV driven; and two-thirds of AIDS-related non-Hodgkin lymphomas are EBV positive.

Very recently the EBV open reading frame BILF1 was demonstrated to encode a functional, constitutively active GPCR that is heavily glycosylated and localizes to the plasma membrane in epithelial cell lines. The BILF1 signals via $G\alpha_i$, but not $G\alpha_{q/11}$, as evidenced by its ability to inhibit forskolin-stimulated CREB activity in a pertussis-sensitive manner in COS-7 cells. Like

KSHV vGPCR, BILF1 appears to be an immediate-early product and can activate NFκB. Interestingly, BILF1 reduces levels of phosphorylated RNA-dependent antiviral protein kinase (PKR). PKR is an interferon-inducible enzyme with a role in intracellular antiviral defense; BILF1 is the first GPCR shown to potentially inhibit this pathway (Williams 1999; Beisser et al. 2005; Paulsen et al. 2005). Further studies of BILF1's role in the EBV life cycle and its influence over host antiviral responses are exciting prospects.

The β-herpesvirus family includes cytomegalovirus (CMV) and human herpesvirus-6 and -7 (HHV-6, -7). In healthy hosts, acute CMV infection is a self-limited flulike syndrome, and by adulthood, 70%–90% of people are latently infected. Infection during pregnancy, however, can lead to serious birth defects including blindness, deafness, seizures, and microencephaly. In the setting of AIDS, CMV causes retinitis, gastrointestinal disease, and various neurologic syndromes. In recipients of solid organ transplants, CMV causes hematologic disorders, hepatitis, pneumonitis, and organ rejection. In bone marrow transplants the more common complications are graft-versus host disease, delayed engraftment, and pneumonitis.

The CMV genome encodes four GPCRs, two of which will be discussed here: US28 and UL33. Like KSHV vGPCR, CMV US28 signals independently of an agonist but can be modified by CC chemokines and fractalkine (CX3CL1). Unlike KSHV vGPCR, US28 is located predominantly endosomally, not on the cell surface (Waldhoer et al. 2003). Expression of US28 results in migration of smooth muscle endothelial cells, and it is therefore postulated to play a role in viral dissemination and CMV-driven vascular disease (Kledal et al. 1998; Streblow et al. 1999; Casarosa et al. 2001). A role in immune evasion is also argued by virtue of the high turnover of US28 and the resultant sequestration of host-produced chemokines (Bodaghi et al. 1998). As discussed for KSHV vGPCR, another important role for US28 may lie in its latent expression in hematopoietic cells, resulting in homing to sites of inflammation rather than lymph nodes (Streblow et al. 2003). CMV UL33 also demonstrates constitutive basal activity but binds no known chemokines. It is incorporated into viral particles and expressed on virus-infected cells (Margulies et al. 1996). UL33 activates CREB-mediated signaling and because many CMV gene promoters contain CREs, it has been postulated that UL33 is involved in establishing viral infection or possibly reactivation (Casarosa et al. 2003). The rat and mouse CMV homologues of UL33 are important for replication in salivary glands and for general virulence, but similar studies are not available for CMV UL33 (Beisser et al. 1998).

Infection with HHV-6 and -7 generally occurs early in life. The former is the usual agent of exanthem subitum (also known as roseola infantum or sixth disease), a self-limited syndrome of fever and rash in children . Primary infection later in life causes an infectious mononucleosis syndrome much like EBV or CMV. HHV-7 is not convincingly associated with disease in the normal host but, like HHV-6, may become reactivated in the immuncompromised patient (Clark et al. 2003). Two ligand-dependent GPCRs with homology to the CC-chemokine receptors are encoded by both HHV-6 and HHV-7. U12 and U51 are positional and structural homologues of CMV UL33 and UL78 (Gompels et al. 1995; Nicholas 1996; Isegawa et al. 1998; Milne et al. 2000). HHV-6 UL12 binds RANTES, macrophage inflammatory protein-1α and -1β (MIP-1α, MIP-1β), and monocyte chemoattractant protein 1 (MCP-1); HHV-7 UL12 binds EBI1-ligand chemokine (ELC), secondary lymphoid-tissue chemokine (SLC), thymus and activation-regulated chemokine (TARC), and macrophage-derived chemokine (MDC) (Nakano et al. 2003; Tadagaki et al. 2005). Both U12 homologues are functional and signal via non-$G\alpha_i$ pathways resulting in transient calcium flux. HHV-7 U12 induces migration of stably transfected Jurkat cells toward ELC and SLC, but not TARC or MDC. This leads the authors to postulate that HHV-7 U12 induces the migration of infected cells toward lymph nodes, perhaps facilitating viral transmission (Tadagaki et al. 2005). Little is known of the function of HHV-6 U12.

In both HHV-6 and HHV-7, UL51 is expressed in the immediate-early and early postinfection stages . Both homologs bind many CC-type chemokines as well as the KSHV-encoded homolog of MIP-II (vMIP-II). HHV-6 UL51 downregulates transcription of RANTES, which may lend it an immunomodulatory role, but this requires further confirmation. Even less is known about the function of HHV-7 UL51 (Milne et al. 2000).

4.2
Nonviral Constitutively Active GPCRs and Disease

Aside from viral GPCRs, there exist several constitutively active GPCR mutants associated with human disease that may arise in germline or somatic cells (Coughlin 1994; Spiegel 1996; Arvanitakis et al. 1998). Familial forms of hyperthyroidism and hypoparathyroidism are due to mutations in the thyroid-stimulating hormone receptor in the former and the extracellular domain of the calcium-sensing receptor in the latter (Van Sande et al. 1995; Chattopadhyay et al. 1996). The retinal degeneration of retinitis pigmentosa is thought due to mutations in the seventh transmembrane domain of rhodopsin, leading to overactivation of photoreceptor cells (Robinson et al. 1992). Mutation of the rhodopsin second transmembrane domain leads to congenital night blindness (Rao et al. 1994). A change in the sixth transmembrane helix of the luteinizing hormone receptor causes familial male precocious puberty, as do other activating mutations (Shenker et al. 1993; Kosugi et al. 1995).

5 Conclusions

Both mammalian and viral constitutively active GPCRs are associated with human disease. The study of viral GPCRs continues to teach us about viral reproduction, immune system evasion, and tumorigenesis, and even about normal GPCR function. The KSHV vGPCR is one of the best-studied and most versatile viral GPCRs. Although KSHV vGPCR alone can cause KS-like tumors in animal models, its overall role in the KSHV life cycle and biology of KSHV-driven endothelial and hematopoietic proliferative disorders is only partly understood. Undoubtedly though, its ability to affect cell growth and differentiation, cytokine production, chemotaxis, and viral transcription make it a promising target for rationally designed anti-KSHV therapies.

References

Ahuja SK, Murphy PM (1993) Molecular piracy of mammalian interleukin-8 receptor type B by herpesvirus saimiri. J Biol Chem 268:20691–20694

Aoki Y, Jaffe ES, Chang Y, Jones K, Teruya-Feldstein J, Moore PS, Tosato G (1999) Angiogenesis and hematopoiesis induced by Kaposi's sarcoma-associated herpesvirus-encoded interleukin-6. Blood 93:4034–4043

Aoki Y, Tosato G (1999) Role of vascular endothelial growth factor/vascular permeability factor in the pathogenesis of Kaposi's sarcoma-associated herpesvirus-infected primary effusion lymphomas. Blood 94:4247–4254

Aoki Y, Tosato G (2001) Vascular endothelial growth factor/vascular permeability factor in the pathogenesis of primary effusion lymphomas. Leuk Lymphoma 41:229–237

Arvanitakis L, Geras-Raaka E, Gershengorn MC (1998) Constitutively signaling G-protein-coupled receptors and human disease. TEM 9:27–31

Arvanitakis L, Geras-Raaka E, Varma A, Gershengorn MC, Cesarman E (1997) Human herpesvirus KSHV encodes a constitutively active G-protein-coupled receptor linked to cell proliferation. Nature 385:347–350

Bais C, Santomasso B, Coso O, Arvanitakis L, Raaka EG, Gutkind JS, Asch AS, Cesarman E, Gershengorn MC, Mesri EA (1998) G-protein-coupled receptor of Kaposi's sarcoma-associated herpesvirus is a viral oncogene and angiogenesis activator. Nature 391:86–89

Bais C, Van Geelen A, Eroles P, Mutlu A, Chiozzini C, Dias S, Silverstein RL, Rafii S, Mesri EA (2003) Kaposi's sarcoma associated herpesvirus G protein-coupled receptor immortalizes human endothelial cells by activation of the VEGF receptor-2/KDR. Cancer Cell 3:131–143

Barillari G, Gendelman R, Gallo RC, Ensoli B (1993) The Tat protein of human immunodeficiency virus type 1, a growth factor for AIDS Kaposi sarcoma and cytokine-activated vascular cells, induces adhesion of the same cell types by using integrin receptors recognizing the RGD amino acid sequence. Proc Natl Acad Sci USA 90:7941–7945

Beisser PS, Verzijl D, Gruijthuijsen YK, Beuken E, Smit MJ, Leurs R, Bruggeman CA, Vink C (2005) The Epstein-Barr virus BILF1 gene encodes a G protein-coupled receptor that inhibits phosphorylation of RNA-dependent protein kinase. J Virol 79:441–449

Beisser PS, Vink C, Van Dam JG, Grauls G, Vanherle SJ, Bruggeman CA (1998) The R33 G protein-coupled receptor gene of rat cytomegalovirus plays an essential role in the pathogenesis of viral infection. J Virol 72:2352–2363

Bodaghi B, Jones TR, Zipeto D, Vita C, Sun L, Laurent L, Arenzana-Seisdedos F, Virelizier JL, Michelson S (1998) Chemokine sequestration by viral chemoreceptors as a novel viral escape strategy: withdrawal of chemokines from the environment of cytomegalovirus-infected cells. J Exp Med 188:855–866

Boshoff C, Schulz TF, Kennedy MM, Graham AK, Fisher C, Thomas A, McGee JO, Weiss RA, JJ OL (1995) Kaposi's sarcoma-associated herpesvirus infects endothelial and spindle cells. Nat Med 1:1274–1278

Cannon M, Cesarman E, Boshoff C (2006) KSHV G protein-coupled receptor inhibits lytic gene transcription in primary-effusion lymphoma cells via p21-mediated inhibition of Cdk2. Blood 107:277–284

Cannon M, Philpott NJ, Cesarman E (2003) The Kaposi's sarcoma-associated herpesvirus G protein-coupled receptor has broad signaling effects in primary effusion lymphoma cells. J Virol 77:57–67

Cannon ML, Cesarman E (2004) The KSHV G protein-coupled receptor signals via multiple pathways to induce transcription factor activation in primary effusion lymphoma cells. Oncogene 23:514–523

Cantaluppi V, Biancone L, Boccellino M, Doublier S, Benelli R, Carlone S, Albini A, Camussi G (2001) HIV type 1 Tat protein is a survival factor for Kaposi's sarcoma and endothelial cells. AIDS Res Hum Retroviruses 17:965–976

Casarosa P, Bakker RA, Verzijl D, Navis M, Timmerman H, Leurs R, Smit MJ (2001) Constitutive signaling of the human cytomegalovirus-encoded chemokine receptor US28. J Biol Chem 276:1133–1137

Casarosa P, Gruijthuijsen YK, Michel D, Beisser PS, Holl J, Fitzsimons CP, Verzijl D, Bruggeman CA, Mertens T, Leurs R, Vink C, Smit MJ (2003) Constitutive signaling of the human cytomegalovirus-encoded receptor UL33 differs from that of its rat cytomegalovirus homolog R33 by promiscuous activation of G proteins of the Gq, Gi, and Gs classes. J Biol Chem 278:50010–50023

Cesarman E, Nador RG, Aozasa K, Delsol G, Said JW, Knowles DM (1996a) Kaposi's sarcoma-associated herpesvirus in non-AIDS related lymphomas occurring in body cavities. Am J Pathol 149:53–57

Cesarman E, Nador RG, Bai F, Bohenzky RA, Russo JJ, Moore PS, Chang Y, Knowles DM (1996b) Kaposi's sarcoma-associated herpesvirus contains G protein-coupled receptor and cyclin D homologs which are expressed in Kaposi's sarcoma and malignant lymphoma. J Virol 70:8218–8223

Chang Y, Cesarman E, Pessin MS, Lee F, Culpepper J, Knowles DM, Moore PS (1994) Identification of herpesvirus-like DNA sequences in AIDS-associated Kaposi's sarcoma. Science 266:1865–1869

Chang Y, Moore PS, Talbot SJ, Boshoff CH, Zarkowska T, Godden K, Paterson H, Weiss RA, Mittnacht S (1996a) Cyclin encoded by KS herpesvirus [letter]. Nature 382:410

Chang Y, Ziegler J, Wabinga H, Katangole-Mbidde E, Boshoff C, Schulz T, Whitby D, Maddalena D, Jaffe HW, Weiss RA, Moore PS (1996b) Kaposi's sarcoma-associated herpesvirus and Kaposi's sarcoma in Africa. Uganda Kaposi's Sarcoma Study Group. Arch Intern Med 156:202–204

Chattopadhyay N, Mithal A, Brown EM (1996) The calcium-sensing receptor: a window into the physiology and pathophysiology of mineral ion metabolism. Endocr Rev 17:289–307

Cheng EH, Nicholas J, Bellows DS, Hayward GS, Guo HG, Reitz MS, Hardwick JM (1997) A Bcl-2 homolog encoded by Kaposi sarcoma-associated virus, human herpesvirus 8, inhibits apoptosis but does not heterodimerize with Bax or Bak. Proc Natl Acad Sci USA 94:690–694

Chiou CJ, Poole LJ, Kim PS, Ciufo DM, Cannon JS, ap Rhys CM, Alcendor DJ, Zong JC, Ambinder RF, Hayward GS (2002) Patterns of gene expression and a transactivation function exhibited by the vGCR (ORF74) chemokine receptor protein of Kaposi's sarcoma- associated herpesvirus. J Virol 76:3421–3439

Clark DA, Emery VC, Griffiths PD (2003) Cytomegalovirus, human herpesvirus-6, and human herpesvirus-7 in hematological patients. Semin Hematol 40:154–162

Coughlin SR (1994) Expanding horizons for receptors coupled to G proteins: diversity and disease. Curr Opin Cell Biol 6:191–197

Couty JP, Geras-Raaka E, Weksler BB, Gershengorn MC (2001) Kaposi's sarcoma-associated herpesvirus G protein-coupled receptor signals through multiple pathways in endothelial cells. J Biol Chem 276:33805–33811

Dezube BJ, Zambela M, Sage DR, Wang JF, Fingeroth JD (2002) Characterization of Kaposi sarcoma-associated herpesvirus/human herpesvirus-8 infection of human vascular endothelial cells: early events. Blood 100:888–896

Dupin N, Diss TL, Kellam P, Tulliez M, Du MQ, Sicard D, Weiss RA, Isaacson PG, Boshoff C (2000) HHV-8 is associated with a plasmablastic variant of Castleman disease that is linked to HHV-8-positive plasmablastic lymphoma. Blood 95:1406–1412

Dupin N, Fisher C, Kellam P, Ariad S, Tulliez M, Franck N, van Marck E, Salmon D, Gorin I, Escande JP, Weiss RA, Alitalo K, Boshoff C (1999) Distribution of human herpesvirus-8 latently infected cells in Kaposi's sarcoma, multicentric Castleman's disease, and primary effusion lymphoma. Proc Natl Acad Sci USA 96:4546–4551

Dupin N, Grandadam M, Calvez V, Gorin I, Aubin JT, Havard S, Lamy F, Leibowitch M, Huraux JM, Escande JP, et al. (1995) Herpesvirus-like DNA sequences in patients with Mediterranean Kaposi's sarcoma. Lancet 345:761–762

Ensoli B, Buonaguro L, Barillari G, Fiorelli V, Gendelman R, Morgan RA, Wingfield P, Gallo RC (1993) Release, uptake, and effects of extracellular human immunodeficiency virus type 1 Tat protein on cell growth and viral transactivation. J Virol 67:277–287

Estep RD, Axthelm MK, Wong SW (2003) A G protein-coupled receptor encoded by rhesus rhadinovirus is similar to ORF74 of Kaposi's sarcoma-associated herpesvirus. J Virol 77:1738–1746

Geras-Raaka E, Arvanitakis L, Bais C, Cesarman E, Mesri EA, Gershengorn MC (1998a) Inhibition of constitutive signaling of Kaposi's sarcoma-associated herpesvirus G protein-coupled receptor by protein kinases in mammalian cells in culture. J Exp Med 187:801–806

Geras-Raaka E, Varma A, Clark-Lewis I, Gershengorn M (1998b) Kaposi's sarcoma-associated herpesvirus (KSHV) chemokine vMIP-II and human SDF-1α inhibit signaling by KSHV G protein-coupled receptor. Biochem Biophys Res Commun 253: p725–727

Geras-Raaka E, Varma A, Ho H, Clark-Lewis I, Gershengorn MC (1998c) Human interferon-γ-inducible protein 10 (IP-10) inhibits constitutive signaling of Kaposi's sarcoma-associated herpesvirus G protein-coupled receptor. J Exp Med 188:405–408

Gershengorn MC, Geras-Raaka E, Varma A, Clark-Lewis I (1998) Chemokines activate Kaposi's sarcoma-associated herpesvirus G protein-coupled receptor in mammalian cells in culture. J Clin Invest 102:1469–1472

Gershengorn MC, Osman R (2001) Minireview: Insights into G protein-coupled receptor function using molecular models. Endocrinology 142: p2–10

Gompels UA, Nicholas J, Lawrence G, Jones M, Thomson BJ, Martin ME, Efstathiou S, Craxton M, Macaulay HA (1995) The DNA sequence of human herpesvirus-6: structure, coding content, and genome evolution. Virology 209:29–51

Guo HG, Browning P, Nicholas J, Hayward GS, Tschachler E, Jiang YW, Sadowska M, Raffeld M, Colombini S, Gallo RC, Reitz MS, Jr. (1997) Characterization of a chemokine receptor-related gene in human herpesvirus 8 and its expression in Kaposi's sarcoma. Virology 228:371–378

Guo HG, Pati S, Sadowska M, Charurat M, Reitz M (2004) Tumorigenesis by human herpesvirus 8 vGPCR is accelerated by human immunodeficiency virus type 1 Tat. J Virol 78:9336–9342

Guo HG, Sadowska M, Reid W, Tschachler E, Hayward G, Reitz M (2003) Kaposi's sarcoma-like tumors in a human herpesvirus 8 ORF74 transgenic mouse. J Virol 77:2631–2639

Ho HH, Du D, Gershengorn MC (1999) The N terminus of Kaposi's sarcoma-associated herpesvirus G protein-coupled receptor is necessary for high affinity chemokine binding but not for constitutive activity. J Biol Chem 274:31327–31332

Ho HH, Ganeshalingam N, Rosenhouse-Dantsker A, Osman R, Gershengorn MC (2001) Charged residues at the intracellular boundary of transmembrane helices 2 and 3 independently affect constitutive activity of Kaposi's sarcoma-associated herpesvirus G protein-coupled receptor. J Biol Chem 276:1376–1382

Isegawa Y, Ping Z, Nakano K, Sugimoto N, Yamanishi K (1998) Human herpesvirus 6 open reading frame U12 encodes a functional beta-chemokine receptor. J Virol 72:6104–6112

Jones KD, Aoki Y, Chang Y, Moore PS, Yarchoan R, Tosato G (1999) Involvement of interleukin-10 (IL-10) and viral IL-6 in the spontaneous growth of Kaposi's sarcoma herpesvirus-associated infected primary effusion lymphoma cells. Blood 94:2871–2879

Kaaya EE, Parravicini C, Ordonez C, Gendelman R, Berti E, Gallo RC, Biberfeld P (1995) Heterogeneity of spindle cells in Kaposi's sarcoma: comparison of cells in lesions and in culture. J Acquir Immune Defic Syndr Hum Retrovirol 10:295–305

Kirshner JR, Staskus K, Haase A, Lagunoff M, Ganem D (1999) Expression of the open reading frame 74 (G-protein-coupled receptor) gene of Kaposi's sarcoma (KS)-associated herpesvirus: implications for KS pathogenesis. J Virol 73:6006–6014

Kledal TN, Rosenkilde MM, Schwartz TW (1998) Selective recognition of the membrane-bound CX3C chemokine, fractalkine, by the human cytomegalovirus-encoded broad-spectrum receptor US28. FEBS Lett 441:209–214

Kosugi S, Van Dop C, Geffner ME, Rabl W, Carel JC, Chaussain JL, Mori T, Merendino JJ, Jr., Shenker A (1995) Characterization of heterogeneous mutations causing constitutive activation of the luteinizing hormone receptor in familial male precocious puberty. Hum Mol Genet 4:183–188

Lagunoff M, Bechtel J, Venetsanakos E, Roy AM, Abbey N, Herndier B, McMahon M, Ganem D (2002) De novo infection and serial transmission of Kaposi's sarcoma-associated herpesvirus in cultured endothelial cells. J Virol 76:2440–2448

Liang Y, Ganem D (2004) RBP-J (CSL) is essential for activation of the K14/vGPCR promoter of Kaposi's sarcoma-associated herpesvirus by the lytic switch protein RTA. J Virol 78:6818–6826

Linderoth J, Rambech E, Dictor M (1999) Dominant human herpesvirus type 8 RNA transcripts in classical and AIDS-related Kaposi's sarcoma. J Pathol 187:582–587

Liu C, Okruzhnov Y, Li H, Nicholas J (2001) Human herpesvirus 8 (HHV-8)-encoded cytokines induce expression of and autocrine signaling by vascular endothelial growth factor (VEGF) in HHV- 8-infected primary-effusion lymphoma cell lines and mediate VEGF-independent antiapoptotic effects. J Virol 75:10933–10940

Macian F, Lopez-Rodriguez C, Rao A (2001) Partners in transcription: NFAT and AP-1. Oncogene 20:2476–2489

Margulies BJ, Browne H, Gibson W (1996) Identification of the human cytomegalovirus G protein-coupled receptor homologue encoded by UL33 in infected cells and enveloped virus particles. Virology 225:111–125

Masood R, Cai J, Zheng T, Smith DL, Naidu Y, Gill PS (1997) Vascular endothelial growth factor/vascular permeability factor is an autocrine growth factor for AIDS-Kaposi sarcoma. Proc Natl Acad Sci USA 94:979–984

Milne RS, Mattick C, Nicholson L, Devaraj P, Alcami A, Gompels UA (2000) RANTES binding and down-regulation by a novel human herpesvirus-6 beta chemokine receptor. J Immunol 164:2396–2404

Montaner S, Sodhi A, Pece S, Mesri EA, Gutkind JS (2001) The Kaposi's sarcoma-associated herpesvirus G protein-coupled receptor promotes endothelial cell survival through the activation of Akt/protein kinase B. Cancer Res 61:2641–2648

Nador RG, Milligan LL, Flore O, Wang X, Arvanitakis L, Knowles DM, Cesarman E (2001) Expression of Kaposi's sarcoma-associated herpesvirus G protein-coupled receptor monocistronic and bicistronic transcripts in primary effusion lymphomas. Virology 287:62–70

Nakano K, Tadagaki K, Isegawa Y, Aye MM, Zou P, Yamanishi K (2003) Human herpesvirus 7 open reading frame U12 encodes a functional beta-chemokine receptor. J Virol 77:8108–8115

Neipel F, Albrecht JC, Fleckenstein B (1997) Cell-homologous genes in the Kaposi's sarcoma-associated rhadinovirus human herpesvirus 8: determinants of its pathogenicity? J Virol 71:4187–4192

Nicholas J (1996) Determination and analysis of the complete nucleotide sequence of human herpesvirus 7. J Virol 70:5975–5989

Nishi J, Maruyama I (2000) Increased expression of vascular endothelial growth factor (VEGF) in Castleman's disease: proposed pathomechanism of vascular proliferation in the affected lymph node. Leuk Lymphoma 38:387–394

Pati S, Cavrois M, Guo HG, Foulke JS, Jr., Kim J, Feldman RA, Reitz M (2001) Activation of NF-κB by the human herpesvirus 8 chemokine receptor ORF74: evidence for a paracrine model of Kaposi's sarcoma pathogenesis. J Virol 75:8660–8673

Pati S, Foulke JS, Jr., Barabitskaya O, Kim J, Nair BC, Hone D, Smart J, Feldman RA, Reitz M (2003) Human herpesvirus 8-encoded vGPCR activates nuclear factor of activated T cells and collaborates with human immunodeficiency virus type 1 Tat. J Virol 77:5759–5773

Paulsen SJ, Rosenkilde MM, Eugen-Olsen J, Kledal TN (2005) Epstein-Barr virus-encoded BILF1 is a constitutively active G protein-coupled receptor. J Virol 79:536–546

Premont RT, Inglese J, Lefkowitz RJ (1995) Protein kinases that phosphorylate activated G protein-coupled receptors. FASEB J 9:175–182

Rabkin CS, Janz S, Lash A, Coleman AE, Musaba E, Liotta L, Biggar RJ, Zhuang Z (1997) Monoclonal origin of multicentric Kaposi's sarcoma lesions. N Engl J Med 336:988–993

Rao VR, Cohen GB, Oprian DD (1994) Rhodopsin mutation G90D and a molecular mechanism for congenital night blindness. Nature 367:639–642

Robinson PR, Cohen GB, Zhukovsky EA, Oprian DD (1992) Constitutively active mutants of rhodopsin. Neuron 9:719–725

Rosenkilde MM, Schwartz TW (2000) Potency of ligands correlates with affinity measured against agonist and inverse agonists but not against neutral ligand in constitutively active chemokine receptor. Mol Pharmacol 57:602–609

Russo J, Bohenzky R, Chien M, Chen J, Yan M, Maddalena D, Parry J, Peruzzi D, Edelman I, Chang Y, Moore P (1996) Nucleotide sequence of the Kaposi sarcoma-associated herpesvirus (HHV8). Proc Natl Acad Sci USA 93: p14862–14867

Salahuddin SZ, Nakamura S, Biberfeld P, Kaplan MH, Markham PD, Larsson L, Gallo RC (1988) Angiogenic properties of Kaposi's sarcoma-derived cells after long-term culture in vitro. Science 242:430–433

Samaniego F, Markham PD, Gendelman R, Watanabe Y, Kao V, Kowalski K, Sonnabend JA, Pintus A, Gallo RC, Ensoli B (1998) Vascular endothelial growth factor and basic fibroblast growth factor present in Kaposi's sarcoma (KS) are induced by inflammatory cytokines and synergize to promote vascular permeability and KS lesion development. Am J Pathol 152:1433–1443

Schwarz M, Murphy PM (2001) Kaposi's sarcoma-associated herpesvirus G protein-coupled receptor constitutively activates NF-κB and induces proinflammatory cytokine and chemokine production via a C-terminal signaling determinant. J Immunol 167:505–513

Shenker A, Laue L, Kosugi S, Merendino JJ, Jr., Minegishi T, Cutler GB, Jr. (1993) A constitutively activating mutation of the luteinizing hormone receptor in familial male precocious puberty. Nature 365:652–654

Shepard LW, Yang M, Xie P, Browning DD, Voyno-Yasenetskaya T, Kozasa T, Ye RD (2001) Constitutive activation of NF-κB and secretion of interleukin-8 induced by the G protein-coupled receptor of Kaposi's sarcoma-associated herpesvirus involve G α_{13} and RhoA. J Biol Chem 276:45979–45987

Smit MJ, Verzijl D, Casarosa P, Navis M, Timmerman H, Leurs R (2002) Kaposi's sarcoma-associated herpesvirus-encoded G protein-coupled receptor ORF74 constitutively activates p44/p42 MAPK and Akt via G_i and phospholipase C-dependent signaling pathways. J Virol 76:1744–1752

Sodhi A, Montaner S, Gutkind JS (2004) Does dysregulated expression of a deregulated viral GPCR trigger Kaposi's sarcomagenesis? FASEB J 18:422–427

Sodhi A, Montaner S, Patel V, Zohar M, Bais C, Mesri EA, Gutkind JS (2000) The Kaposi's sarcoma-associated herpes virus G protein-coupled receptor up-regulates vascular endothelial growth factor expression and secretion through mitogen-activated protein kinase and p38 pathways acting on hypoxia-inducible factor 1α. Cancer Res 60:4873–4880

Soulier J, Grollet L, Oksenhendler E, Cacoub P, Cazals-Hatem D, Babinet P, d'Agay MF, Clauvel JP, Raphael M, Degos L, et al. (1995) Kaposi's sarcoma-associated herpesvirus-like DNA sequences in multicentric Castleman's disease. Blood 86:1276–1280

Spiegel AM (1996) Mutations in G proteins and G protein-coupled receptors in endocrine disease. J Clin Endocrinol Metab 81:2434–2442

Staskus KA, Zhong W, Gebhard K, Herndier B, Wang H, Renne R, Beneke J, Pudney J, Anderson DJ, Ganem D, Haase AT (1997) Kaposi's sarcoma-associated herpesvirus gene expression in endothelial (spindle) tumor cells. J Virol 71:715–719

Streblow DN, Soderberg-Naucler C, Vieira J, Smith P, Wakabayashi E, Ruchti F, Mattison K, Altschuler Y, Nelson JA (1999) The human cytomegalovirus chemokine receptor US28 mediates vascular smooth muscle cell migration. Cell 99:511–520

Streblow DN, Vomaske J, Smith P, Melnychuk R, Hall L, Pancheva D, Smit M, Casarosa P, Schlaepfer DD, Nelson JA (2003) Human cytomegalovirus chemokine receptor US28-induced smooth muscle cell migration is mediated by focal adhesion kinase and Src. J Biol Chem 278:50456–50465

Strieter RM, Polverini PJ, Kunkel SL, Arenberg DA, Burdick MD, Kasper J, Dzuiba J, Van Damme J, Walz A, Marriott D, et al. (1995) The functional role of the ELR motif in CXC chemokine-mediated angiogenesis. J Biol Chem 270:27348–27357

Swanton C, Mann DJ, Fleckenstein B, Neipel F, Peters G, Jones N (1997) Herpes viral cyclin/Cdk6 complexes evade inhibition by CDK inhibitor proteins. Nature 390:184–187

Tadagaki K, Nakano K, Yamanishi K (2005) Human herpesvirus 7 open reading frames U12 and U51 encode functional beta-chemokine receptors. J Virol 79:7068–7076

Teruya-Feldstein J, Zauber P, Setsuda JE, Berman EL, Sorbara L, Raffeld M, Tosato G, Jaffe ES (1998) Expression of human herpesvirus-8 oncogene and cytokine homologues in an HIV-seronegative patient with multicentric Castleman's disease and primary effusion lymphoma. Lab Invest 78:1637–1642

Van Sande J, Parma J, Tonacchera M, Swillens S, Dumont J, Vassart G (1995) Somatic and germline mutations of the TSH receptor gene in thyroid diseases. J Clin Endocrinol Metab 80:2577–2585

Wakeling MN, Roy DJ, Nash AA, Stewart JP (2001) Characterization of the murine gammaherpesvirus 68 ORF74 product: a novel oncogenic G protein-coupled receptor. J Gen Virol 82:1187–1197

Waldhoer M, Casarosa P, Rosenkilde MM, Smit MJ, Leurs R, Whistler JL, Schwartz TW (2003) The carboxyl terminus of human cytomegalovirus-encoded 7 transmembrane receptor US28 camouflages agonism by mediating constitutive endocytosis. J Biol Chem 278:19473–19482

Wang HW, Trotter MW, Lagos D, Bourboulia D, Henderson S, Makinen T, Elliman S, Flanagan AM, Alitalo K, Boshoff C (2004) Kaposi sarcoma herpesvirus-induced cellular reprogramming contributes to the lymphatic endothelial gene expression in Kaposi sarcoma. Nat Genet 36:687–693

Whitby D, Howard MR, Tenant-Flowers M, Brink NS, Copas A, Boshoff C, Hatzioannou T, Suggett FE, Aldam DM, Denton AS (1995) Detection of Kaposi sarcoma associated herpesvirus in peripheral blood of HIV-infected individuals and progression to Kaposi's sarcoma. Lancet 346:799–802

Williams BR (1999) PKR; a sentinel kinase for cellular stress. Oncogene 18:6112–6120

Wong WW (1998) ICE family proteases in inflammation and apoptosis. Agents Actions Suppl 49:5-13

Yang T, Chen S, Leach M, Manfra D, Homey B, Wiekowski M, Sullivan L, Jenh C, Narula S, Chensue S, Lira S (2000) Transgenic expression of the chemokine receptor encoded by human herpesvirus 8 induces an angioproliferative disease resembling Kaposi's sarcoma [see comments]. J Exp Med 191: p445–454

Yen-Moore A, Hudnall SD, Rady PL, Wagner RF, Jr., Moore TO, Memar O, Hughes TK, Tyring SK (2000) Differential expression of the HHV-8 vGCR cellular homolog gene in AIDS-associated and classic Kaposi's sarcoma: potential role of HIV-1 Tat. Virology 267:247–251

Zoeteweij JP, Moses AV, Rinderknecht AS, Davis DA, Overwijk WW, Yarchoan R, Orenstein JM, Blauvelt A (2001) Targeted inhibition of calcineurin signaling blocks calcium-dependent reactivation of Kaposi sarcoma-associated herpesvirus. Blood 97:2374–2380

CTMI (2006) 312:157–183

Regulation of KSHV Lytic Gene Expression

H. Deng[1,2] · Y. Liang[3] · R. Sun[4] (✉)

[1]Center for Infection and Immunity, National Laboratory of Biomacromolecules, Institute of Biophysics, Chinese Academy of Sciences, 100101 Beijing, P.R. China

[2]School of Dentistry, University of California at Los Angeles, Los Angeles, CA 90095, USA

[3]Department of Pathology and Laboratory Medicine, Emory University, Atlanta, GA 30322, USA

[4]Department of Molecular and Medical Pharmacology, University of California at Los Angeles, Los Angeles, CA 90095, USA
RSun@mednet.ucla.edu

Abstract The life cycle of KSHV, latency versus lytic replication, is mainly determined at the transcriptional regulation level. A viral immediate-early gene product, replication and transcription activator (RTA), has been identified as the molecular switch for initiation of the lytic gene expression program from latency. Here we review progress on two key questions: how RTA gene expression is controlled by viral proteins and cellular signals and how RTA regulates the expression of downstream viral genes. We summarize the interactions of RTA with cellular and other viral proteins. We also discuss critical issues that must be addressed in the near future.

1 Introduction

Kaposi sarcoma-associated herpesvirus (KSHV), also known as human herpesvirus-8 (HHV-8), was discovered by its association with Kaposi sarcoma (KS), the most common form of cancer in human immunodeficiency virus type 1 (HIV-1)-infected patients (Chang et al. 1994). Subsequently, KSHV was found to be associated with two lymphoproliferative diseases related to HIV-1 infection: primary effusion lymphoma (PEL, a non-Hodgkin lymphoma) and multicentric Castleman disease (MCD) (Cesarman et al. 1995a; Soulier et al. 1995; Knowles and Cesarman 1997). The genomic DNA of KSHV (Russo et al. 1996) consists of a 140.5-kb-long unique coding region flanked by multiple 801-bp GC-rich terminal repeat sequences. The long unique coding region encodes over 80 open reading frames (ORFs), including many cellular gene homologues implicated in viral pathogenesis (Russo et al. 1996). Based on its genomic organization and other properties, KSHV has been classified as a member of the gamma 2-herpesvirus subfamily (Neipel and Fleckenstein 1999; Schulz and Moore 1999).

Like other herpesviruses, KSHV has two distinct phases in its life cycle, latency and lytic replication. Latency is characterized by persistence of the viral genome with expression of a limited set of viral genes. Once the virus is reactivated from latency and enters the lytic cycle, most viral genes are expressed in an orderly fashion (immediate-early, early, and late), leading to the production of infectious virions. KSHV gene expression patterns in latency and lytic phase have been studied with biopsies from KS/MCD/PEL, cell lines derived from PEL, and de novo infection of cultured cells in vitro. In situ hybridization studies of KS biopsies showed that the majority of tumor cells in the lesions contain KSHV genomic DNA and express viral latent transcripts, but in a subpopulation of tumor cells (1%–3%), viruses spontaneously enter the lytic cycle as evidenced by the expression of lytic transcripts (Staskus et al. 1997; Sun et al. 1999). Notably, these transcripts include those encoding for viral macrophage inflammatory protein-I (vMIP-I), viral interleukin-6 (vIL-6), viral Bcl-2 homologue, as well as an unusual polyadenylated nuclear RNA (PAN). Colocalization experiments have also shown that it is the same subpopulation of cells in KS lesions that spontaneously express both the early (e.g., PAN RNA and vMIP-I) and late [e.g., major capsid protein (MCP) and small viral capsid protein (sVCA)] viral transcripts. These observations suggest a novel paradigm for pathogenesis and tumorigenesis by an oncogenic herpesvirus. Whereas transforming function of certain viral latent gene products expressed in a majority of the tumor cells may play direct roles in tumorigenesis, production of viral and cellular cytokines by adjacent infected cells in which KSHV undergoes lytic replication may create a favorable microenvironment to enhance the growth of latently infected cells and hence also contribute significantly to tumor development.

A number of human B cell lines derived from PEL have greatly facilitated the study of KSHV gene expression, gene function, and viral pathogenesis. Similar to KS biopsies, these PEL cell lines predominantly carry the virus in a latent state, with a small subset of cells in which the virus undergoes spontaneous reactivation (Cesarman et al. 1995b; Arvanitakis et al. 1996; Renne et al. 1996; Said et al. 1996; Staskus et al. 1997; Boshoff et al. 1998; Sun et al. 1999). Viral lytic replication in the whole cell population can be further activated by treating cells with inducing chemical agents such as phorbol esters or sodium butyrate (NaB) (Renne et al. 1996; Miller et al. 1997; Nicholas et al. 1997a; Sarid et al. 1998). With these cell lines (e.g., BC-1, BCBL-1, and BC-3), KSHV gene expression patterns have been more extensively examined through Northern blot analysis (Zhong et al. 1996; Sarid et al. 1998; Sun et al. 1999). These studies showed that the kinetics of KSHV gene expression is similar to those of other herpesviruses, that is, viral gene expression is restricted in latency, and lytic genes are expressed in a cascade fashion (immediate-early, early, and late), leading to the production of infectious virions (Renne et al. 1996; Vieira et al. 1997; Gradoville et al. 2000).

More recently, the temporal pattern of KSHV gene expression has been examined on a genomic scale with custom-built DNA microarrays (Jenner et al. 2001; Paulose-Murphy et al. 2001). These studies confirmed previous results on the kinetics of viral gene expression and identified a correlation between stages of gene expression and the function of the gene in viral replication. For instance, early genes such as those encoding proteins involved in viral DNA replication are expressed at relatively early time points, whereas those involved in virion assembly are expressed at later time points. Gene expression regulation of Epstein-Barr virus (EBV) has been extensively studied, providing information that is instructive for KSHV studies. In this review, we mainly discuss some of the regulatory steps in KSHV lytic gene expression and their implications in viral replication and pathogenesis.

2
Transcriptional Regulation of Immediate-Early Genes

2.1
Overview

Immediate-early genes are the first group of genes expressed during herpesvirus lytic replication (de novo infection or reactivation) and are usually defined by their transcription without requiring de novo protein synthesis. They generally encode regulatory proteins that either activate the downstream viral lytic gene expression cascade and/or modulate host cellular environment to facilitate viral replication. In early studies, because of a lack of an efficient permissive cell system for de novo infection, KSHV lytic gene expression was generally studied by treating latently infected PEL cell lines with chemical inducers, such as TPA or NaB, to activate the viral lytic cycle. Recently, studies on viral transcription in infected TIME, HFF cells, and 293 cells have been initiated, which provide additional information (Moses et al. 1999; Krishnan et al. 2004; Sharma-Walia et al. 2005). With chemical inducers and the protein synthesis inhibitor cycloheximide, several immediate-early genes of KSHV have been identified (Sarid et al. 1998; Sun et al. 1998; Lukac et al. 1999; Zhu et al. 1999; Gradoville et al. 2000; Haque et al. 2000; Saveliev et al. 2002). These include ORF50, ORF45, ORF K4.2, and a 4.5-kb mRNA species that corresponds to the viral genomic region between nt. 49419 and 54688. ORF K5 has also been reported as an immediate-early gene, even though its transcription is only resistant to cycloheximide at a low concentration (10 μg/ml) and its promoter is moderately activated (~5-fold) by another immediate-early protein, RTA (Haque et al. 2000).

The best-characterized immediate-early gene is RTA (replication and transcription activator, a homologue of EBV RTA, also called regulator of transcription activation, ART, Lyta, or ORF50). The major transcript is a 3.6-kb multiply spliced bicistronic message containing ORF50 and K8 (Sun et al. 1998; Zhu et al. 1999). The RTA protein is mainly encoded within ORF50. A splicing event upstream of ORF50 introduces a new methionine initiation codon and adds a coding region for an additional 60 amino acids (aa) to the N terminus of ORF50. The resulting RTA protein of 691 aa has a N-terminal DNA binding and dimerization domain, a C-terminal activation domain, and two nuclear localization signals (Lukac et al. 1998; Sun et al. 1998). The splicing event as well as the domain organization are highly conserved among the RTA homologues of gamma-herpesviruses including EBV and herpesvirus saimiri (HVS). The strongest homologous sequence with other RTA is found in between residues 103 and 202 within the DNA binding and dimerization domain (Sun et al. 1998). The activation domain also shows limited sequence conservation with other RTA homologues as well as several cellular and viral transcription factors such as the herpes simplex virus-1 VP16 (Lukac et al. 1998). RTA protein has a predicted molecular mass of 73.7 kDa; however, when it is expressed in mammalian cells its migration pattern on Western blot reveals an apparent molecular mass of approximately 110 kDa, suggesting posttranslational modifications (Lukac et al. 1999; Song et al. 2002). Indeed, the RTA protein sequence contains numerous potential sites for phosphorylation. However, whether and how these sites are utilized and what other modification mechanisms are responsible for the observed increase of RTA molecular mass remain to be investigated.

Several lines of evidences have demonstrated that RTA serves as the "molecular switch" for KSHV life cycle. Expression of RTA alone in latently infected PEL cells disrupted latency and activated the expression of viral lytic genes such as vIL-6, PAN RNA, and ORF59; induction of viral late protein synthesis, ORF65 and K8.1, by RTA indicated that RTA drives the lytic cycle to completion (Lukac et al. 1998; Sun et al. 1998). Moreover, ectopic expression of RTA increased the production of DNase-resistant encapsidated viral DNA, providing the ultimate proof that RTA is capable of initiating and driving the complete viral lytic cycle, leading to the release of newly produced viral particles (Gradoville et al. 2000). On the other hand, introduction of dominant-negative mutant RTA proteins that lacked the C-terminal activation domain into latently infected cells reduced spontaneous viral reactivation, suggesting that RTA function is necessary for lytic reactivation (Lukac et al. 1999). A similar conclusion was reached by using the ribozyme approach to inhibit RTA expression (Zhu et al. 2004). Finally, deletion of RTA generated by the KSHV BAC system further confirmed the essential roles of RTA in viral lytic reactivation (Xu et al. 2005). Taken together, these data have demonstrated that RTA is both necessary and sufficient to mediate the switch from latency to lytic replication of KSHV in vitro. Because of the lack of an animal model for KSHV infection, several groups utilized a murine gammaherpesvirus-68 (MHV-68) to study the functions of RTA in vivo. The results suggested that RTA of MHV-68 also plays a central role in initiating the lytic replication cycle during infection of mice (Wu et al. 2001; Pavlova et al. 2003; Boname et al. 2004; Rickabaugh et al. 2004).

Because RTA serves as the molecular switch for viral life cycle, activation of the RTA promoter becomes the key to understanding the mechanisms controlling KSHV latency and reactivation. The activity of the RTA promoter on the viral genome should be determined by a number of factors: activators and repressors (both viral and cellular) as well as chromatin structures of the viral genome. The remainder of this section will be devoted to discussing the recent progress in understanding the transcriptional regulation of this key molecule.

2.2 Regulation of RTA Gene Transcription by KSHV Proteins

As a powerful transcriptional activator, RTA activates the expression of many viral genes, including itself (Seaman et al. 1999; Deng et al. 2000; Gradoville et al. 2000). Autoactivation of the immediate-early RTA gene represents an important strategy for KSHV to effectively respond to environmental stimuli and maximally activate the virus lytic cycle. A 0.5-kb RTA promoter sequence was activated 2.7-fold by RTA expression (Seaman et al. 1999). With a construct with a much longer upstream sequence, it was shown that a 3-kb RTA promoter is highly responsive to TPA and NaB. In addition, RTA autoactivated the 3-kb promoter reporter construct up to 144-fold, independent of other viral factors or B cell-specific factors. Furthermore, ectopic expression of RTA in latently infected cells activated the expression of the RTA gene from endogenous viral genomes as demonstrated by either ribonuclease protection assay of the 5′ untranslated region (Deng et al. 2000) or Northern blot analysis of the RTA bicistronic transcripts using a probe derived from the ORF K8 region (Gradoville et al. 2000).

Several studies were conducted to map the RTA response element (RRE) in RTA promoter. A luciferase reporter with the 950-bp fragment upstream of the RTA coding region was weakly activated four- to eightfold by RTA, and the response element was mapped to an octamer-binding site (Sakakibara et al. 2001). Electrophoretic mobility shift assay (EMSA) showed that cellular Oct1 protein binds to an octamer-binding site; however, an RTA-Oct-1 complex was not observed, suggesting that the RTA protein may not associate with the octamer-binding site strongly and therefore an indirect mechanism may be involved. Nevertheless, this 950-bp promoter supports only mild activation by RTA (up to 8-fold) in comparison to a 3-kb promoter (Deng et al. 2000) that can be highly activated (144-fold), indicating that other unidentified RREs may exist further upstream. Sequence analysis of this region reveals multiple binding sites for the cellular transcription factor RBPJ. Using a luciferase construct with 3-kb promoter sequences, Liang et al. revealed that autoactivation of this RTA promoter is significantly lowered in RBPJ-null cells (10-fold) but much stronger in wild-type cells (up to 70-fold) (Liang and Ganem 2003). These observations suggested that RBPJ, a known cellular partner of RTA (Liang et al. 2002) (see more details below), is involved in the autoactivation of RTA promoter. Indeed, Chang et al. have shown that a DNA-binding-defective RTA mutant is still competent to induce the ORF50 transcription in vivo and that the RTA promoter with multiple RBPJ-binding sites was autoactivated at a higher level than that lacking RBPJ-binding sites (Chang et al. 2005c). These lines of evidence suggested that autostimulation of RTA promoter is mediated mainly through indirect non-RTA-binding mechanisms and involves at least two cellular proteins, RBPJ and Oct-1.

Several viral proteins have been reported to inhibit the expression of RTA (Cannon et al. 2006). Interestingly, RBPJ-binding sites are also involved in the repression of RTA expression by latency-associated nuclear antigen (LANA). LANA is a multifunctional protein that is essential for latency establishment and maintenance. LANA mediates viral episomal DNA persistence during latency by tethering the viral episomes to cellular mitotic chromosomes (Ballestas et al. 1999; Hu et al. 2002; Fejer et al. 2003; Grundhoff and Ganem 2003). It also promotes cell survival and regulates cell cycle progression by modulating various cellular targets, including p53, Rb, and Wnt pathways (Friborg et al. 1999; Fujimuro and Hayward 2003; An et al. 2005). More recently, it has been shown that LANA suppresses lytic reactivation by repressing not only the basal level expression of the RTA promoter but also RTA-mediated autoactivation (Lan et al. 2004). Intriguingly, LANA-mediated repression of RTA promoter depends on the RBPJ-binding sites (Lan et al. 2005) that are also involved in RTA autoactivation (see above). By directly interacting with RBPJ protein, LANA may not only recruit additional corepressors to suppress the transcription of the RTA gene but also repress RTA autoactivation activity by competing with RTA in RBPJ-binding. The fact that both positive and negative regulators of RTA gene expression use the same RBPJ-dependent mechanism suggests that the switch between latency and lytic reactivation is finely controlled by the levels of LANA and RTA proteins in virus-infected cells. Furthermore, RTA also induces LANA expression (Lan et al. 2005; Matsumura et al. 2005), providing a negative feedback in keeping viral lytic reactivation under check.

Although RTA expression is strictly controlled in latency, RTA and a subset of lytic genes are transiently expressed very early after viral entry and quickly replaced by latent gene expression (Krishnan et al. 2004). The mechanisms and physiological significance for such transient expression and repression are still unknown. One possibility is that some viral factors are directly brought into the infected cells in the form of virion proteins to effectively modulate the intracellular environment to facilitate viral replication, as suggested by data from different perspectives (Lu et al. 2005). An advantage of bringing viral regulatory factors into infected cells as virion-associated proteins is that these factors can directly interact with the host cellular environment as soon as they are delivered into the cells by virus entry, independent of transcription and translation. Alphaherpesviruses and betaherpesviruses are known to incorporate a number of regulatory gene products as tegument proteins. For example, herpes simplex virus-1 virion carries transcription factor VP-16 and virion host shut-off (VHS) proteins. However, little is known about the physiological functions of any gammaherpesvirus virion proteins immediately after infection of the cell. RTA turns out to be one of the KSHV virion-associated proteins (Bechtel et al. 2005; Lan et al. 2005). It may be involved in triggering transient expression of some lytic genes that modulate the host cell for the benefit of viral infection and/or induce the expression of latency proteins (Lan et al. 2005; Matsumura et al. 2005) that eventually shut down the lytic gene expression and establish latency in some types of cells. Further studies are needed to address the roles of KSHV virion proteins in establishing viral infection.

2.3
Cellular Signaling Pathways That Control RTA Expression

A critical question that must be addressed in KSHV biology is what host cellular signals control RTA expression and thus the balance between viral latency and lytic replication. Chemical agents such as phorbol esters, NaB, 5′-azacytidine, and glycyrrhizic acid can reactivate KSHV in latently infected cells in culture (Cesarman et al. 1995b; Miller et al. 1996; Renne et al. 1996; Chen et al. 2001; Curreli et al. 2005). These agents, although not necessarily natural physiological inducers, provide valuable tools for studying the mechanism of KSHV reactivation and may reveal potential cellular pathways that may be involved. Phorbol esters such as 12-*O*-tetradecanoyl-phorbol-13-acetate (TPA) can activate protein kinase C (PKC) (Castagna et al. 1982), suggesting that signaling pathways initiated by PKC may contribute to physiologically relevant KSHV reactivation. Experiments of overexpression and selective inhibition indicated an essential but not sufficient role for PKCδ isoform in KSHV reactivation (Deutsch et al. 2004). In addition, activation of Ap-1 pathway has also been suggested to be involved in TPA-induced KSHV reactivation (Wang et al. 2004). A different group of chemical inducers such as NaB and 5′-azacytidine implicates the roles of chromosomal modulation in KSHV reactivation, which is discussed in Sect. 2.4.

Another approach to define the mechanism controlling the reactivation is to explore suspected physiological stimuli, such as inflammatory cytokines (Chang et al. 2000; Mercader et al. 2000; Wang et al. 2005) and hypoxia (Haque et al. 2003). Hypoxia has been proposed to be at the location where KS occurs frequently. Several neuron transmitters associated with stress responses have also been found to reactivate KSHV. One example is that epinephrine and norepinephrine efficiently reactivated lytic replication of KSHV in latently infected PEL cells via β-adrenergic activation of the cellular cyclic AMP/protein kinase A (PKA) signaling pathway (Chang et al. 2005b).

Many cellular signaling pathways have been found to positively regulate herpesvirus reactivation, but negative regulators have been less defined. It has been found that the cellular transcription factor NF-κB, which is highly active in lymphocytes, is required for maintaining viral latency of KSHV and EBV (Brown et al. 2003, 2005; Chang et al. 2005b). Downregulation of NF-kB resulted in viral reactivation in latently infected B cell lines. On the other hand, overexpression of NF-κB in epithelial or fibroblast cells inhibited MHV-68 lytic replication and allowed the virus to establish persistent infection. NF-κB also inhibited the activation of viral lytic genes by RTA in reporter assays. The inhibition was also reversible by I-κB and regulated by the relative amount of RTA versus NF-κB in the cell. These data suggest that high levels of NF-κB activity in lymphocytes play a direct role in the establishment and maintenance of viral latency in these cells. NF-κB can be upregulated by a viral latent protein, vFLIP (An et al. 2003; Guasparri et al. 2004). In addition, MTX was found to inhibit the lytic cycle of KSHV (Curreli et al. 2002). Interestingly, MTX, a well-known DHFR inhibitor, acts in a DHFR-independent fashion in this case. The underlying mechanism is not clear, but potentially bears novelty. The potential link between transcriptional regulation and signaling originating from cellular metabolism is a topic to be explored.

2.4
Effect of Chromatin Architecture on RTA Gene Expression

It is well known that both DNA methylation and histone acetylation play critical roles in gene regulation through chromatin remodeling (Wu and Grunstein 2000). Not surprisingly, expression of the RTA gene on the KSHV genome is also regulated at the chromatin level. Treatment of BCBL-1 cells with 5-azacytidine, a DNA methyltransferase inhibitor, induced KSHV reactivation. Bisulfite genomic sequencing analysis confirmed that the RTA promoter region is methylated and the promoter of a latent gene LANA lacks methylation in latently infected cells. These results suggested that methylation status is critical for controlling the RTA promoter activity, and hence viral reactivation (Chen et al. 2001).

In addition to DNA methylation, the chromatin structure of the RTA promoter is also regulated by histone acetylation. Positively charged histones bind tightly to the phosphate backbone of DNA, keeping chromatin in a transcriptionally silent state. Acetylation of histones by histone acetylase (HAT) neutralizes the positive charges on histones and therefore disrupts the higher-order chromatin structures for easy access by transcription factors and RNA polymerase complex to initiate transcription. Histone deacetylation mediated by histone deacetylases (HDACs) restores a positive charge on histones, leading to a tightly supercoiled chromatin structure that is associated with transcription repression. It has been known that inhibitors of HDACs such as

NaB and trichostatin A can induce lytic reactivation of the KSHV in latently infected PEL cell lines (Riggs et al. 1977; Cousens et al. 1979; Miller et al. 1996). The responsive element was mapped to a GC box that binds Sp1/Sp3, over which transcriptional initiation site a nucleosome is located (Lu et al. 2003). In latently infected cells, the RTA promoter is associated with multiple HDACs (including HDACs 1, 5, and 7), whereas NaB treatment resulted in the rapid binding of the SWI/SNF chromatin remodeling complex. Cyclic AMP-response element binding protein (CREB)-binding protein (CBP) HAT was also shown to stimulate RTA transcription from a plasmid (Lu et al. 2003). Taken together, these studies suggest that chromatin remodeling of the RTA promoter region, including histone acetylation and DNA demethylation, is a critical step in the switch from latency to lytic reactivation.

3 Transcriptional Regulation of Early Genes

3.1 Overview

KSHV early genes generally encode proteins that are involved in nucleic acid metabolism and modulation of cellular functions. They are usually under the control of the immediate-early proteins. A number of KSHV early lytic genes have been shown to be activated by RTA through direct or indirect mechanisms. They include PAN RNA (also called nut-1 or T1.1), Kaposin (Kpsn; also called K12), ORF57 (a posttranscriptional activator), K-bZIP (the KSHV homologue of ZEBRA encoded by EBV; also called K8), vIL-6, K5, K9 [viral interferon (IFN) regulatory factor or vIRF], K14 (viral OX-2), K15, ORF6 (single-stranded DNA binding protein), ORF59 (DNA polymerase-associated processivity factor), ORF21 (viral thymidine kinase or vTK), ORF74 (viral G protein -coupled receptor) (Zhang et al. 1998; Chang et al. 2000; Chen et al. 2000; Haque et al. 2000; Duan et al. 2001; Jeong et al. 2001; Lukac et al. 2001; Song et al. 2001; Deng et al. 2002b; Wong and Damania 2006). Interactions of RTA with some of these viral promoters have been characterized in detail. In this section, we will use examples to illustrate how RTA differentially activates the downstream gene expression through various mechanisms.

3.2 RTA Activation of PAN and Kaposin Genes

PAN RNA is a novel 1.1- to 1.2-kb noncoding polyadenylated transcript, forming a speckled pattern in the nucleus and colocalizing with cellular Sm protein. Therefore, PAN RNA possesses features of both U snRNA and mRNA (Sun et al. 1996; Zhong and Ganem 1997). It is the most abundant transcript expressed during KSHV lytic phase, comprising approximately 80% of the total polyadenylated RNA in infected cells. Although its function in KSHV replication and pathogenesis has remained a mystery, PAN RNA serves as a good model to define the mechanism of RTA activation. With a transient transfection reporter assay, it was demonstrated that RTA activates the PAN promoter up to 7,000-fold and that this activation is independent of other KSHV or B cell-specific factors (Song et al. 2001). Deletion analysis further mapped the RTA-responsive element (RRE) to a 31-bp region of the PAN promoter. EMSAs showed that RTA directly binds to the PAN RRE and forms a highly stable complex (Song et al. 2001). Independently, Chang et al. also mapped PAN RRE to a 25-bp region (Chang et al. 2002), which is contained within the 31-bp RRE mapped by Song et al. previously (Song et al. 2001). A detailed analysis of specific interactions between RTA and the PAN promoter showed that RTA has a strong affinity for the RRE in the PAN promoter, which is reflected in the K_d at the nanomolar range. In addition, the minimal length for RTA binding was mapped to a 30-bp region spanning from −74 to −45 of the PAN promoter (with transcription initiation site of PAN RNA at nt 28667 on KSHV genome set as +1). Results from methylation interference assay, deletion analysis, and extensive mutagenesis study using both reporter assays and EMSAs correlated with one another and revealed base pairs critical for both RTA binding in vitro and RTA transactivation in vivo (Song et al. 2002). These studies were performed with the N-terminal half of RTA protein purified from *Escherichia coli* because the full-length protein was less efficient in binding, a problem that hindered many investigators. Only recently has the Miller lab overcome the hurdle by discovering that the C-terminal end of RTA inhibits the DNA binding activity (Chang and Miller 2004).

A sequence analysis of the KSHV genome revealed that a region in Kaposin promoter bears remarkable homology to the PAN RRE sequence. The Kaposin gene is abundantly expressed during latency and strongly induced during the viral lytic cycle. Kaposin A (K12) was reported to possess cellular transformation ability (Muralidhar et al. 1998, 2000; Kliche et al. 2001). Kaposin B increases the production of cytokines via activating the p38/MK2 pathway that is known to stabilize the AU-rich element-containing transcripts as seen for most cytokines (Sadler et al. 1999; McCormick and Ganem 2005). This 25-bp homologous region in the Kaposin promoter contains a consecutive stretch of 16-bp matches and additional 5-bp matches downstream. When tested in EMSA, this region conferred binding to RTA, although at a lower affinity. Mutation of two nucleotides (CC→TG) so that the mutated Kaposin promoter sequence (Kpsn/TG) has 19 consecutive matches plus additional

4-bp matches showed that its binding affinity for RTA is higher than that of the Kaposin promoter but still lower than that of the PAN promoter (Song et al. 2003). This result indicated that the sequences downstream of the 16-bp homology are also important for binding by RTA, consistent with the previous analysis of the PAN promoter (Song et al. 2003). Deletion analyses of the Kaposin promoter sequences have further confirmed that the 16-bp homologous region is essential for RTA binding and *trans* -activation and that the 5′ and 3′ flanking sequences also play important roles (Chang et al. 2002).

3.3 RTA Activation of ORF57 and vMIP1 Genes

ORF57 encodes a posttranscriptional regulator that is conserved in herpesviruses. Its expression is upregulated by RTA expression (Kirshner et al. 1999, 2000; Duan et al. 2001). Using luciferase reporter assays, Lukac et al. identified a core 25-bp region in the ORF57 promoter that is responsive for RTA activation (Lukac et al. 2001). When Liang et al. first discovered the protein-protein interactions between RTA and a cellular DNA-binding factor RBPJ, they noticed an authentic RBPJ-binding site GTGGGAA within this 25-bp ORF57 RRE and showed that mutation of this site almost completely abolished RTA activation (Liang et al. 2002). Furthermore, RTA fails to activate ORF57 promoter in RBPJ-null cells while such activation can be restored by cotransfection of an RBPJ-encoding vector (Liang et al. 2002). These observations clearly demonstrated that RTA activation of ORF57 gene mainly depends on the cellular protein RBPJ bound to the RBPJ-binding site within the identified 25-bp RRE. This conclusion was further supported by a recent study from the Miller lab (Chang et al. 2005c), which showed that RBPJ rather than RTA binds ORF57 promoter in EMSA and that DNA-binding-deficient mutants of RTA can still activate ORF57 promoter but not promoters of PAN and Kaposin. Moreover, an RRE with similarity to that of ORF57 was identified in the promoter of vMIP-I (a virus-encoded chemokine homologue), and studies using EMSA and mutational analysis demonstrated that it too contains a real RBPJ-binding site and that RTA activates vMIP-I through protein-protein interaction with the bound RBPJ protein (Chang et al. 2005c).

3.4 RTA Activation of IL-6 Genes

KSHV encodes homologues of several cellular cytokines and chemokines, one of which is viral IL-6 (vIL-6). vIL-6 is encoded by ORF K2 and shares 25% amino acid identity with human IL-6. Similar to its human counterpart, vIL-6 promotes growth and proliferation of IL-6-dependent human and murine hybridoma B cell lines (Moore et al. 1996; Nicholas et al. 1997b). vIL-6 also activates multiple signal transduction pathways, including JAK/STAT and Ras-MAP kinase pathways (Molden et al. 1997; Osborne et al. 1999). The vIL-6 transcript is one of the most abundant viral mRNAs expressed during lytic phase in PEL-derived cells (Sun et al. 1999), and its expression is under the control of RTA as well (Deng et al. 2002b). Through reporter assays and EMSAs, the vIL-6 RRE has been mapped to a 26-bp region, which, however, bears no similarities to either the PAN RRE or the ORF57 RRE.

Defining the vIL-6 transcription unit has also revealed the presence of two transcriptional initiation sites, and hence two promoters (Deng et al. 2002b). The existence of two promoters is intriguing, especially in light of the variable vIL-6 levels observed in KS, PEL, and MCD samples. In situ hybridization and immunohistochemistry studies demonstrated that vIL-6 is expressed at higher levels in PEL and MCD than in KS, and the average levels of vIL-6 expressed in individual infected PEL and MCD cells are greater by at least an order of magnitude than those in KS cells (Staskus et al. 1999). The presence of two promoters allows for differential regulation of vIL-6 gene expression in cell- and tissue-specific environments, which may lead to different manifestations in distinct malignancies associated with KSHV. In this aspect, it is interesting to note that the human cellular IL-6 (hIL-6) gene is also strongly upregulated by RTA (Deng et al. 2002a). It is remarkable that the virus takes multiple approaches to upregulate and maintain IL-6 levels during latency and lytic replication, which involve the roles from v-FLIP, RTA, and Kaposin, as well as encoding a viral homologue of IL-6. This strongly implicates the essential roles of IL-6-related functions in KSHV life cycle.

3.5 RTA Activation of Other KSHV Early Lytic Genes

A virally encoded G protein-coupled receptor (vGPCR) plays important roles in KSHV-induced angiogenesis. The most abundant vGPCR-containing transcripts are bicistronic RNAs with K14 (Nador et al. 2001) at the 5′ end and vGPCR at the 3′ end. This K14/vGPCR transcript is strongly induced during lytic reactivation (Kirshner et al. 1999, 2000; Nador et al. 2001; Chiou et al. 2002). The promoter governing this transcript is highly responsive to RTA activation, in which three putative RREs (sites A, B, and C) were identified through deletion and mutation mapping (Liang and Ganem 2004). None of these sites binds directly to RTA; rather, both sites A and C bind to RBP-J, whereas site B binds to yet-unknown cellular factors. The importance of RBP-J in the transcription of K14/vGPCR mRNAs is demonstrated by the observation that RTA activation of K14/vGPCR promoter is significantly inhibited in cells lacking RBP-J (Liang and Ganem 2004). Interestingly, a recent study (Matsumura et al. 2005) has shown that RTA uses these same RREs to activate the latency transcript (LT) locus, encoding LANA (ORF73), v-cyclin (ORF72), and v-FLIP (ORF71), which are transcribed in the opposite direction of K14/vGPCR. Another study (Zhang et al. 2005), however, has shown that RTA can directly bind to and activate through IFN-stimulated response element (ISRE) and that one such ISRE is present within the K14/vGPCR promoter, partially overlapping site A (Liang and Ganem 2004). RTA was further shown to selectively induce some cellular IFN-responsive genes (IRGs) including ISG-54, MxA, and STAF50 (Zhang et al. 2005), whose in vivo functions in viral replication and pathogenesis wait further studies.

The KSHV K-bZIP (KSHV basic leucine zipper) or K8 protein is a homologue of the EBV ZEBRA (Zta), one of the two immediate-early proteins that control the life cycle of EBV. However, the KSHV K-bZIP is an early gene, whose expression is activated by RTA (Sun et al. 1998; Lin et al. 1999) and has been shown to play an essential role in viral DNA replication (Lin et al. 2003) and cell cycle regulation. Wang et al. reported that K8 gene is differentially transcribed during immediate-early (IE) and delayed-early stages with different promoters (Wang et al. 2004). K8 IE promoter can be activated by butyrate, possibly through Sp1-binding sites. K8 delayed-early promoter is activated by RTA, the mechanism of which has been characterized in detail by several groups (Lukac et al. 2001; Seaman and Quinlivan 2003; Wang et al. 2003, 2004). A total of three RREs were identified. RRE-I, via cooperative binding of RTA, RAP (K-bZIP), and cellular factor C/EBPα, has a minor effect in B cells but more in 293 cells. RRE-II plays a major role via a non-DNA-binding mechanism in both 293 and B cells, the details of which are yet to be characterized. RRE-III contains a standard RBPJ-binding site, mutation of which causes a 2.5-fold transcriptional reduction in both cell types.

KSHV vIRF (K9) has differential transcription patterns during latency and lytic phase (Chen et al. 2000). Its lytic promoter is highly responsive to RTA activation, and the RREs were finely mapped to two regions that have no sequence homology and cannot directly bind RTA in vitro (Ueda et al. 2002). Multiple cellular factors were found to bind to these elements, and their binding correlated with the RTA responsiveness, suggesting an indirect mechanism for RTA to act through cellular factor(s) (Ueda et al. 2002). The response of the viral *tk* promoter to RTA appears to require Sp1-binding sites (Zhang et al. 1998) and also the RBPJ protein (Liang unpublished data).

4 Viral and Cellular Factors That Interact with RTA Protein

4.1 Overview

RTA must function through interactions via other proteins. Multiple viral and cellular proteins have been identified to interact with RTA, which is consistent with the observation that RTA is a multifunctional protein. However, the challenge is to define the biological relevance of such interactions. Current progress is summarized here.

4.2 RBP-J as a Coactivator of RTA

One of the best-characterized RTA-interacting cellular proteins is the DNA-binding transcription factor RBP-J (Liang et al. 2002). RBP-J is a downstream target of the cellular Notch signal pathway. It recognizes the specific DNA sequence GTGGGAA and acts as transcriptional repressor by recruiting a corepressor complex. On activation, the Notch molecule spanning the cell membrane is cleaved to release the intracellular domain, which translocates to the nucleus, where it binds RBP-J, replaces the corepressor complex, and activates the target promoters using its intrinsic activation domain. Viral proteins such as EBV EBNA2 and KSHV RTA have been shown to pirate this cellular pathway in order to activate target promoters containing RBPJ-binding sites. Like Notch molecule and EBNA2, RTA also binds to the central repressor domain of RBP-J, and, by competing off the corepressor complex, activates target promoters using its strong activation domain. So far the mechanism of RBPJ-dependent RTA activation has been utilized by multiple RTA target promoters, such as ORF57, TK, K14/vGPCR, RTA, K3, K5, vIL-6, and vMIP-1 (Liang et al. 2002; Liang and Ganem 2003, 2004; Chang et al. 2005a, 2005c; Lan et al. 2005). The biological functions of RBP-J in the KSHV life cycle have been evaluated with a mouse RBPJ-knockout cell line (Liang and Ganem 2003), in which KSHV can still establish latency but RTA-induced lytic reactivation is completely abolished. This is likely caused by the defective expression of multiple lytic genes that depend on RBPJ for activation by RTA.

4.3 Other Cellular Proteins Interacting with RTA

Other RTA-interacting cellular transcription factors include CCAAT/enhancer-binding protein-α (C/EBPα) (Wang et al. 2003) and STAT3 (Gwack et al. 2002). C/EBPα is a member of the leucine zipper family of transcription factors. Potential C/EBPα-binding sites are found in the promoters of K-bZIP, PAN, ORF57 (MTA), and RTA, whose activations can be cooperatively activated by RTA and C/EBPα (Wang et al. 2003). STATs are a family of proteins that is phosphorylated and activated by signaling pathways of various cytokines and growth factors. Activated STATs form homo- or heterodimers and translocate to the nucleus, where they bind to specific DNA response elements. Gwack et al. showed that interaction between RTA and STAT3 leads to STAT3 dimerization and nuclear translocation in the absence of phosphorylation and that RTA is able to induce STAT3-mediated transcription (Gwack et al. 2002). However, the biological roles of STAT3 in KSHV replication and pathogenesis have not yet been characterized.

Using GST pull-down combined with mass spectrometry, Gwack et al. identified multiple cellular transcription cofactors that can interact with RTA, including the CBP HAT, SWI/SNF complex, and TRAP/Mediator complex (Gwack et al. 2003). These cofactors interact with the RTA activation domain and presumably function to modulate chromatin structures of viral promoters and recruit general transcription factors to initiate viral gene expression. RTA also interacts with the cellular poly(ADP-ribose) polymerase I (PARP-1) and a human homologue of Ste20-related kinase from chicken (hKFC) (Gwack et al. 2003) through its serine/threonine-rich region. It was shown that PARP-1 can poly(ADP)-ribosylate RTA and hKFC can phosphorylate RTA (Gwack et al. 2003). Both PARP-1 and hKFC repress RTA transcriptional activity in vitro and suppress RTA-mediated lytic reactivation in vivo. The KSHV RTA-binding protein (K-RBP, or MGC2663) contains potential zinc finger domains and was shown to cooperate with RTA in activation of target genes such as ORF57, K8, and vMIP-1 (Wang et al. 2001). Using nuclear extracts for EMSA, Chang et al. showed that cellular proteins such as YY1 and Sp1 also bind to the sequences within the identified RREs in the promoters of PAN, Kpsn, ORF57, and vMIP-I genes (Chang et al. 2002, 2005c; Chang and Miller 2004). However, whether these cellular proteins contribute to RTA activation of the respective promoters and, if so, whether they act by protein-protein interactions with RTA remain to be studied.

4.4 Viral Proteins Interacting with RTA

In addition to cellular factors, RTA also interacts with many viral proteins that function to modulate its activity. We have described that viral latency protein LANA interacts with RTA and antagonizes its autoactivation activity and thus keeps lytic reactivation under check (Sect. 2.2). A viral early protein KbZIP was shown to interact directly with RTA and to repress RTA activation in promoter-dependent manner, that is, it suppresses RTA activation of promoters of ORF57 and KbZIP but not of PAN. There is evidence suggesting (Izumiya et al. 2005) that sumoylation plays an important role in KbZIP's transcriptional repression activity. KbZIP represses RTA activation of a subset of promoters, which was suggested to contribute to the highly controlled expression of viral lytic genes. However, the significance of such control during viral infection is not yet understood. Another viral early protein, ORF57 (MTA), recently shown to interact with RTA in vitro and in vivo (Malik et al. 2004), cooperates with RTA to enhance the expression of target genes at a posttranscriptional level (Kirshner et al. 2000). The recent finding that RTA contains ubiquitin E3 ligase activity and targets interferon regulatory factor 7 (IRF-7) suggested a novel role of RTA in downregulating immune responses (Yu et al. 2005).

4.5 Other Regulators of RTA Activity

In addition to binding RTA directly, some regulators of RTA activity may indirectly modulate RTA transcriptional activity. Song et al. has shown that HMGB1 (also called HMG-1) protein, a highly conserved nonhistone chromatin protein, promotes RTA binding to different RTA target sites in vitro and stimulates RTA activation of target genes in vivo. The importance of HMGB1 in viral gene expression and replication of gammaherpesvirus was highlighted by a study using MHV-68, whose viral gene expression, as well as viral replication, was significantly reduced in HMGB1-deficient cells but can be partially restored by HMGB1 transfection (Song et al. 2004). A recent study by Wang et al showed that cellular IRF-7, an essential gene for IFN-α and -β response, competes with RTA for binding to the RTA response element and thus negatively regulates RTA activation of target genes such as ORF 57 (Wang et al. 2005). This implicates that host interferon responses, on KSHV infection, may play a role in suppressing lytic gene expression by attenuating RTA transcriptional activity by IRF-7. In addition, the stability and the DNA binding activity of the RTA protein are autoregulated through its distinct domains (Chang et al. 2005c).

5 Perspectives

After the virus enters the cell, viral gene expression control is the main determinant of viral replication, especially in the case of reactivation. Identification of RTA as the key molecular switch for KSHV gene expression serves as the starting point for defining the viral gene expression cascade. Despite significant progress during the last decade, we are still faced with gaps in our knowledge that are critical in understanding the viral gene expression program and its underlying mechanisms. Among these issues, we would like to discuss the following five challenges.

First, we know very little about signaling upstream of RTA that controls the initiation of the lytic cycle, in comparison to the events immediately downstream of RTA. Cellular signals that positively or negatively regulate RTA expression or activity have not been systematically identified. With the available genomic approaches, this question can be addressed in the near future. It is expected that multiple cellular signaling pathways affect the balance between latency and lytic replication. The next challenge is how to define the combinatory effects of these multiple signals that simultaneously act on the infected cells. The current approaches in molecular or cellular biology usually examine one or two signaling pathways at a time. New methods must be introduced to define the biological effects of multiple inputs. This advance will help us to understand the mechanism coupling viral reactivation with cellular differentiation and associated pathogenesis (Young et al. 1991; Longworth and Laimins 2004; Johnson et al. 2005). On a related issue, we expect to see significant progress in defining the effects of virion proteins on RTA expression during de novo infection and the effects of chromatin structure on reactivation from latency.

Second, the activation of early genes by RTA was mostly studied with reporter assays. It requires direct evidence for the interactions between RTA or its partners with their DNA targets. The temporal relationship among the activation of the viral downstream genes has not been defined, although viral gene array data provide some hints at a low resolution. Chromosomal immunoprecipitation should facilitate the construction of gene expression cascade with a clearer molecular definition. A similar approach should be applied to cellular genes regulated by RTA.

Third, some lytic genes are differentially expressed among the KSHV-associated diseases (KS, PEL, and MCD). Such differential expression may contribute to the tissue tropism and the diverse disease outcomes of KSHV infection. We expect to see more studies addressing how different cellular and microenvironmental signals affect the expression of viral genes and how these varied expressions may alter disease pathogenesis.

Fourth, a gap in the current understanding of KSHV gene expression cascade is how late gene expression is controlled. Because some early genes and most late genes are not directly activated by RTA, the intermediator(s) functioning downstream of RTA must be identified. A current assumption is that the late genes will be expressed once the viral DNA is replicated. How viral late gene expression is coupled to viral DNA replication is a long-standing question. Our recent study with MHV-68 suggested that additional viral factors are required in addition to the viral DNA replication in *cis*. The mechanism controlling late gene expression is a less well-defined area in the herpesvirus field.

Finally, the biological relevance of the majority of the studies reviewed here must be validated in the context of viral replication. The availability of a BAC system to manipulate the KSHV genome will greatly facilitate such work (Zhou et al. 2002; Gao et al. 2003). The next challenge is to apply the knowledge to clinical applications for both tumor therapy and prevention. Based on the concept that KSHV lytic replication plays a direct role in tumor development by providing a favorable microenvironment, inhibition of lytic gene expression may have therapeutic values. On the other hand, intentional induction of lytic gene expression in the presence of a gancyclovir-type drug may drive the destruction of tumor lesions because of the elevated immune responses and the bystander killing effect of gancyclovir.

Acknowledgements The authors appreciate the careful review and editing by Dr. Hinh Ly and Ms. Joyce X. Wu.

References

An FQ, Compitello N, Horwitz E, Sramkoski M, Knudsen ES, Renne R (2005) The latency-associated nuclear antigen of Kaposi's sarcoma-associated herpesvirus modulates cellular gene expression and protects lymphoid cells from p16 INK4A-induced cell cycle arrest. J Biol Chem 280:3862–3874

An J, Sun Y, Sun R, Rettig MB (2003) Kaposi's sarcoma-associated herpesvirus encoded vFLIP induces cellular IL-6 expression: the role of the NF-κB and JNK/AP1 pathways. Oncogene 22:3371–3385

Arvanitakis L, Mesri EA, Nador RG, Said JW, Asch AS, Knowles DM, Cesarman E (1996) Establishment and characterization of a primary effusion (body cavity-based) lymphoma cell line (BC-3) harboring Kaposi's sarcoma-associated herpesvirus (KSHV/HHV-8) in the absence of Epstein-Barr virus. Blood 88:2648–2654

Ballestas ME, Chatis PA, Kaye KM (1999) Efficient persistence of extrachromosomal KSHV DNA mediated by latency-associated nuclear antigen. Science 284:641–644

Bechtel JT, Winant RC, Ganem D (2005) Host and viral proteins in the virion of Kaposi's sarcoma-associated herpesvirus. J Virol 79:4952–4964

Boname JM, Coleman HM, May JS, Stevenson PG (2004) Protection against wild-type murine gammaherpesvirus-68 latency by a latency-deficient mutant. J Gen Virol 85:131–135

Boshoff C, Gao SJ, Healy LE, Matthews S, Thomas AJ, Coignet L, Warnke RA, Strauchen JA, Matutes E, Kamel OW, Moore PS, Weiss RA, Chang Y (1998) Establishing a KSHV+ cell line (BCP-1) from peripheral blood and characterizing its growth in Nod/SCID mice. Blood 91:1671–1679

Brown HJ, McBride WH, Zack JA, Sun R (2005) Prostratin and bortezomib are novel inducers of latent Kaposi's sarcoma-associated herpesvirus. Antivir Ther 10:745–751

Brown HJ, Song MJ, Deng H, Wu TT, Cheng G, Sun R (2003) NF-κB inhibits gammaherpesvirus lytic replication. J Virol 77:8532–8540

Cannon M, Cesarman E, Boshoff C (2006) KSHV G protein-coupled receptor inhibits lytic gene transcription in primary-effusion lymphoma cells via p21-mediated inhibition of Cdk2. Blood 107:277–284

Castagna M, Takai Y, Kaibuchi K, Sano K, Kikkawa U, Nishizuka Y (1982) Direct activation of calcium-activated, phospholipid-dependent protein kinase by tumorpromoting phorbol esters. J Biol Chem 257:7847–7851

Cesarman E, Chang Y, Moore PS, Said JW, Knowles DM (1995a) Kaposi's sarcomaassociated herpesvirus-like DNA sequences in AIDS-related body-cavity-based lymphomas. N Engl J Med 332:1186–1191

Cesarman E, Moore PS, Rao PH, Inghirami G, Knowles DM, Chang Y (1995b) In vitro establishment and characterization of two acquired immunodeficiency syndrome-related lymphoma cell lines (BC-1 and BC-2) containing Kaposi's sarcoma-associated herpesvirus-like (KSHV) DNA sequences. Blood 86: 2708–2714

Chang H, Dittmer DP, Chul SY, Hong Y, Jung JU (2005a) Role of Notch signal transduction in Kaposi's sarcoma-associated herpesvirus gene expression. J Virol 79:14371–14382

Chang J, Renne R, Dittmer D, Ganem D (2000) Inflammatory cytokines and the reactivation of Kaposi's sarcoma-associated herpesvirus lytic replication. Virology 266:17–25

Chang M, Brown HJ, Collado-Hidalgo A, Arevalo JM, Galic Z, Symensma TL, Tanaka L, Deng H, Zack JA, Sun R, Cole SW (2005b) β-Adrenoreceptors reactivate Kaposi's sarcoma-associated herpesvirus lytic replication via PKA-dependent control of viral RTA. J Virol 79:13538–13547

Chang PJ, Miller G (2004) Autoregulation of DNA binding and protein stability of Kaposi's sarcoma-associated herpesvirus ORF50 protein. J Virol 78:10657–10673

Chang PJ, Shedd D, Gradoville L, Cho MS, Chen LW, Chang J, Miller G (2002) Open reading frame 50 protein of Kaposi's sarcoma-associated herpesvirus directly activates the viral PAN and K12 genes by binding to related response elements. J Virol 76:3168–3178

Chang PJ, Shedd D, Miller G (2005c) Two subclasses of Kaposi's sarcoma-associated herpesvirus lytic cycle promoters distinguished by open reading frame 50 mutant proteins that are deficient in binding to DNA. J Virol 79:8750–8763

Chang Y, Cesarman E, Pessin MS, Lee F, Culpepper J, Knowles DM, Moore PS (1994) Identification of herpesvirus-like DNA sequences in AIDS-associated Kaposi's sarcoma. Science 266:1865–1869

Chen J, Ueda K, Sakakibara S, Okuno T, Parravicini C, Corbellino M, Yamanishi K (2001) Activation of latent Kaposi's sarcoma-associated herpesvirus by demethylation of the promoter of the lytic transactivator. Proc Natl Acad Sci USA 98:4119–4124

Chen J, Ueda K, Sakakibara S, Okuno T, Yamanishi K (2000) Transcriptional regulation of the Kaposi's sarcoma-associated herpesvirus viral interferon regulatory factor gene. J Virol 74:8623–8634

Chiou CJ, Poole LJ, Kim PS, Ciufo DM, Cannon JS, ap Rhys CM, Alcendor DJ, Zong JC, Ambinder RF, Hayward GS (2002) Patterns of gene expression and a transactivation function exhibited by the vGCR (ORF74) chemokine receptor protein of Kaposi's sarcoma-associated herpesvirus. J Virol 76:3421–3439

Cousens LS, Gallwitz D, Alberts BM (1979) Different accessibilities in chromatin to histone acetylase. J Biol Chem 254:1716–1723

Curreli F, Cerimele F, Muralidhar S, Rosenthal LJ, Cesarman E, Friedman-Kien AE, Flore O (2002) Transcriptional downregulation of ORF50/Rta by methotrexate inhibits the switch of Kaposi's sarcoma-associated herpesvirus/human herpesvirus 8 from latency to lytic replication. J Virol 76:5208–5219

Curreli F, Friedman-Kien AE, Flore O (2005) Glycyrrhizic acid alters Kaposi sarcoma-associated herpesvirus latency, triggering p53-mediated apoptosis in transformed B lymphocytes. J Clin Invest 115:642–652

Deng H, Chu JT, Rettig MB, Martinez-Maza O, Sun R (2002a) Rta of the human herpesvirus 8/Kaposi sarcoma-associated herpesvirus up-regulates human interleukin-6 gene expression. Blood 100:1919–1921

Deng H, Song MJ, Chu JT, Sun R (2002b) Transcriptional regulation of the interleukin-6 gene of human herpesvirus 8 (Kaposi's sarcoma-associated herpesvirus). J Virol 76:8252–8264

Deng H, Young A, Sun R (2000) Auto-activation of the rta gene of human herpesvirus-8/Kaposi's sarcoma-associated herpesvirus. J Gen Virol 81:3043–3048

Deutsch E, Cohen A, Kazimirsky G, Dovrat S, Rubinfeld H, Brodie C, Sarid R (2004) Role of protein kinase Cδin reactivation of Kaposi's sarcoma-associated herpesvirus. J Virol 78:10187–10192

Duan W, Wang S, Liu S, Wood C (2001) Characterization of Kaposi's sarcoma-associated herpesvirus/human herpesvirus-8 ORF57 promoter. Arch Virol 146:403–413

Fejer G, Medveczky MM, Horvath E, Lane B, Chang Y, Medveczky PG (2003) The latency-associated nuclear antigen of Kaposi's sarcoma-associated herpesvirus interacts preferentially with the terminal repeats of the genome in vivo and this complex is sufficient for episomal DNA replication. J Gen Virol 84:1451–1462

Friborg J, Jr., Kong W, Hottiger MO, Nabel GJ (1999) p53 inhibition by the LANA protein of KSHV protects against cell death. Nature 402:889–894

Fujimuro M, Hayward SD (2003) The latency-associated nuclear antigen of Kaposi's sarcoma-associated herpesvirus manipulates the activity of glycogen synthase kinase-3beta. J Virol 77:8019–8030

Gao SJ, Deng JH, Zhou FC (2003) Productive lytic replication of a recombinant Kaposi's sarcoma-associated herpesvirus in efficient primary infection of primary human endothelial cells. J Virol 77:9738–9749

Gradoville L, Gerlach J, Grogan E, Shedd D, Nikiforow S, Metroka C, Miller G (2000) Kaposi's sarcoma-associated herpesvirus open reading frame 50/Rta protein activates the entire viral lytic cycle in the HH-B2 primary effusion lymphoma cell line. J Virol 74:6207–6212

Grundhoff A, Ganem D (2003) The latency-associated nuclear antigen of Kaposi's sarcoma-associated herpesvirus permits replication of terminal repeat-containing plasmids. J Virol 77:2779–2783

Guasparri I, Keller SA, Cesarman E (2004) KSHV vFLIP is essential for the survival of infected lymphoma cells. J Exp Med 199:993–1003

Gwack Y, Hwang S, Lim C, Won YS, Lee CH, Choe J (2002) Kaposi's Sarcoma-associated herpesvirus open reading frame 50 stimulates the transcriptional activity of STAT3. J Biol Chem 277:6438–6442

Gwack Y, Nakamura H, Lee SH, Souvlis J, Yustein JT, Gygi S, Kung HJ, Jung JU (2003) Poly(ADP-ribose) polymerase 1 and Ste20-like kinase hKFC act as transcriptional repressors for gamma-2 herpesvirus lytic replication. Mol Cell Biol 23:8282–8294

Haque M, Chen J, Ueda K, Mori Y, Nakano K, Hirata Y, Kanamori S, Uchiyama Y, Inagi R, Okuno T, Yamanishi K (2000) Identification and analysis of the K5 gene of Kaposi's sarcoma-associated herpesvirus. J Virol 74:2867–2875

Haque M, Davis DA, Wang V, Widmer I, Yarchoan R (2003) Kaposi's sarcoma-associated herpesvirus (human herpesvirus 8) contains hypoxia response elements: relevance to lytic induction by hypoxia. J Virol 77:6761–6768

Hu J, Garber AC, Renne R (2002) The latency-associated nuclear antigen of Kaposi's sarcoma-associated herpesvirus supports latent DNA replication in dividing cells. J Virol 76:11677–11687

Izumiya Y, Ellison TJ, Yeh ET, Jung JU, Luciw PA, Kung HJ (2005) Kaposi's sarcoma-associated herpesvirus K-bZIP represses gene transcription via SUMO modification. J Virol 79:9912–9925

Jenner RG, Alba MM, Boshoff C, Kellam P (2001) Kaposi's sarcoma-associated herpesvirus latent and lytic gene expression as revealed by DNA arrays. J Virol 75:891–902

Jeong J, Papin J, Dittmer D (2001) Differential regulation of the overlapping Kaposi's sarcoma-associated herpesvirus vGCR (orf74) and LANA (orf73) promoters. J Virol 75:1798–1807

Johnson AS, Maronian N, Vieira J (2005) Activation of Kaposi's sarcoma-associated herpesvirus lytic gene expression during epithelial differentiation. J Virol 79:13769–13777

Kirshner JR, Lukac DM, Chang J, Ganem D (2000) Kaposi's sarcoma-associated herpesvirus open reading frame 57 encodes a posttranscriptional regulator with multiple distinct activities. J Virol 74:3586–3597

Kirshner JR, Staskus K, Haase A, Lagunoff M, Ganem D (1999) Expression of the open reading frame 74 (G-protein-coupled receptor) gene of Kaposi's sarcoma (KS)-associated herpesvirus: implications for KS pathogenesis. J Virol 73:6006–6014

Kliche S, Nagel W, Kremmer E, Atzler C, Ege A, Knorr T, Koszinowski U, Kolanus W, Haas J (2001) Signaling by human herpesvirus 8 kaposin A through direct membrane recruitment of cytohesin-1. Mol Cell 7:833–843

Knowles DM, Cesarman E (1997) The Kaposi's sarcoma-associated herpesvirus (human herpesvirus-8) in Kaposi's sarcoma, malignant lymphoma, and other diseases. Ann Oncol 8 Suppl 2:123–129

Krishnan HH, Naranatt PP, Smith MS, Zeng L, Bloomer C, Chandran B (2004) Concurrent expression of latent and a limited number of lytic genes with immune modulation and antiapoptotic function by Kaposi's sarcoma-associated herpesvirus early during infection of primary endothelial and fibroblast cells and subsequent decline of lytic gene expression. J Virol 78:3601–3620

Lan K, Kuppers DA, Robertson ES (2005) Kaposi's sarcoma-associated herpesvirus reactivation is regulated by interaction of latency-associated nuclear antigen with recombination signal sequence-binding protein Jκ, the major downstream effector of the Notch signaling pathway. J Virol 79:3468–3478

Lan K, Kuppers DA, Verma SC, Robertson ES (2004) Kaposi's sarcoma-associated herpesvirus-encoded latency-associated nuclear antigen inhibits lytic replication by targeting Rta: a potential mechanism for virus-mediated control of latency. J Virol 78:6585–6594

Liang Y, Chang J, Lynch SJ, Lukac DM, Ganem D (2002) The lytic switch protein of KSHV activates gene expression via functional interaction with RBP-Jκ (CSL), the target of the Notch signaling pathway. Genes Dev 16:1977–1989

Liang Y, Ganem D (2003) Lytic but not latent infection by Kaposi's sarcoma-associated herpesvirus requires host CSL protein, the mediator of Notch signaling. Proc Natl Acad Sci USA 100:8490–8495

Liang Y, Ganem D (2004) RBP-J (CSL) is essential for activation of the K14/vGPCR promoter of Kaposi's sarcoma-associated herpesvirus by the lytic switch protein RTA. J Virol 78:6818–6826

Lin CL, Li H, Wang Y, Zhu FX, Kudchodkar S, Yuan Y (2003) Kaposi's sarcoma-associated herpesvirus lytic origin (ori-Lyt)-dependent DNA replication: identification of the ori-Lyt and association of K8 bZip protein with the origin. J Virol 77:5578–5588

Lin SF, Robinson DR, Miller G, Kung HJ (1999) Kaposi's sarcoma-associated herpesvirus encodes a bZIP protein with homology to BZLF1 of Epstein-Barr virus. J Virol 73:1909–1917

Longworth MS, Laimins LA (2004) Pathogenesis of human papillomaviruses in differentiating epithelia. Microbiol Mol Biol Rev 68:362–372

Lu F, Day L, Lieberman PM (2005) Kaposi's sarcoma-associated herpesvirus virion-induced transcription activation of the ORF50 immediate-early promoter. J Virol 79:13180–13185

Lu F, Zhou J, Wiedmer A, Madden K, Yuan Y, Lieberman PM (2003) Chromatin remodeling of the Kaposi's sarcoma-associated herpesvirus ORF50 promoter correlates with reactivation from latency. J Virol 77:11425–11435

Lukac DM, Garibyan L, Kirshner JR, Palmeri D, Ganem D (2001) DNA binding by Kaposi's sarcoma-associated herpesvirus lytic switch protein is necessary for transcriptional activation of two viral delayed early promoters. J Virol 75:6786–6799

Lukac DM, Kirshner JR, Ganem D (1999) Transcriptional activation by the product of open reading frame 50 of Kaposi's sarcoma-associated herpesvirus is required for lytic viral reactivation in B cells. J Virol 73:9348–9361

Lukac DM, Renne R, Kirshner JR, Ganem D (1998) Reactivation of Kaposi's sarcoma-associated herpesvirus infection from latency by expression of the ORF 50 transactivator, a homolog of the EBV R protein. Virology 252:304–312

Malik P, Blackbourn DJ, Cheng MF, Hayward GS, Clements JB (2004) Functional co-operation between the Kaposi's sarcoma-associated herpesvirus ORF57 and ORF50 regulatory proteins. J Gen Virol 85:2155–2166

Matsumura S, Fujita Y, Gomez E, Tanese N, Wilson AC (2005) Activation of the Kaposi's sarcoma-associated herpesvirus major latency locus by the lytic switch protein RTA (ORF50). J Virol 79:8493–8505

McCormick C, Ganem D (2005) The kaposin B protein of KSHV activates the p38/MK2 pathway and stabilizes cytokine mRNAs. Science 307:739–741

Mercader M, Taddeo B, Panella JR, Chandran B, Nickoloff BJ, Foreman KE (2000) Induction of HHV-8 lytic cycle replication by inflammatory cytokines produced by HIV-1-infected T cells. Am J Pathol 156:1961–1971

Miller G, Heston L, Grogan E, Gradoville L, Rigsby M, Sun R, Shedd D, Kushnaryov VM, Grossberg S, Chang Y (1997) Selective switch between latency and lytic replication of Kaposi's sarcoma herpesvirus and Epstein-Barr virus in dually infected body cavity lymphoma cells. J Virol 71:314–324

Miller G, Rigsby MO, Heston L, Grogan E, Sun R, Metroka C, Levy JA, Gao SJ, Chang Y, Moore P (1996) Antibodies to butyrate-inducible antigens of Kaposi's sarcoma-associated herpesvirus in patients with HIV-1 infection. N Engl J Med 334:1292–1297

Molden J, Chang Y, You Y, Moore PS, Goldsmith MA (1997) A Kaposi's sarcoma-associated herpesvirus-encoded cytokine homolog (vIL-6) activates signaling through the shared gp130 receptor subunit. J Biol Chem 272:19625–19631

Moore PS, Boshoff C, Weiss RA, Chang Y (1996) Molecular mimicry of human cytokine and cytokine response pathway genes by KSHV. Science 274:1739–1744

Moses AV, Fish KN, Ruhl R, Smith PP, Strussenberg JG, Zhu L, Chandran B, Nelson JA (1999) Long-term infection and transformation of dermal microvascular endothelial cells by human herpesvirus 8. J Virol 73:6892–6902

Muralidhar S, Pumfery AM, Hassani M, Sadaie MR, Kishishita M, Brady JN, Doniger J, Medveczky P, Rosenthal LJ (1998) Identification of kaposin (open reading frame K12) as a human herpesvirus 8 (Kaposi's sarcoma-associated herpesvirus) transforming gene. J Virol 72:4980–4988

Muralidhar S, Veytsmann G, Chandran B, Ablashi D, Doniger J, Rosenthal LJ (2000) Characterization of the human herpesvirus 8 (Kaposi's sarcoma-associated herpesvirus) oncogene, kaposin (ORF K12). J Clin Virol 16:203–213

Nador RG, Milligan LL, Flore O, Wang X, Arvanitakis L, Knowles DM, Cesarman E (2001) Expression of Kaposi's sarcoma-associated herpesvirus G protein-coupled receptor monocistronic and bicistronic transcripts in primary effusion lymphomas. Virology 287:62–70

Neipel F, Fleckenstein B (1999) The role of HHV-8 in Kaposi's sarcoma. Semin Cancer Biol 9:151–164

Nicholas J, Ruvolo V, Zong J, Ciufo D, Guo HG, Reitz MS, Hayward GS (1997a) A single 13-kilobase divergent locus in the Kaposi sarcoma-associated herpesvirus (human herpesvirus 8) genome contains nine open reading frames that are homologous to or related to cellular proteins. J Virol 71:1963–1974

Nicholas J, Ruvolo VR, Burns WH, Sandford G, Wan X, Ciufo D, Hendrickson SB, Guo HG, Hayward GS, Reitz MS (1997b) Kaposi's sarcoma-associated human herpesvirus-8 encodes homologues of macrophage inflammatory protein-1 and interleukin-6. Nat Med 3:287–292

Osborne J, Moore PS, Chang Y (1999) KSHV-encoded viral IL-6 activates multiple human IL-6 signaling pathways. Hum Immunol 60:921–927

Paulose-Murphy M, Ha NK, Xiang C, Chen Y, Gillim L, Yarchoan R, Meltzer P, Bittner M, Trent J, Zeichner S (2001) Transcription program of human herpesvirus 8 (Kaposi's sarcoma-associated herpesvirus). J Virol 75:4843–4853

Pavlova IV, Virgin HWt, Speck SH (2003) Disruption of gammaherpesvirus 68 gene 50 demonstrates that Rta is essential for virus replication. J Virol 77:5731–5739

Renne R, Zhong W, Herndier B, McGrath M, Abbey N, Kedes D, Ganem D (1996) Lytic growth of Kaposi's sarcoma-associated herpesvirus (human herpesvirus 8) in culture. Nat Med 2:342–346

Rickabaugh TM, Brown HJ, Martinez-Guzman D, Wu TT, Tong L, Yu F, Cole S, Sun R (2004) Generation of a latency-deficient gammaherpesvirus that is protective against secondary infection. J Virol 78:9215–9223

Riggs MG, Whittaker RG, Neumann JR, Ingram VM (1977) n-Butyrate causes histone modification in HeLa and Friend erythroleukaemia cells. Nature 268:462–464

Russo JJ, Bohenzky RA, Chien MC, Chen J, Yan M, Maddalena D, Parry JP, Peruzzi D, Edelman IS, Chang Y, Moore PS (1996) Nucleotide sequence of the Kaposi sarcoma-associated herpesvirus (HHV8). Proc Natl Acad Sci USA 93:14862–14867

Sadler R, Wu L, Forghani B, Renne R, Zhong W, Herndier B, Ganem D (1999) A complex translational program generates multiple novel proteins from the latently expressed kaposin (K12) locus of Kaposi's sarcoma-associated herpesvirus. J Virol 73:5722–5730

Said W, Chien K, Takeuchi S, Tasaka T, Asou H, Cho SK, de Vos S, Cesarman E, Knowles DM, Koeffler HP (1996) Kaposi's sarcoma-associated herpesvirus (KSHV or HHV8) in primary effusion lymphoma: ultrastructural demonstration of herpesvirus in lymphoma cells. Blood 87:4937–4943

Sakakibara S, Ueda K, Chen J, Okuno T, Yamanishi K (2001) Octamer-binding sequence is a key element for the autoregulation of Kaposi's sarcoma-associated herpesvirus ORF50/Lyta gene expression. J Virol 75:6894–6900

Sarid R, Flore O, Bohenzky RA, Chang Y, Moore PS (1998) Transcription mapping of the Kaposi's sarcoma-associated herpesvirus (human herpesvirus 8) genome in a body cavity-based lymphoma cell line (BC-1). J Virol 72:1005–1012

Saveliev A, Zhu F, Yuan Y (2002) Transcription mapping and expression patterns of genes in the major immediate-early region of Kaposi's sarcoma-associated herpesvirus. Virology 299:301–314

Schulz TF, Moore PS (1999) Kaposi's sarcoma-associated herpesvirus: a new human tumor virus, but how? Trends Microbiol 7:196–200

Seaman WT, Quinlivan EB (2003) Lytic switch protein (ORF50) response element in the Kaposi's sarcoma-associated herpesvirus K8 promoter is located within but does not require a palindromic structure. Virology 310:72–84

Seaman WT, Ye D, Wang RX, Hale EE, Weisse M, Quinlivan EB (1999) Gene expression from the ORF50/K8 region of Kaposi's sarcoma-associated herpesvirus. Virology 263:436–449

Sharma-Walia N, Krishnan HH, Naranatt PP, Zeng L, Smith MS, Chandran B (2005) ERK1/2 and MEK1/2 induced by Kaposi's sarcoma-associated herpesvirus (human herpesvirus 8) early during infection of target cells are essential for expression of viral genes and for establishment of infection. J Virol 79:10308–10329

Song MJ, Brown HJ, Wu TT, Sun R (2001) Transcription activation of polyadenylated nuclear RNA by RTA in human herpesvirus 8/Kaposi's sarcoma-associated herpesvirus. J Virol 75:3129–3140

Song MJ, Deng H, Sun R (2003) Comparative study of regulation of RTA-responsive genes in Kaposi's sarcoma-associated herpesvirus/human herpesvirus 8. J Virol 77: 9451–9462

Song MJ, Hwang S, Wong W, Round J, Martinez-Guzman D, Turpaz Y, Liang J, Wong B, Johnson RC, Carey M, Sun R (2004) The DNA architectural protein HMGB1 facilitates RTA-mediated viral gene expression in gamma-2 herpesviruses. J Virol 78:12940–12950

Song MJ, Li X, Brown HJ, Sun R (2002) Characterization of interactions between RTA and the promoter of polyadenylated nuclear RNA in Kaposi's sarcoma-associated herpesvirus/human herpesvirus 8. J Virol 76:5000–5013

Soulier J, Grollet L, Oksenhendler E, Cacoub P, Cazals-Hatem D, Babinet P, d'Agay MF, Clauvel JP, Raphael M, Degos L, et al. (1995) Kaposi's sarcoma-associated herpesvirus-like DNA sequences in multicentric Castleman's disease. Blood 86:1276–1280

Staskus KA, Sun R, Miller G, Racz P, Jaslowski A, Metroka C, Brett-Smith H, Haase AT (1999) Cellular tropism and viral interleukin-6 expression distinguish human herpesvirus 8 involvement in Kaposi's sarcoma, primary effusion lymphoma, and multicentric Castleman's disease. J Virol 73:4181–4187

Staskus KA, Zhong W, Gebhard K, Herndier B, Wang H, Renne R, Beneke J, Pudney J, Anderson DJ, Ganem D, Haase AT (1997) Kaposi's sarcoma-associated herpesvirus gene expression in endothelial (spindle) tumor cells. J Virol 71:715–719

Sun R, Lin SF, Gradoville L, Miller G (1996) Polyadenylylated nuclear RNA encoded by Kaposi sarcoma-associated herpesvirus. Proc Natl Acad Sci USA 93:11883–11888

Sun R, Lin SF, Gradoville L, Yuan Y, Zhu F, Miller G (1998) A viral gene that activates lytic cycle expression of Kaposi's sarcoma-associated herpesvirus. Proc Natl Acad Sci USA 95:10866–10871

Sun R, Lin SF, Staskus K, Gradoville L, Grogan E, Haase A, Miller G (1999) Kinetics of Kaposi's sarcoma-associated herpesvirus gene expression. J Virol 73:2232–2242

Ueda K, Ishikawa K, Nishimura K, Sakakibara S, Do E, Yamanishi K (2002) Kaposi's sarcoma-associated herpesvirus (human herpesvirus 8) replication and transcription factor activates the K9 (vIRF) gene through two distinct *cis* elements by a non-DNA-binding mechanism. J Virol 76:12044–12054

Vieira J, Huang ML, Koelle DM, Corey L (1997) Transmissible Kaposi's sarcoma-associated herpesvirus (human herpesvirus 8) in saliva of men with a history of Kaposi's sarcoma. J Virol 71:7083–7087

Wang J, Zhang J, Zhang L, Harrington W, Jr., West JT, Wood C (2005) Modulation of human herpesvirus 8/Kaposi's sarcoma-associated herpesvirus replication and transcription activator transactivation by interferon regulatory factor 7. J Virol 79:2420–2431

Wang S, Liu S, Wu MH, Geng Y, Wood C (2001) Identification of a cellular protein that interacts and synergizes with the RTA (ORF50) protein of Kaposi's sarcoma-associated herpesvirus in transcriptional activation. J Virol 75:11961–11973

Wang SE, Wu FY, Yu Y, Hayward GS (2003) CCAAT/enhancer-binding protein-α is induced during the early stages of Kaposi's sarcoma-associated herpesvirus (KSHV) lytic cycle reactivation and together with the KSHV replication and transcription activator (RTA) cooperatively stimulates the viral RTA, MTA, and PAN promoters. J Virol 77:9590–9612

Wang Y, Chong OT, Yuan Y (2004) Differential regulation of K8 gene expression in immediate-early and delayed-early stages of Kaposi's sarcoma-associated herpesvirus. Virology 325:149–163

Wong EL, Damania B (2006) Transcriptional regulation of the Kaposi's sarcoma-associated herpesvirus K15 gene. J Virol 80:1385–1392

Wu J, Grunstein M (2000) 25 years after the nucleosome model: chromatin modifications. Trends Biochem Sci 25:619–623

Wu TT, Tong L, Rickabaugh T, Speck S, Sun R (2001) Function of Rta is essential for lytic replication of murine gammaherpesvirus 68. J Virol 75:9262–9273

Xu Y, AuCoin DP, Huete AR, Cei SA, Hanson LJ, Pari GS (2005) A Kaposi's sarcoma-associated herpesvirus/human herpesvirus 8 ORF50 deletion mutant is defective for reactivation of latent virus and DNA replication. J Virol 79:3479–3487

Young LS, Lau R, Rowe M, Niedobitek G, Packham G, Shanahan F, Rowe DT, Greenspan D, Greenspan JS, Rickinson AB, et al. (1991) Differentiation-associated expression of the Epstein-Barr virus BZLF1 transactivator protein in oral hairy leukoplakia. J Virol 65:2868–2874

Yu Y, Wang SE, Hayward GS (2005) The KSHV immediate-early transcription factor RTA encodes ubiquitin E3 ligase activity that targets IRF7 for proteosome-mediated degradation. Immunity 22:59–70

Zhang J, Wang J, Wood C, Xu D, Zhang L (2005) Kaposi's sarcoma-associated herpesvirus/human herpesvirus 8 replication and transcription activator regulates viral and cellular genes via interferon-stimulated response elements. J Virol 79:5640–5652

Zhang L, Chiu J, Lin JC (1998) Activation of human herpesvirus 8 (HHV-8) thymidine kinase (TK) TATAA-less promoter by HHV-8 ORF50 gene product is SP1 dependent. DNA Cell Biol 17:735–742

Zhong W, Ganem D (1997) Characterization of ribonucleoprotein complexes containing an abundant polyadenylated nuclear RNA encoded by Kaposi's sarcoma-associated herpesvirus (human herpesvirus 8). J Virol 71:1207–1212

Zhong W, Wang H, Herndier B, Ganem D (1996) Restricted expression of Kaposi sarcoma-associated herpesvirus (human herpesvirus 8) genes in Kaposi sarcoma. Proc Natl Acad Sci USA 93:6641–6646

Zhou FC, Zhang YJ, Deng JH, Wang XP, Pan HY, Hettler E, Gao SJ (2002) Efficient infection by a recombinant Kaposi's sarcoma-associated herpesvirus cloned in a bacterial artificial chromosome: application for genetic analysis. J Virol 76:6185–6196

Zhu FX, Cusano T, Yuan Y (1999) Identification of the immediate-early transcripts of Kaposi's sarcoma-associated herpesvirus. J Virol 73:5556-5567

Zhu J, Trang P, Kim K, Zhou T, Deng H, Liu F (2004) Effective inhibition of Rta expression and lytic replication of Kaposi's sarcoma-associated herpesvirus by human RNase P. Proc Natl Acad Sci USA 101:9073–9078

CTMI (2006) 312:185–209

Kaposi Sarcoma Herpesvirus-Encoded Interferon Regulator Factors

M. K. Offermann (✉)

Winship Cancer Institute, 1365-B Clifton Rd NE, Atlanta, GA 30322, USA
mofferm@emory.edu

Abstract The Kaposi sarcoma herpesvirus (KSHV) encodes multiple proteins that disrupt host antiviral responses, including four viral proteins that have homology to the interferon regulatory factor (IRF) family of transcription factors. At least three of the KSHV vIRFs (vIRFs 1–3) alter responses to cellular IRFs and to interferons (IFNs), whereas functional changes resulting from the fourth vIRF (vIRF-4) have not been reported. The vIRFs also affect other important regulatory proteins in the cell, including responses to transforming growth factor β (TGF-β) and the tumor suppressor protein p53. This review examines the expression of the vIRFs during the life cycle of KSHV and the functional consequences of their expression.

1 Consequences of Infection with KSHV

Kaposi sarcoma herpesvirus (KSHV), also known as human herpesvirus 8 (HHV-8), is a large double-stranded DNA virus belonging to the gammaherpesvirus subfamily (Boshoff and Weiss 1998; Moore and Chang 2001). As with all herpesviruses, once an individual becomes infected with KSHV, the virus persists throughout the lifetime of the host. This lifelong persistence occurs because KSHV enters a latent state that is not well recognized by innate and acquired host antiviral defenses (Moore and Chang 2003). During latency, viral replication is coupled to cellular replication without producing infectious virus (Rainbow et al. 1997; Ballestas et al. 1999; Hu et al. 2002), and viral gene products are expressed that subvert cellular antiviral defenses so that infected cells are not destroyed (Friborg et al. 1999; Brander et al. 2000; Radkov et al. 2000; Guasparri et al. 2004; An et al. 2005). Production of infectious virus can occur when KSHV shifts from latent to lytic replication (Renne et al. 1996; Lukac et al. 1998). KSHV expresses proteins during the lytic cascade that disrupt components of the cellular antiviral machinery and the host response, thereby enhancing the likelihood of successful production of infectious virus (Gao et al. 1997; Sarid et al. 1997; Zimring et al. 1998; Coscoy and Ganem 2000; Means et al. 2002; Pozharskaya et al. 2004b).

Although most individuals who are latently infected with KSHV remain asymptomatic and never experience KSHV-associated diseases, the lifelong persistence of KSHV puts the host at risk for the development of diseases associated with KSHV, including Kaposi sarcoma (KS), primary effusion lymphoma (PEL), and a plasmacytic variant of multicentric Castleman disease (Moore and Chang 2003). Tumor cells in both KS and PEL contain KSHV that is primarily latent (Parravicini et al. 2000; Dittmer 2003), but a few cells in KS lesions support lytic replication and express lytic viral proteins that contribute to the disease process (Staskus et al. 1997; Cannon et al. 2003). Many of the KSHV-infected cells in multicentric Castleman disease support lytic replication of KSHV, with viral proteins expressed during the lytic cascade playing a critical role in the disease process (Dupin et al. 1999; Parravicini et al. 2000; Waterston and Bower 2004).

2
KSHV-Encoded Proteins with Homology to the IRF Family of Transcription Factors

KSHV has significant sequence homology with other herpesviruses, with conservation of many open reading frames (ORFs) that encode proteins that are critical for herpesvirus replication and packaging of virions. KSHV ORFs were named after their homologues in herpesvirus saimiri (ORFs 1–75), whereas ORFs without recognizable homologues in herpesvirus saimiri were numbered separately as K1–K15 (Russo et al. 1996; Neipel et al. 1997). ORFs or Ks with decimals refer to ORFs that were not identified before the initial numbering (Neipel et al. 1997). Many of the proteins encoded by the Ks have homology to mammalian proteins, including four viral proteins that have homology to the interferon (IFN) regulatory factor (IRF) family of transcription factors. All four vIRFs are located between ORF57 and ORF58 on the KSHV genome (Sarid et al. 1998; Lubyova and Pitha 2000; Rivas et al. 2001; Cunningham et al. 2003). The splicing of vIRF-2 and vIRF-3 has led to several different reported locations on the KSHV genome (Table 1). vIRF-1 is located at K9, vIRF-2 at K11.1, vIRF-3, also called latency-associated nuclear antigen 2 (LANA2) at K10.5, and vIRF-4 at K10.

The location of the vIRFs and their sequence homologies suggest that they arose from capture of cellular sequences followed by gene duplication events. Each gene possesses its own promoter and polyadenylation site, and all except vIRF-1 are spliced from two exons. There may be some alternative splicing that leads to the minor differences in sizes of transcripts that have been detected, but there is no evidence of splicing between the different vIRFs (Cunningham et al. 2003). The only other herpesvirus that is known to encode vIRFs is the rhesus rhadinovirus (RRV), a close relative to KSHV (Searles et al. 1999; Alexander et al. 2000). The RRV encodes eight tandem vIRFs, but, unlike KSHV, none of them are spliced.

3
Subversion of Host Antiviral Defenses by vIRFs

Nine cellular IRF genes have been identified that function in the regulation of expression of IFN and other genes (Pitha et al. 1998), and at least three of the KSHV vIRFs (vIRFs 1–3) alter responses to cellular IRFs and to IFNs (Gao et al. 1997; Flowers et al. 1998; Zimring et al. 1998; Burysek et al. 1999a; Burysek et al. 1999b; Burysek and Pitha 2001; Lin et al. 2001; Nakamura et al. 2001; Lubyova et al. 2004) (Table 2). The functional changes resulting from

Table 1 vIRFs located between ORF 57 and ORF 58

	Reported genomic locations	Splicing pattern	Expression pattern	Cellular location
vIRF-1	K9 (Russo et al. 1996)	Not spliced (Cunningham et al. 2003)	Low constitutive expression during latency (Dittmer 2003; Pozharskaya et al. 2004b); strongly induced during lytic cascade (Pozharskaya et al. 2004b)	Predominantly in nucleus and sometimes in cytoplasm (Inagi et al. 1999; Parravicini et al. 2000; Pozharskaya et al. 2004b)
vIRF-2	K11.1 spliced with K11 (Jenner et al. 2001; Cunningham et al. 2003)	Spliced (Cunningham et al. 2003)	Cell type variation; some expression during latency (Burysek et al. 1999b; Burysek and Pitha 2001; Cunningham et al. 2003); sometimes strongly induced during lytic cascade (Cunningham et al. 2003)	GFP-vIRF-2 fusion protein expressed in nucleus (Burysek and Pitha 2001)
vIRF-3/ LANA 2	K10.1, K10.5, K10.6 (Lubyova and Pitha 2000; Rivas et al. 2001)	Spliced (Lubyova and Pitha 2000; Rivas et al. 2001; Cunningham et al. 2003)	Constitutively expressed during latency (Rivas et al. 2001; Fakhari and Dittmer 2002); some increase at the mRNA but not protein level in response to TPA (Lubyova and Pitha 2000; Lubyova et al. 2004)	Nucleus (Rivas et al. 2001; Munoz-Fontela et al. 2003)
VIRF-4	K10	Spliced (Cunningham et al. 2003)	Strongly induced during lytic cascade (Cunningham et al. 2003)	Not reported

vIRF-4 have not been reported. The cellular IRFs play a critical role in the response to viral infection (Pitha et al. 1998; Barnes et al. 2002). All of the IRFs share homology in the amino-terminal region, including five tryptophan repeats that regulate binding to DNA (Taniguchi et al. 2001; Barnes et al. 2002). Some of the IRFs serve as transcriptional activators, whereas others serve as transcriptional repressors. The vIRFs do not retain the amino acids that are utilized by the nine known cellular IRFs for binding to DNA. It is thus not

Table 2 Changes in gene expression and protein activation resulting from vIRFs

	vIRF-1	vIRF-2	vIRF-3
Virus-induced type I IFN expression	Inhibited (Lin et al. 2001)	Inhibited (Burysek et al. 1999b)	Inhibited in L929 cells (Lubyova and Pitha 2000); Induced in BJAB cells (Lubyova et al. 2004)
Gene expression induced by IFNα, β, or γ	Inhibited (Flowers et al. 1998; Li et al. 1998; Zimring et al. 1998)	Not reported	Not reported
Activation of IFN-induced proteins	Not reported	Activation of PKR blocked (Burysek and Pitha 2001)	Activation of PKR blocked, whereas activation of 2′,5′-OAS not affected (Esteban et al. 2003)
NF-κB-induced gene expression	Not inhibited (Zimring et al. 1998)	Inhibited (Burysek et al. 1999b)	Inhibited (Seo et al. 2004)
p53-induced apoptosis and gene expression	Inhibited (Seo et al. 2001)	Not reported	Inhibited (Rivas et al. 2001)
IRF-1-induced gene expression	Inhibited (Flowers et al. 1998; Zimring et al. 1998)	Inhibited (Burysek et al. 1999b)	Not reported
IRF-3-induced gene expression	Inhibited (Lin et al. 2001)	Inhibited (Burysek et al. 1999b)	Inhibited in L929 cells (Lubyova and Pitha 2000); induced in BJAB cells (Lubyova et al. 2004)
IRF-7-induced gene expression	Not inhibited (Lin et al. 2001)	Not reported	Inhibited in L929 cell (Lubyova and Pitha 2000); induced in BJAB cells (Lubyova et al. 2004)

surprising that none of the vIRFs binds to the sites in DNA to which the cellular IRFs bind (Flowers et al. 1998; Zimring et al. 1998; Lin et al. 2001). They also do not prevent the cellular IRFs from binding to DNA. They nonetheless can disrupt the function of multiple cellular IRFs.

Most studies on the functions of vIRFs have used expression vectors to constitutively express vIRFs in non-KSHV-infected cells, thereby characterizing the consequence of vIRF expression in the absence of other viral proteins. In addition, studies have been done in KSHV-infected PEL cells (Li et al.

1998, 2000; Nakamura et al. 2001; Zhou et al. 2002; Pozharskaya et al. 2004b). These studies demonstrate that vIRFs affect multiple antiviral pathways, with some shared and some distinct functions occurring in response to the different vIRFs.

All three vIRFs that have been characterized alter IFN expression and/or signaling. IFNs play an important role in the host response to viral infection (Samuel 2001; Malmgaard 2004). The type I IFNs consist of multiple forms of IFNα and a single IFNβ, whereas type II IFN consists exclusively of IFNγ (Malmgaard 2004). There are distinct receptors for type I and type II IFNs, but they share some components of the Jak-Stat signaling pathway and induce expression of some shared genes, including proteins that play an important role in cellular and host defenses against viral infection (Kotenko and Pestka 2000; Pozharskaya et al. 2004a). Proteins such as the double-stranded RNA activated protein kinase (PKR) and 2′,5′-oligoadenylate synthetase (2′5′-OAS) that are induced by both types of IFN display very little enzymatic activity unless they are activated, a process that often occurs in response to molecular motifs encountered during viral infection (Rebouillat and Hovanessian 1999; Justesen et al. 2000; Vattem et al. 2001; Chawla-Sarkar et al. 2003; Donze et al. 2004). The activated enzymes then inhibit protein synthesis, thereby impairing the ability of viruses to survive and replicate.

For viruses to persist within the host, the antiviral effects of IFNs generally must be subverted. vIRF-1 and vIRF-2 block transcription of type I IFN, whereas vIRF-3 can either block or enhance transcription of type I IFN, depending on cellular context (Burysek et al. 1999a, 1999b; Li et al. 2000; Lubyova and Pitha 2000; Lin et al. 2001; Lubyova et al. 2004). Both vIRF-2 and vIRF-3 block the activation of the IFN-inducible antiviral protein PKR, and they also block activation of NF-κB (Burysek et al. 1999b; Esteban et al. 2003; Seo et al. 2004). In addition, vIRF-1 and vIRF-3 bind to the tumor suppressor protein p53, blocking p53-induced transcription and apoptosis (Nakamura et al. 2001; Rivas et al. 2001; Seo et al. 2001; Munoz-Fontela et al. 2003). Thus the vIRFs affect the function of multiple important regulatory pathways in a way that generally favors the survival of virally infected cells.

4
Alterations in IFN Induction and Responsiveness Resulting from vIRF-1

The expression of vIRF-1 blocks transcription of type I IFNs as well as responses induced by IFNα, IFNβ, and IFNγ (Li et al. 1998; Zimring et al. 1998; Lin et al. 2001). The inhibition results in large part from the binding of vIRF-1 to the coactivator proteins CBP and p300, thereby interfering with coactivator binding and function (Burysek et al. 1999a; Li et al. 2000; Seo et al. 2000; Lin et al. 2001). CBP and p300 are structurally and functionally related proteins that possess histone acetyltransferase (HAT) activity, and they associate with CBP/p300-associated factor (pCAF), another protein that has HAT activity (Chan and La Thangue 2001). The interaction between vIRF-1 and p300 displaces p/CAF from p300, thereby inhibiting the HAT activity of p300 (Li et al. 2000). In addition, vIRF-1 can displace CBP/p300 from IRF-3, thereby inhibiting transcriptional activation by IRF-3 (Lin et al. 2001). HAT activity is important for altering chromatin structure and making the DNA accessible for transcription (Chan and La Thangue 2001). The disruption of CBP/p300 binding and function inhibits transcriptional activation by IRF-3, whereas transcriptional activation by IRF-7 is less affected (Suhara et al. 2000; Lin et al. 2001; Yang et al. 2002, 2003, 2004).

Both cellular IRF-3 and IRF-7 reside in the cytoplasm, and activation requires phosphorylation followed by nuclear translocation. Activated IRF-3 binds to CBP/p300 in order to become transcriptionally active (Suhara et al. 2000; Yang et al. 2002). IRF-3 and IRF-7 continue to undergo phosphorylation and bind to DNA when vIRF-1 is expressed, but IRF-3 is no longer able to enhance transcription (Lin et al. 2001). Because IRF-3 is required for the transcription of IFNα and IFNβ, the disruption of IRF-3 function that results from vIRF-1 blocks expression of all forms of type I IFN (Lin et al. 2001). The IRF-7 transcription factor contributes to type I IFN expression (Yang et al. 2003, 2004). vIRF-1 binds to cellular IRF-7, but IRF-7 continues to enhance transcription because it is less dependent on CBP/p300 than IRF-3 (Lin et al. 1998, 2000a, 2000b, 2001; Li et al. 2000; Seo et al. 2000; Morin et al. 2002; Servant et al. 2002; Sharma et al. 2003).

The expression of vIRF-1 also blocks the ability of IRF-1 to enhance transcription (Zimring et al. 1998; Burysek et al. 1999a). IRF-1 is generally induced to higher levels of expression by both type I and type II IFN, and it serves as a transcription factor for many cellular proteins involved in regulating proliferation and immune responses (Coccia et al. 1999; Taniguchi et al. 2001; Kroger et al. 2002; Romeo et al. 2002; Lee et al. 2003; Dornan et al. 2004). IRF-1 also has activities that complement the activities of other transcription factors such as p53. For example, IRF-1 is important in the acetylation of p53 through its recruitment of p300 to the p21 promoter, and this enhances transcriptional activation by p53 (Dornan et al. 2004). Thus the disruption of IRF-1 function affects the expression of many cellular genes.

Because vIRF-1 blocks the transcriptional activity of several IRF family members, it has profound effects on induction of IFN-regulated, immunoregulatory and growth regulatory gene products. Microarray analysis of cells expressing vIRF-1 demonstrate that approximately 7% of the 4,500 transcripts examined changed by at least twofold (Li et al. 2000). The expression of vIRF-1 inhibits apoptosis (Burysek et al. 1999a; Seo et al. 2001) and induces malignant transformation of several different cell types (Gao et al. 1997; Li et al. 1998).

The inhibition of HAT activity that results from vIRF-1 causes some global changes in the chromatin. Cells expressing vIRF-1 have reduced histone acetylation and condensation of the chromatin, thereby reducing the ability of propidium iodide to bind stoichiometrically to DNA (Li et al. 2000). This can lead to falsely low quantitation of cellular DNA content, whereas binding of propidium iodide to DNA can be restored to normal levels by incubation with the histone deacetylase inhibitor butyrate.

4.1 vIRF-1 Expression

Studies in the primary effusion lymphoma cell line BCBL-1 demonstrate that vIRF-1 is expressed during both latent and lytic replication (Pozharskaya et al. 2004b) (Table 1). The levels of vIRF-1 that are achieved during lytic replication are much higher than during latency, and hence the latent expression is often overlooked (Parravicini et al. 2000; Dittmer 2003). The amount of vIRF-1 that is expressed during latency is insufficient to block responses to exogenous IFNs in BCBL-1 cells, whereas the high levels expressed during lytic replication block gene induction by exogenous IFNα (Pozharskaya et al. 2004b). The vIRF-1 expressed during latency is associated with PML bodies, but the functional consequence of this association is not known. When vIRF-1 is expressed at the high levels that occur during lytic replication of KSHV, it is often detected in the cytoplasm as well as in the nucleus (Inagi et al. 1999; Parravicini et al. 2000; Pozharskaya et al. 2004b). Immunoprecipitation studies reveal that 5%–10% of vIRF-1 in TPA-stimulated BCBL-1 cells is bound to p300 (Li et al. 2000), providing evidence that much of the vIRF-1 that is expressed during the course of KSHV infection is complexed with p300. When KSHV in BCBL-1 cells is induced to enter the lytic cascade, the percentage of cells that express high levels of vIRF-1 is considerably lower than the percentage that express other lytic markers, such as the product of ORF 59 (Pozharskaya et al. 2004b). This is because vIRF-1 is transiently expressed and has a short half-life, whereas some other lytic viral proteins are more stable. Thus vIRF-1 is able to inhibit cellular antiviral defenses for only a limited portion of the lytic cascade.

The level of expression of vIRF-1 in latently infected cells is beneath the threshold for detection unless a method that is more sensitive than immunohistochemistry is used. vIRF-1 protein is readily detectible in multifocal Castleman disease and in cultured PEL cells supporting lytic replication, whereas vIRF-1 is not detected by immunohistochemical analysis in freshly excised KS biopsies or in primary effusion lymphomas (Parravicini et al. 2000). It is nonetheless expressed in latently infected KS biopsies as revealed by microarray analysis, and its expression profile in KS biopsies corresponds with latent rather than lytic expression (Dittmer 2003).

vIRF-1 is transcribed in both uninduced and induced cells from a single initiating site proceeded by a TATA box (Chen et al. 2000; Cunningham et al. 2003). An additional TATA box and initiation site is used to initiate vIRF-1 transcription during latency. vIRF-1 transcription is induced by replication and transcription activator (Rta), the viral gene product that induces lytic replication of KSHV (Ueda et al. 2002). vIRF-1 has a transactivation domain that can drive transcription (Roan et al. 1999), and vIRF-1 autoactivates its own expression (Wang and Gao 2003).

5
Alteration in IFN Induction and Responsiveness Resulting from vIRF-2

In transient transfection assays, the expression of vIRF-2 disrupts the ability of viruses to induce type I IFNs (Burysek et al. 1999b) (Table 2). vIRF-2 inhibits NF-κB and IRF-1-dependent transcription (Burysek et al. 1999b). This blocks IFNβ transcription because the coordinate activation of both IRF family members and NF-κB are required for IFNβ transcription (Parekh and Maniatis 1999; Han et al. 2004).

The expression of vIRF-2 blocks the activation of PKR (Burysek and Pitha 2001), an antiviral protein that is induced by IFN and activated by molecular motifs such as double-stranded RNA that are encountered during viral infection (Clemens 1997; Tan and Katze 1999; Justesen et al. 2000). PKR catalyzes the phosphorylation of eIF2α, a translation factor that blocks the initiation of protein synthesis when it becomes phosphorylated (Clemens and Elia 1997; Samuel et al. 1997). The prevention of activation of PKR by vIRF-2 decreases the ability of cells to respond to viral infection with reduced protein synthesis as part of their antiviral strategy. Thus vIRF-2 not only blocks expression of IFNβ but also blocks the functional consequences of IFNs by blocking the activation of an IFN-induced protein.

5.1
vIRF-2 Expression

Initial studies classified vIRF-2 as a latent viral gene because its mRNA and protein levels did not change in response to agents that induce lytic replication (Burysek et al. 1999b; Burysek and Pitha 2001) (Table 1). Other investigators have reported that vIRF-2 levels increase in response to TPA, thereby reclassifying it as a lytic transcript (Cunningham et al. 2003). When different types of KSHV-infected cells were examined for vIRF-2 mRNA expression with reverse transcriptase-PCR, some KSHV-infected cells showed considerable expression of vIRF-2 in unstimulated cells, whereas other cell lines showed minimal expression of vIRF-2 before induction of lytic replication with TPA (Cunningham et al. 2003). This suggests that vIRF-2 might be expressed during latent and lytic replication of KSHV, with some cell type differences. Expression of green fluorescent protein-tagged vIRF-2 in non-KSHV-infected cells provides evidence that vIRF-2 is a nuclear protein (Burysek and Pitha 2001). The percentage of KSHV-infected cells that express vIRF-2 is not known, and viral proteins that are coexpressed with vIRF-2 have not been reported.

6 Alterations in IFN Induction and Responsiveness Resulting from vIRF-3

Some studies report that IFNα transcription and expression increase in response to vIRF-3 (Lubyova et al. 2004), whereas other studies show vIRF-3 inhibits IFNα expression (Lubyova and Pitha 2000) (Table 2). When L929 cells were transfected with an expression vector for vIRF-3, it decreased virus-induced transcriptional activation of an IFNα-reporter construct and decreased the amount of biologically active IFNα that was produced in response to Newcastle disease virus (NDV) infection (Lubyova and Pitha 2000). In contrast, vIRF-3 increased virus-induced transcription and production of IFNα in BJAB cells (Lubyova et al. 2004). vIRF-3 binds to activated cellular IRF-3, whereas vIRF-3 binds to cellular IRF-7 irrespective of its activation state (Lubyova et al. 2004). vIRF-3 binds to a site in p300 that is distinct from the site that binds to vIRF-1, and the interaction between vIRF-3 and CBP/p300 enhances HAT activity. The transcriptional complexes that contain vIRF-3 are recruited to the IFNα promoter and more effectively enhance transcription than in the absence of vIRF-3. This increases the amount of IFNα produced by BJAB cells infected with NDV by approximately fourfold compared to control cells lacking vIRF-3 (Lubyova et al. 2004). The increase in IFN production that occurs when vIRF-3 is expressed in BJAB cells differs from the reduction that occurs when vIRF-3 is expressed in L929 cells (Lubyova and Pitha 2000), suggesting there might be cell type differences in the functional consequences of vIRF-3 expression.

6.1
vIRF-3 Expression

Initial studies suggested that vIRF-3 was expressed during lytic replication because mRNA levels increased when BCBL-1 cells were incubated with TPA (Lubyova and Pitha 2000), but vIRF-3 protein levels do not increase when BCBL-1 cells are incubated with TPA (Lubyova et al. 2004). Expression of vIRF-3 is detected at the protein level in the nuclei of all latently infected BCBL-1 cells with a speckled pattern that is distinct from the pattern that results from LANA (Rivas et al. 2001) (Table 2). Its constitutive expression during latency is responsible for vIRF-3 having an alternative name, LANA2 (Rivas et al. 2001; Esteban et al. 2003; Munoz-Fontela et al. 2003). vIRF-3 is detected in nearly all PEL tumor cells and the many KSHV-associated Castleman disease cells, but vIRF-3 is not expressed in KS tumors (Rivas et al. 2001; Fakhari and Dittmer 2002, 2003). Thus there are tissue-specific differences in the expression of vIRF-3, with expression occurring in cells of B lymphocyte origin that are latently infected with KSHV but not in cells of endothelial origin. vIRF-3 is encoded by a spliced mRNA and can exist in two different sizes, 2.2 or 1.8 kb (Lubyova and Pitha 2000; Rivas et al. 2001). The spliced transcript for vIRF-3 originates at ORFK10.5 and is composed of a 455-bp 5′ exon that is joined to the 1,339-bp 3′ exon previously called K10.1. The promoter of vIRF-3 is the only member of the vIRF family that lacks an obvious TATA box (Cunningham et al. 2003).

7
vIRF-4 Expression

Functional studies on vIRF-4 have not been reported. vIRF-4 is a spliced transcript that is induced by TPA in multiple KSHV-infected cell types (Sarid et al. 1998; Jenner et al. 2001; Fakhari and Dittmer 2002; Cunningham et al. 2003).

8
Inhibition of p53 by vIRFs

Expression of the tumor suppressor gene p53 is one of the mechanisms used by cells to prevent the survival and replication of virally infected cells (Collot-Teixeira et al. 2004; Meek 2004; Michalak et al. 2005; O'Shea 2005). p53 is a transcription factor that induces expression of cellular proteins that prevent replication of cells containing abnormal DNA by inhibiting proliferation and/or inducing apoptosis (Oren 2003; Meek 2004; Slee et al. 2004; Harris and Levine 2005; Michalak et al. 2005; Sengupta and Harris 2005). Viruses often express proteins that disrupt the action of p53 in order to overcome its growth-suppressive and proapoptotic actions (Friborg et al. 1999; Moore and Chang 2003; Collot-Teixeira et al. 2004). Binding of vIRF-1 to p53 suppresses phosphorylation and acetylation of p53 and inhibits the transcriptional activation and apoptosis driven by p53 (Nakamura et al. 2001; Seo et al. 2001) (Table 2). vIRF-3 also binds to p53 and blocks its ability to enhance transcription and induce apoptosis (Nakamura et al. 2001; Rivas et al. 2001; Seo et al. 2001; Munoz-Fontela et al. 2003). Thus both vIRF-1 and vIRF-3 join other KSHV-encoded proteins such as LANA and K-bZIP that repress p53 activity (Friborg et al. 1999; Park et al. 2000). Immunofluorescent microscopy demonstrates that vIRF-1 and p53 colocalize in some KSHV-infected BCBL-1 cells, but sometimes the amount of p53 exceeds the amount that is bound by vIRF-1 (Nakamura et al. 2001; Pozharskaya et al. 2004a). Immunoprecipitation studies provide additional evidence that vIRF-1 and p53 interact in KSHV-infected cells (Nakamura et al. 2001). In contrast, vIRF-3 does not coprecipitate with p53 in KSHV infected cells, and thus it might not associate with p53 when other viral proteins are present (Rivas et al. 2001). It is not known whether the other KSHV-encoded proteins that associate with p53 compete for binding or coassociate with p53, and the relative importance of specific KSHV proteins in disrupting p53 function is not known.

9
Inhibition of NF-κB by vIRFs

The transcription factor NF-κB plays an important role in the cellular response to viral infection (Offermann et al. 1995; Alexopoulou et al. 2001; Harcourt and Offermann 2001; Kaiser and Offermann 2005), and it is also involved in controlling the balance between latent and lytic replication (Brown et al. 2003). NF-κB is often activated as part of the cellular antiviral response, leading to expression of proinflammatory proteins that can enhance the host response to viral infection (Offermann et al. 1995; Alexopoulou et al. 2001; Harcourt and Offermann 2001; Kaiser et al. 2004). The activation of NF-κB by viral proteins can also be beneficial to the virus by enhancing expression of antiapoptotic proteins that help virally infected cells escape cellular defenses (Stehlik et al. 1998; Kreuz et al. 2001; Li et al. 2001; Mak and Yeh 2002). Both vIRF-2 and vIRF-3 decrease activation of NF-κB (Burysek et al. 1999b; Seo et al. 2004) (Table 2). vIRF-2 binds to the κB consensus element in DNA, blocking the ability of NF-κB to enhance transcription (Burysek et al. 1999b).

vIRF-3 inhibits the kinase activity of IκB kinase β (IKKβ) but not of IKKα (Seo et al. 2004). This blocks the nuclear translocation of NF-κB and sensitizes cells to TNF-induced apoptosis.

The functional consequences of the alterations in NF-κB that occur in response to vIRF-2 and vIRF-3 are unclear. The studies showing an interaction between vIRF-3 and IKKβ were done with transfection of expression vectors. Immunoprecipitation studies using extracts from BCBL-1 cells did not show evidence of vIRF-3 and IKKβ interactions (Seo et al. 2004), suggesting that the interaction might not occur when other viral proteins that interact with IKKβ are present. NF-κB is activated in latently infected PEL cells despite expression of vIRF-3, and inhibition of NF-κB activation leads to apoptosis of PEL cells (Keller et al. 2000). This indicates that vIRF-3 is insufficient to inhibit NF-κB activation in KSHV-infected BCBL-1 cells, but it might be modulating the level of activation or the function of the NF-κB that results. The activation of NF-κB that occurs during latency is likely to result from vFLIP, a viral protein expressed during latency that activates NF-κB through interactions with multiple components of the IKK (Matta et al. 2003). The activated NF-κB not only plays an important role in the survival of KSHV-infected cells (Keller et al. 2000; Belanger et al. 2001; Ghosh et al. 2003) but also helps maintain latency by inhibiting the function of Rta (Roan et al. 2002; Brown et al. 2003).

10
Inhibition of Transforming Growth Factor-β Signaling by vIRF-1

vIRF-1 also inhibits transforming growth factor-β (TGF-β) signaling, thereby suppressing the ability of TGF-β to inhibit proliferation (Seo et al. 2005). TGF-β initiates signaling by assembling a receptor kinase that activates members of the Smad family of transcription factors (Shen et al. 1998; Seoane et al. 2001; Feng and Derynck 2005; Massague et al. 2005). The activated Smad proteins reduce proliferation by inducing transcription of inhibitors of cyclin-dependent kinases, including p15 and p21, and by reducing the transcription of c-myc (Seoane et al. 2001; Massague et al. 2005). vIRF-1 binds to both Smad3 and Smad4, thereby interfering with their ability to bind to each other and to bind DNA, and this disrupts the transcriptional responses that are normally induced by TGF-β (Seo et al. 2005). Because TGF-β plays an important role in growth regulation and differentiation, its disruption by vIRF-1 is likely to further contribute to the deregulation of growth control that occurs in KSHV-infected cells.

11
Expression of IRF-Regulated Gene Products in KSHV-Infected Cells and Responses to Exogenous IFNs

When microvascular endothelial cells (MECs) are infected with KSHV, they express elevated levels of PKR and 2′5′-OAS (Krug et al. 2004). Transcription of these genes can be induced by IRF family members or by IFN (Wang and Floyd-Smith 1998; Coccia et al. 1999; Floyd-Smith et al. 1999; Yu et al. 1999; Nakaya et al. 2001). The expression of PKR is higher in MECs that are latently infected with KSHV than in surrounding uninfected cells, providing evidence that gene products expressed in virally infected cells are responsible for the increase rather than secreted IFNs (Krug et al. 2004). The increased expression of the PKR and 2′5′-OAS suggests that IRF transcription factors are activated in KSHV-infected MECs, leading to the enhanced expression of these gene products. This provides evidence that the vIRFs are not sufficient to fully block the ability of IRF family members to induce transcription of cellular antiviral gene products. The PKR and 2′5′-OAS are probably not fully activated because sufficient protein synthesis occurs in latently infected MECs for replication to continue, so that the latently infected cells have a growth advantage over uninfected cells (Krug et al. 2004). vIRF-3 is not expressed in KSHV-infected MECs (Cunningham et al. 2003), indicating that expression of vIRF-3 is not responsible for the induction of these genes or for the disruption of PKR activation.

Both type I and type II IFNs induce enhanced gene expression in KSHV-infected cells despite expression of viral proteins that disrupt responses to IRFs and to IFNs (Krug et al. 2004; Pozharskaya et al. 2004a, 2004b). Incubation of KSHV-infected MECs with IFNα for 24 h enhances expression of PKR in latently infected cells without reducing cell number, demonstrating that KSHV-infected cells respond to IFNα with increased PKR expression yet cells continue to survive (Krug et al. 2004). BCBL-1 cells that are latently infected with KSHV respond to IFNα and IFNγ with increased expression of the antiviral proteins PKR and 2′5′-OAS, yet very little cell death is observed unless KSHV enters the lytic phase of viral replication (Pozharskaya et al. 2004a, 2004b). This suggests that either these antiviral proteins are not activated or viral proteins expressed during latency block the functional consequences of activated PKR and 2′5′-OAS. Both vIRF-2 and vIRF-3 can block the activation of PKR (Burysek and Pitha 2001; Esteban et al. 2003), but vIRF-3 does not prevent activation of 2′5′-OAS (Esteban et al. 2003), and vIRF-3 is not expressed in KSHV-infected MECs (Cunningham et al. 2003). The vFLIP that is expressed during latency blocks apoptosis (Sturzl et al. 1999; Belanger et al. 2001; Sun et al. 2003; Guasparri et al. 2004), and this probably inhibits the apoptosis that occurs when protein synthesis decreases in response to activated PKR and 2′5′-OAS (Justesen et al. 2000; Ghosh et al. 2001; Vorburger et al. 2002; Chawla-Sarkar et al. 2003; Donze et al. 2004).

The host response to viral infection is generally greater during lytic replication when multiple viral gene products are expressed, increasing the targets that can be recognized by innate antiviral defenses (Samuel 2001; Chawla-Sarkar et al. 2003; Malmgaard 2004). It is thus not surprising that most of the vIRFs and other viral gene products that block components of the cellular antiviral response are expressed during lytic replication. Apoptosis occurs in most BCBL-1 cells supporting lytic replication of KSHV (D'Agostino et al. 1999), but it generally does not prevent the release of infectious virus unless the process is accelerated by incubation with either IFNα or IFNγ (D'Agostino et al. 1999; Pozharskaya et al. 2004a, 2004b; Klass et al. 2005). The apoptosis that occurs during lytic replication is not dependent on viral DNA replication or viral late gene expression because it continues to occur when virus production is blocked with either ganciclovir or phosphoformic acid (Klass et al. 2005). Thus the role of expression of vIRFs and other viral gene products that disrupt the innate antiviral response may be to delay and/or attenuate the cellular response until production of infectious virus is complete, but they are not sufficient to allow production of infectious virus when the host response is augmented by exogenous IFN encountered early in the lytic cascade (Pozharskaya et al. 2004b).

The majority of the cells that are in the lytic cascade do not express high levels of vIRF-1 because vIRF-1 transcription is transiently induced and vIRF-1 protein has a short half-life (Pozharskaya et al. 2004b). The percentage of cells within the lytic cascade that express vIRF-2 or vIRF-4 is not known, and the duration of their expression during the lytic cascade is also not known. When cells supporting lytic replication are incubated with either IFNα or IFNγ, increased expression of PKR, 2′5′-OAS, and IRF-1 occurs, and the production of infectious virus is reduced, especially when the IFNs are added early in the lytic cascade (Pozharskaya et al. 2004a, 2004b). Thus none of the vIRFs is sufficient to fully protect KSHV-infected MECs from the antiviral effects of IFNα and IFNγ.

12
Other KSHV-Encoded Proteins That Block Responsiveness to IFNs

A structurally unrelated KSHV protein encoded by ORF45 binds cellular IRF-7 and blocks its phosphorylation and nuclear translocation, thereby blocking the induction of IFNα during KSHV infection (Zhu et al. 2002). The disruption of IRF-7 function should also reduce expression of IFNβ (Yang et al. 2003). The ORF45 protein is expressed early in the lytic cascade, and the expression persists throughout the lytic cascade so that ORF45 protein is found in KSHV virions as a tegument protein (Zhu and Yuan 2003). Unlike the vIRFs that are expressed in a subset of gammaherpesviruses, ORF45 is conserved among members of the gammaherpesvirus subfamily (Jia et al. 2005). A murine gammaherpesvirus 68 (MHV68) mutant lacking ORF45 loses infectivity, whereas it is restored by expression of KSHV ORF45. This provides evidence that disruption of IRF-7 function and type I IFN expression must occur for herpesviruses to infect and replicate in mammalian cells. ORF45 is expressed early and throughout the lytic cascade (Zhu et al. 2002; Zhu and Yuan 2003), yet virally infected cells supporting lytic replication continue to respond to IFNs (Monini et al. 1999; Krug et al. 2004; Pozharskaya et al. 2004a, 2004b). Thus the vIRFs and ORF45 are unable to block responses to high levels of exogenous IFN, suggesting that their role is to disrupt responses to endogenous IRFs, IFNs, and other components of the cellular antiviral response, thereby enhancing production of infectious virus.

13 Concluding Remarks

It is likely that the vIRFs are not essential for viral replication because their incorporation into herpesviruses is relatively recent and is not found in the majority of gammaherpesviruses. They nonetheless are likely to provide some protection from innate and acquired antiviral defenses. Each of the characterized vIRFs affects the transcription of type I IFNs, and they also affect responses induced by IFNs. vIRFs also block activation of PKR, activation of p53, and activation of the NF-κB pathway. They alter the expression of a wide range of cellular and viral gene products by altering the responses to multiple transcription factors including IRF-1, IRF-3, IRF-7, p53, NF-κB, and Smads. Several vIRFs block apoptosis, and at least one of the vIRFs can induce malignant transformation. Thus their expression is likely to have a significant impact on the phenotype of KSHV-infected cells.

Very little is known about the function of the vIRFs in the context of KSHV infection. KSHV with deletion of vIRF-1 maintains its ability to infect cells and produce infectious virus (Zhou et al. 2002), but it is not known whether the deletion of vIRF-1 alters the amount of infectious virus that is produced, the cellular response to the virus, or the ability of IFNs to modulate the cellular response. There are no reports on the consequences of deletion of the other vIRFs. vIRF-3 is constitutively expressed during latent infection with

KSHV, but its expression is limited to cells of B lymphocyte origin and hence is not essential for protecting latently infected cells from cellular defenses. Expression of both vIRF-2 and vIRF-4 increases in response to agents that induce lytic replication, but it is not known when in the lytic cascade these proteins are expressed and whether they are expressed at sufficient levels to modulate the processes that are affected when they are overexpressed by transfection. There are no reports on the functional consequences of vIRF-4 expression, and no studies have been reported using KSHV that is defective in the expression of vIRF-2, vIRF-3, or vIRF-4. The functional consequence of the vIRF-1 that colocalizes with PML bodies in all KSHV-infected cells is currently unknown. Thus future studies will need to explore the consequences of de novo infection using KSHV with targeted deletion of the vIRFs compared to wild-type virus, with special attention to pathways that have been shown to be affected by transfected vIRFs.

References

Alexander L, Denekamp L, Knapp A, Auerbach MR, Damania B, Desrosiers RC (2000) The primary sequence of rhesus monkey rhadinovirus isolate 26–95: sequence similarities to Kaposi's sarcoma-associated herpesvirus and rhesus monkey rhadinovirus isolate 17577. J Virol 74:3388–3398

Alexopoulou L, Holt AC, Medzhitov R, Flavell RA (2001) Recognition of double-stranded RNA and activation of NF-κB by Toll- like receptor 3. Nature 413:732–738.

An FQ, Compitello N, Horwitz E, Sramkoski M, Knudsen ES, Renne R (2005) The latency-associated nuclear antigen of Kaposi's sarcoma-associated herpesvirus modulates cellular gene expression and protects lymphoid cells from p16 INK4A-induced cell cycle arrest. J Biol Chem 280:3862–3874

Ballestas ME, Chatis PA, Kaye KM (1999) Efficient persistence of extrachromosomal KSHV DNA mediated by latency-associated nuclear antigen. Science 284:641–644

Barnes B, Lubyova B, Pitha PM (2002) On the role of IRF in host defense. J Interferon Cytokine Res 22:59–71

Belanger C, Gravel A, Tomoiu A, Janelle ME, Gosselin J, Tremblay MJ, Flamand L (2001) Human herpesvirus 8 viral FLICE-inhibitory protein inhibits Fas-mediated apoptosis through binding and prevention of procaspase-8 maturation. J Hum Virol 4:62–73

Boshoff C, Weiss RA (1998) Kaposi's sarcoma-associated herpesvirus. Adv Cancer Res 75:57–86

Brander C, Suscovich T, Lee Y, Nguyen PT, O'Connor P, Seebach J, Jones NG, van Gorder M, Walker BD, Scadden DT (2000) Impaired CTL recognition of cells latently infected with Kaposi's sarcoma-associated herpes virus. J Immunol 165:2077–2083

Brown HJ, Song MJ, Deng H, Wu TT, Cheng G, Sun R (2003) NF-κB inhibits gammaherpesvirus lytic replication. J Virol 77:8532–8540

Burysek L, Pitha PM (2001) Latently expressed human herpesvirus 8-encoded interferon regulatory factor 2 inhibits double-stranded RNA-activated protein kinase. J Virol 75:2345–2352

Burysek L, Yeow WS, Lubyova B, Kellum M, Schafer SL, Huang YQ, Pitha PM (1999a) Functional analysis of human herpesvirus 8-encoded viral interferon regulatory factor 1 and its association with cellular interferon regulatory factors and p300. J Virol 73:7334–7342

Burysek L, Yeow WS, Pitha PM (1999b) Unique properties of a second human herpesvirus 8-encoded interferon regulatory factor (vIRF-2). J Hum Virol 2:19–32

Cannon M, Philpott NJ, Cesarman E (2003) The Kaposi's sarcoma-associated herpesvirus G protein-coupled receptor has broad signaling effects in primary effusion lymphoma cells. J Virol 77:57–67

Chan HM, La Thangue NB (2001) p300/CBP proteins: HATs for transcriptional bridges and scaffolds. J Cell Sci 114:2363–2373

Chawla-Sarkar M, Lindner DJ, Liu YF, Williams BR, Sen GC, Silverman RH, Borden EC (2003) Apoptosis and interferons: role of interferon-stimulated genes as mediators of apoptosis. Apoptosis 8:237–249

Chen J, Ueda K, Sakakibara S, Okuno T, Yamanishi K (2000) Transcriptional regulation of the Kaposi's sarcoma-associated herpesvirus viral interferon regulatory factor gene. J Virol 74:8623–8634

Clemens MJ (1997) PKR–a protein kinase regulated by double-stranded RNA. Int J Biochem Cell Biol 29:945–949

Clemens MJ, Elia A (1997) The double-stranded RNA-dependent protein kinase PKR: structure and function. J Interferon Cytokine Res 17:503–524.

Coccia EM, Del Russo N, Stellacci E, Orsatti R, Benedetti E, Marziali G, Hiscott J, Battistini A (1999) Activation and repression of the 2-5A synthetase and p21 gene promoters by IRF-1 and IRF-2. Oncogene 18:2129–2137

Collot-Teixeira S, Bass J, Denis F, Ranger-Rogez S (2004) Human tumor suppressor p53 and DNA viruses. Rev Med Virol 14:301–319

Coscoy L, Ganem D (2000) Kaposi's sarcoma-associated herpesvirus encodes two proteins that block cell surface display of MHC class I chains by enhancing their endocytosis. Proc Natl Acad Sci USA

Cunningham C, Barnard S, Blackbourn DJ, Davison AJ (2003) Transcription mapping of human herpesvirus 8 genes encoding viral interferon regulatory factors. J Gen Virol 84:1471–1483

D'Agostino G, Arico E, Santodonato L, Venditti M, Sestili P, Masuelli L, Coletti A, Modesti A, Picchio G, Mosier DE, Ferrantini M, Belardelli F (1999) Type I consensus IFN (IFN-con1) gene transfer into KSHV/HHV-8-infected BCBL-1 cells causes inhibition of viral lytic cycle activation via induction of apoptosis and abrogates tumorigenicity in sCID mice. J Interferon Cytokine Res 19:1305–1316

Dittmer DP (2003) Transcription profile of Kaposi's sarcoma-associated herpesvirus in primary Kaposi's sarcoma lesions as determined by real-time PCR arrays. Cancer Res 63:2010–2015

Donze O, Deng J, Curran J, Sladek R, Picard D, Sonenberg N (2004) The protein kinase PKR: a molecular clock that sequentially activates survival and death programs. EMBO J 23:564–571

Dornan D, Eckert M, Wallace M, Shimizu H, Ramsay E, Hupp TR, Ball KL (2004) Interferon regulatory factor 1 binding to p300 stimulates DNA-dependent acetylation of p53. Mol Cell Biol 24:10083–10098

Dupin N, Fisher C, Kellam P, Ariad S, Tulliez M, Franck N, van Marck E, Salmon D, Gorin I, Escande JP, Weiss RA, Alitalo K, Boshoff C (1999) Distribution of human herpesvirus-8 latently infected cells in Kaposi's sarcoma, multicentric Castleman's disease, and primary effusion lymphoma. Proc Natl Acad Sci USA 96:4546–4551

Esteban M, Garcia MA, Domingo-Gil E, Arroyo J, Nombela C, Rivas C (2003) The latency protein LANA2 from Kaposi's sarcoma-associated herpesvirus inhibits apoptosis induced by dsRNA-activated protein kinase but not RNase L activation. J Gen Virol 84:1463–1470

Fakhari FD, Dittmer DP (2002) Charting latency transcripts in Kaposi's sarcoma-associated herpesvirus by whole-genome real-time quantitative PCR. J Virol 76:6213–6223

Feng XH, Derynck R (2005) Specificity and versatility in Tgf-β signaling through Smads. Annu Rev Cell Dev Biol 21:659–693

Flowers CC, Flowers SP, Nabel GJ (1998) Kaposi's sarcoma-associated herpesvirus viral interferon regulatory factor confers resistance to the antiproliferative effect of interferon-α. Mol Med 4:402–412

Floyd-Smith G, Wang Q, Sen GC (1999) Transcriptional induction of the p69 isoform of 2′,5′-oligoadenylate synthetase by interferon-β and interferon-γinvolves three regulatory elements and interferon-stimulated gene factor 3. Exp Cell Res 246:138–147

Friborg J, Jr., Kong W, Hottiger MO, Nabel GJ (1999) p53 inhibition by the LANA protein of KSHV protects against cell death. Nature 402:889–894

Gao SJ, Boshoff C, Jayachandra S, Weiss RA, Chang Y, Moore PS (1997) KSHV ORF K9 (vIRF) is an oncogene which inhibits the interferon signaling pathway. Oncogene 15:1979–1985

Ghosh A, Sarkar SN, Rowe TM, Sen GC (2001) A specific isozyme of 2′-5′ oligoadenylate synthetase is a dual function proapoptotic protein of the Bcl-2 family. J Biol Chem 276:25447–25455.

Ghosh SK, Wood C, Boise LH, Mian AM, Deyev VV, Feuer G, Toomey NL, Shank NC, Cabral L, Barber GN, Harrington WJ, Jr. (2003) Potentiation of TRAIL-induced apoptosis in primary effusion lymphoma through azidothymidine-mediated inhibition of NF-κB. Blood 101:2321–2327

Guasparri I, Keller SA, Cesarman E (2004) KSHV vFLIP is essential for the survival of infected lymphoma cells. J Exp Med 199:993–1003

Han KJ, Su X, Xu LG, Bin LH, Zhang J, Shu HB (2004) Mechanisms of the TRIF-induced interferon-stimulated response element and NF-κB activation and apoptosis pathways. J Biol Chem 279:15652–15661

Harcourt JL, Offermann MK (2001) Multiple signaling cascades are differentially involved in gene induction by double stranded RNA in interferon-α-primed cells. Eur J Biochem 268:1373–1381.

Harris SL, Levine AJ (2005) The p53 pathway: positive and negative feedback loops. Oncogene 24:2899–2908

Hu J, Garber AC, Renne R (2002) The latency-associated nuclear antigen of Kaposi's sarcoma-associated herpesvirus supports latent DNA replication in dividing cells. J Virol 76:11677–11687

Inagi R, Okuno T, Ito M, Chen J, Mori Y, Haque M, Zou P, Yagi H, Kiniwa S, Saida T, Ueyama Y, Hayashi K, Yamanishi K (1999) Identification and characterization of human herpesvirus 8 open reading frame K9 viral interferon regulatory factor by a monoclonal antibody. J Hum Virol 2:63–71

Jenner RG, Alba MM, Boshoff C, Kellam P (2001) Kaposi's sarcoma-associated herpesvirus latent and lytic gene expression as revealed by DNA arrays. J Virol 75:891–902

Jia Q, Chernishof V, Bortz E, McHardy I, Wu TT, Liao HI, Sun R (2005) Murine gammaherpesvirus 68 open reading frame 45 plays an essential role during the immediate-early phase of viral replication. J Virol 79:5129–5141

Justesen J, Hartmann R, Kjeldgaard NO (2000) Gene structure and function of the 2′-5′-oligoadenylate synthetase family. Cell Mol Life Sci 57:1593–1612.

Kaiser WJ, Kaufman JL, Offermann MK (2004) IFN-α sensitizes human umbilical vein endothelial cells to apoptosis induced by double-stranded RNA. J Immunol 172:1699–1710

Kaiser WJ, Offermann MK (2005) Apoptosis Induced by the Toll-like receptor adaptor TRIF Is dependent on its receptor interacting protein homotypic interaction motif. J Immunol 174:4942–4952

Keller SA, Schattner EJ, Cesarman E (2000) Inhibition of NF-κB induces apoptosis of KSHV-infected primary effusion lymphoma cells. Blood 96:2537–2542

Klass CM, Krug LT, Pozharskaya VP, Offermann MK (2005) The targeting of primary effusion lymphoma cells for apoptosis by inducing lytic replication of human herpesvirus 8 while blocking virus production. Blood 105:4028–4034

Kotenko SV, Pestka S (2000) Jak-Stat signal transduction pathway through the eyes of cytokine class II receptor complexes. Oncogene 19:2557–2565

Kreuz S, Siegmund D, Scheurich P, Wajant H (2001) NF-κB inducers upregulate cFLIP, a cycloheximide-sensitive inhibitor of death receptor signaling. Mol Cell Biol 21:3964–3973.

Kroger A, Koster M, Schroeder K, Hauser H, Mueller PP (2002) Activities of IRF-1. J Interferon Cytokine Res 22:5-14

Krug LT, Pozharskaya VP, Yu Y, Inoue N, Offermann MK (2004) Inhibition of infection and replication of human herpesvirus 8 in microvascular endothelial cells by αinterferon and phosphonoformic acid. J Virol 78:8359–8371

Lee SH, Kim JW, Lee HW, Cho YS, Oh SH, Kim YJ, Jung CH, Zhang W, Lee JH (2003) Interferon regulatory factor-1 (IRF-1) is a mediator for interferon-γ induced attenuation of telomerase activity and human telomerase reverse transcriptase (hTERT) expression. Oncogene 22:381–391

Li M, Damania B, Alvarez X, Ogryzko V, Ozato K, Jung JU (2000) Inhibition of p300 histone acetyltransferase by viral interferon regulatory factor. Mol Cell Biol 20:8254–8263

Li M, Lee H, Guo J, Neipel F, Fleckenstein B, Ozato K, Jung JU (1998) Kaposi's sarcoma-associated herpesvirus viral interferon regulatory factor. J Virol 72:5433–5440

Li M, Shillinglaw W, Henzel WJ, Beg AA (2001) The Rela(p65) subunit of NF-κB is essential for inhibiting double- stranded RNA-induced cytotoxicity. J Biol Chem 276:1185–1194.

Lin R, Genin P, Mamane Y, Hiscott J (2000a) Selective DNA binding and association with the CREB binding protein coactivator contribute to differential activation of α/β interferon genes by interferon regulatory factors 3 and 7. Mol Cell Biol 20:6342–6353
Lin R, Genin P, Mamane Y, Sgarbanti M, Battistini A, Harrington WJ, Jr., Barber GN, Hiscott J (2001) HHV-8 encoded vIRF-1 represses the interferon antiviral response by blocking IRF-3 recruitment of the CBP/p300 coactivators. Oncogene 20:800–811
Lin R, Heylbroeck C, Pitha PM, Hiscott J (1998) Virus-dependent phosphorylation of the IRF-3 transcription factor regulates nuclear translocation, transactivation potential, and proteasome-mediated degradation. Mol Cell Biol 18:2986–2996
Lin R, Mamane Y, Hiscott J (2000b) Multiple regulatory domains control IRF-7 activity in response to virus infection. J Biol Chem 275:34320–34327
Lubyova B, Kellum MJ, Frisancho AJ, Pitha PM (2004) Kaposi's sarcoma-associated herpesvirus-encoded vIRF-3 stimulates the transcriptional activity of cellular IRF-3 and IRF-7. J Biol Chem 279:7643–7654
Lubyova B, Pitha PM (2000) Characterization of a novel human herpesvirus 8-encoded protein, vIRF-3, that shows homology to viral and cellular interferon regulatory factors. J Virol 74:8194–8201
Lukac DM, Renne R, Kirshner JR, Ganem D (1998) Reactivation of Kaposi's sarcoma-associated herpesvirus infection from latency by expression of the ORF 50 transactivator, a homolog of the EBV R protein. Virology 252:304–312
Mak TW, Yeh WC (2002) Signaling for survival and apoptosis in the immune system. Arthritis Res 4 Suppl 3: S243–252
Malmgaard L (2004) Induction and regulation of IFNs during viral infections. J Interferon Cytokine Res 24:439–454
Massague J, Seoane J, Wotton D (2005) Smad transcription factors. Genes Dev 19:2783–2810
Matta H, Sun Q, Moses G, Chaudhary PM (2003) Molecular genetic analysis of human herpes virus 8-encoded viral FLICE inhibitory protein-induced NF-κB activation. J Biol Chem 278:52406–52411
Means RE, Choi JK, Nakamura H, Chung YH, Ishido S, Jung JU (2002) Immune evasion strategies of Kaposi's sarcoma-associated herpesvirus. Curr Top Microbiol Immunol 269:187–201
Meek DW (2004) The p53 response to DNA damage. DNA Repair (Amst) 3:1049–1056
Michalak E, Villunger A, Erlacher M, Strasser A (2005) Death squads enlisted by the tumour suppressor p53. Biochem Biophys Res Commun 331:786–798
Monini P, Carlini F, Sturzl M, Rimessi P, Superti F, Franco M, Melucci-Vigo G, Cafaro A, Goletti D, Sgadari C, Butto S, Leone P, Chiozzini C, Barresi C, Tinari A, Bonaccorsi A, Capobianchi MR, Giuliani M, di Carlo A, Andreoni M, Rezza G, Ensoli B (1999) Alpha interferon inhibits human herpesvirus 8 (HHV-8) reactivation in primary effusion lymphoma cells and reduces HHV-8 load in cultured peripheral blood mononuclear cells. J Virol 73:4029–4041
Moore PS, Chang Y (2001) Molecular virology of Kaposi's sarcoma-associated herpesvirus. Philos Trans R Soc Lond B Biol Sci 356:499–516
Moore PS, Chang Y (2003) Kaposi's sarcoma-associated herpesvirus immunoevasion and tumorigenesis: two sides of the same coin? Annu Rev Microbiol 57:609–639

Morin P, Braganca J, Bandu MT, Lin R, Hiscott J, Doly J, Civas A (2002) Preferential binding sites for interferon regulatory factors 3 and 7 involved in interferon-A gene transcription. J Mol Biol 316:1009–1022

Munoz-Fontela C, Rodriguez E, Nombela C, Arroyo J, Rivas C (2003) Characterization of the bipartite nuclear localization signal of protein LANA2 from Kaposi's sarcoma-associated herpesvirus. Biochem J 374:545–550

Nakamura H, Li M, Zarycki J, Jung JU (2001) Inhibition of p53 tumor suppressor by viral interferon regulatory factor. J Virol 75:7572–7582

Nakaya T, Sato M, Hata N, Asagiri M, Suemori H, Noguchi S, Tanaka N, Taniguchi T (2001) Gene induction pathways mediated by distinct IRFs during viral infection. Biochem Biophys Res Commun 283:1150–1156

Neipel F, Albrecht JC, Fleckenstein B (1997) Cell-homologous genes in the Kaposi's sarcoma-associated rhadinovirus human herpesvirus 8: determinants of its pathogenicity? J Virol 71:4187–4192

O'Shea CC (2005) DNA tumor viruses—the spies who lyse us. Curr Opin Genet Dev 15:18–26

Offermann MK, Zimring J, Mellits KH, Hagan MK, Shaw R, Medford RM, Mathews MB, Goodbourn S, Jagus R (1995) Activation of the double-stranded-RNA-activated protein kinase and induction of vascular cell adhesion molecule-1 by poly (I).poly (C) in endothelial cells. Eur J Biochem 232:28–36

Oren M (2003) Decision making by p53: life, death and cancer. Cell Death Differ 10:431–442

Parekh BS, Maniatis T (1999) Virus infection leads to localized hyperacetylation of histones H3 and H4 at the IFN-β promoter. Mol Cell 3: 125–129

Park J, Seo T, Hwang S, Lee D, Gwack Y, Choe J (2000) The K-bZIP protein from Kaposi's sarcoma-associated herpesvirus interacts with p53 and represses its transcriptional activity. J Virol 74:11977–11982

Parravicini C, Chandran B, Corbellino M, Berti E, Paulli M, Moore PS, Chang Y (2000) Differential viral protein expression in Kaposi's sarcoma-associated herpesvirus-infected diseases: Kaposi's sarcoma, primary effusion lymphoma, and multicentric Castleman's disease. Am J Pathol 156: 743–749

Pitha PM, Au WC, Lowther W, Juang YT, Schafer SL, Burysek L, Hiscott J, Moore PA (1998) Role of the interferon regulatory factors (IRFs) in virus-mediated signaling and regulation of cell growth. Biochimie 80:651–658

Pozharskaya VP, Weakland LL, Offermann MK (2004a) Inhibition of infectious human herpesvirus 8 production by gamma interferon and alpha interferon in BCBL-1 cells. J Gen Virol 85:2779–2787

Pozharskaya VP, Weakland LL, Zimring JC, Krug LT, Unger ER, Neisch A, Joshi H, Inoue N, Offermann MK (2004b) Short duration of elevated vIRF-1 expression during lytic replication of human herpesvirus 8 limits its ability to block antiviral responses induced by alpha interferon in BCBL-1 cells. J Virol 78:6621–6635

Radkov SA, Kellam P, Boshoff C (2000) The latent nuclear antigen of Kaposi sarcoma-associated herpesvirus targets the retinoblastoma-E2F pathway and with the oncogene Hras transforms primary rat cells. Nat Med 6:1121–1127

Rainbow L, Platt GM, Simpson GR, Sarid R, Gao SJ, Stoiber H, Herrington CS, Moore PS, Schulz TF (1997) The 222- to 234-kilodalton latent nuclear protein (LNA) of Kaposi's sarcoma-associated herpesvirus (human herpesvirus 8) is encoded by orf73 and is a component of the latency-associated nuclear antigen. J Virol 71:5915–5921
Rebouillat D, Hovanessian AG (1999) The human 2′,5′-oligoadenylate synthetase family: interferon-induced proteins with unique enzymatic properties. J Interferon Cytokine Res 19:295–308
Renne R, Zhong W, Herndier B, McGrath M, Abbey N, Kedes D, Ganem D (1996) Lytic growth of Kaposi's sarcoma-associated herpesvirus (human herpesvirus 8) in culture. Nature Medicine 2:342–346
Rivas C, Thlick AE, Parravicini C, Moore PS, Chang Y (2001) Kaposi's sarcoma-associated herpesvirus LANA2 is a B-cell-specific latent viral protein that inhibits p53. J Virol 75:429–438
Roan F, Inoue N, Offermann MK (2002) Activation of cellular and heterologous promoters by the human herpesvirus 8 replication and transcription activator. Virology 301:293–304.
Roan F, Zimring JC, Goodbourn S, Offermann MK (1999) Transcriptional activation by the human herpesvirus-8-encoded interferon regulatory factor. J Gen Virol 80 (Pt 8): 2205–2209
Romeo G, Fiorucci G, Chiantore MV, Percario ZA, Vannucchi S, Affabris E (2002) IRF-1 as a negative regulator of cell proliferation. J Interferon Cytokine Res 22:39–47
Russo JJ, Bohenzky RA, Chien MC, Chen J, Yan M, Maddalena D, Parry JP, Peruzzi D, Edelman IS, Chang Y, Moore PS (1996) Nucleotide sequence of the Kaposi sarcoma-associated herpesvirus (HHV8). Proc Natl Acad Sci USA 93:14862–14867
Samuel CE (2001) Antiviral actions of interferons. Clin Microbiol Rev 14:778–809, table of contents.
Samuel CE, Kuhen KL, George CX, Ortega LG, Rende-Fournier R, Tanaka H (1997) The PKR protein kinase–an interferon-inducible regulator of cell growth and differentiation. Int J Hematol 65:227–237
Sarid R, Flore O, Bohenzky RA, Chang Y, Moore PS (1998) Transcription mapping of the Kaposi's sarcoma-associated herpesvirus (human herpesvirus 8) genome in a body cavity-based lymphoma cell line (BC-1). J Virol 72:1005–1012
Sarid R, Sato T, Bohenzky RA, Russo JJ, Chang Y (1997) Kaposi's sarcoma-associated herpesvirus encodes a functional bcl-2 homologue. Nat Medi 3:293–298
Searles RP, Bergquam EP, Axthelm MK, Wong SW (1999) Sequence and genomic analysis of a Rhesus macaque rhadinovirus with similarity to Kaposi's sarcoma-associated herpesvirus/human herpesvirus 8. J Virol 73:3040–3053
Sengupta S, Harris CC (2005) p53: traffic cop at the crossroads of DNA repair and recombination. Nat Rev Mol Cell Biol 6:44–55
Seo T, Lee D, Lee B, Chung JH, Choe J (2000) Viral interferon regulatory factor 1 of Kaposi's sarcoma-associated herpesvirus (human herpesvirus 8) binds to, and inhibits transactivation of, CREB-binding protein. Biochem Biophys Res Commun 270:23–27
Seo T, Park J, Choe J (2005) Kaposi's sarcoma-associated herpesvirus viral IFN regulatory factor 1 inhibits transforming growth factor-β signaling. Cancer Res 65:1738–1747

Seo T, Park J, Lee D, Hwang SG, Choe J (2001) Viral interferon regulatory factor 1 of Kaposi's sarcoma-associated herpesvirus binds to p53 and represses p53-dependent transcription and apoptosis. J Virol 75:6193–6198

Seo T, Park J, Lim C, Choe J (2004) Inhibition of nuclear factor κB activity by viral interferon regulatory factor 3 of Kaposi's sarcoma-associated herpesvirus. Oncogene 23:6146–6155

Seoane J, Pouponnot C, Staller P, Schader M, Eilers M, Massague J (2001) TGFβ influences Myc, Miz-1 and Smad to control the CDK inhibitor p15INK4b. Nat Cell Biol 3:400–408

Servant MJ, Tenoever B, Lin R (2002) Overlapping and distinct mechanisms regulating IRF-3 and IRF-7 function. J Interferon Cytokine Res 22:49–58

Sharma S, tenOever BR, Grandvaux N, Zhou GP, Lin R, Hiscott J (2003) Triggering the interferon antiviral response through an IKK-related pathway. Science 300:1148–1151

Shen X, Hu PP, Liberati NT, Datto MB, Frederick JP, Wang XF (1998) TGF-β-induced phosphorylation of Smad3 regulates its interaction with coactivator p300/CREB-binding protein. Mol Biol Cell 9:3309–3319

Slee EA, O'Connor DJ, Lu X (2004) To die or not to die: how does p53 decide? Oncogene 23:2809–2818

Staskus KA, Zhong W, Gebhard K, Herndier B, Wang H, Renne R, Beneke J, Pudney J, Anderson DJ, Ganem D, Haase AT (1997) Kaposi's sarcoma-associated herpesvirus gene expression in endothelial (spindle) tumor cells. J Virol 71:715–719

Stehlik C, de Martin R, Kumabashiri I, Schmid JA, Binder BR, Lipp J (1998) Nuclear factor (NF)-κB-regulated X-chromosome-linked iap gene expression protects endothelial cells from tumor necrosis factor α-induced apoptosis. J Exp Med 188:211–216

Sturzl M, Hohenadl C, Zietz C, Castanos-Velez E, Wunderlich A, Ascherl G, Biberfeld P, Monini P, Browning PJ, Ensoli B (1999) Expression of K13/v-FLIP gene of human herpesvirus 8 and apoptosis in Kaposi's sarcoma spindle cells. J Natl Cancer Inst 91:1725–1733

Suhara W, Yoneyama M, Iwamura T, Yoshimura S, Tamura K, Namiki H, Aimoto S, Fujita T (2000) Analyses of virus-induced homomeric and heteromeric protein associations between IRF-3 and coactivator CBP/p300. J Biochem (Tokyo) 128:301–307

Sun Q, Zachariah S, Chaudhary PM (2003) The human herpes virus 8-encoded viral FLICE-inhibitory protein induces cellular transformation via NF-κB activation. J Biol Chem 278:52437–52445

Tan SL, Katze MG (1999) The emerging role of the interferon-induced PKR protein kinase as an apoptotic effector: a new face of death? J Interferon Cytokine Res 19:543–554.

Taniguchi T, Ogasawara K, Takaoka A, Tanaka N (2001) IRF family of transcription factors as regulators of host defense. Annu Rev Immunol 19:623–655

Ueda K, Ishikawa K, Nishimura K, Sakakibara S, Do E, Yamanishi K (2002) Kaposi's sarcoma-associated herpesvirus (human herpesvirus 8) replication and transcription factor activates the K9 (vIRF) gene through two distinct cis elements by a non-DNA-binding mechanism. J Virol 76:12044–12054

Vattem KM, Staschke KA, Wek RC (2001) Mechanism of activation of the double-stranded-RNA-dependent protein kinase, PKR: role of dimerization and cellular localization in the stimulation of PKR phosphorylation of eukaryotic initiation factor-2 (eIF2). Eur J Biochem 268:3674–3684

Vorburger SA, Pataer A, Yoshida K, Barber GN, Xia W, Chiao P, Ellis LM, Hung MC, Swisher SG, Hunt KK (2002) Role for the double-stranded RNA activated protein kinase PKR in E2F-1-induced apoptosis. Oncogene 21:6278–6288.

Wang Q, Floyd-Smith G (1998) Maximal induction of p69 2′, 5′-oligoadenylate synthetase in Daudi cells requires cooperation between an ISRE and two IRF-1-like elements. Gene 222:83–90

Wang XP, Gao SJ (2003) Auto-activation of the transforming viral interferon regulatory factor encoded by Kaposi's sarcoma-associated herpesvirus (human herpesvirus-8). J Gen Virol 84:329–336

Waterston A, Bower M (2004) Fifty years of multicentric Castleman's disease. Acta Oncol 43:698–704

Yang H, Lin CH, Ma G, Baffi MO, Wathelet MG (2003) Interferon regulatory factor-7 synergizes with other transcription factors through multiple interactions with p300/CBP coactivators. J Biol Chem 278:15495–15504

Yang H, Lin CH, Ma G, Orr M, Baffi MO, Wathelet MG (2002) Transcriptional activity of interferon regulatory factor (IRF)-3 depends on multiple protein-protein interactions. Eur J Biochem 269:6142–6151

Yang H, Ma G, Lin CH, Orr M, Wathelet MG (2004) Mechanism for transcriptional synergy between interferon regulatory factor (IRF)-3 and IRF-7 in activation of the interferon-β gene promoter. Eur J Biochem 271:3693–3703

Yu F, Wang Q, Floyd-Smith G (1999) Transcriptional induction of p69 2'-5'-oligoadenylate synthetase by interferon-alpha is stimulated by 12-*O*-tetradecanoyl phorbol-13-acetate through IRF/ISRE binding motifs. Gene 237:177–184

Zhou FC, Zhang YJ, Deng JH, Wang XP, Pan HY, Hettler E, Gao SJ (2002) Efficient infection by a recombinant Kaposi's sarcoma-associated herpesvirus cloned in a bacterial artificial chromosome: application for genetic analysis. J Virol 76:6185–6196

Zhu FX, King SM, Smith EJ, Levy DE, Yuan Y (2002) A Kaposi's sarcoma-associated herpesviral protein inhibits virus-mediated induction of type I interferon by blocking IRF-7 phosphorylation and nuclear accumulation. Proc Natl Acad Sci USA 99:5573–5578

Zhu FX, Yuan Y (2003) The ORF45 protein of Kaposi's sarcoma-associated herpesvirus is associated with purified virions. J Virol 77:4221–4230

Zimring JC, Goodbourn S, Offermann MK (1998) Human herpesvirus 8 encodes an interferon regulatory factor (IRF) homolog that represses IRF-1-mediated transcription. J Virol 72:701–707

CTMI (2006) 312:211–244

Endothelial Cell- and Lymphocyte-Based In Vitro Systems for Understanding KSHV Biology

S. C. McAllister · A. V. Moses (✉)

Vaccine and Gene Therapy Institute, Oregon Health and Science University, Beaverton, OR 97006, USA
mosesa@ohsu.edu

Abstract Kaposi sarcoma (KS), the most common AIDS-associated malignancy, is a multifocal tumor characterized by deregulated angiogenesis, proliferation of spindle cells, and extravasation of inflammatory cells and erythrocytes. Kaposi sarcoma-associated herpesvirus (KSHV; also human herpesvirus-8) is implicated in all clinical forms of KS. Endothelial cells (EC) harbor the KSHV genome in vivo, are permissive for virus infection in vitro, and are thought to be the precursors of KS spindle cells. Spindle cells are rare in early patch-stage KS lesions but become the predominant cell type in later plaque- and nodular-stage lesions. Alterations in endothelial/spindle cell physiology that promote proliferation and survival are thus thought to be important in disease progression and may represent potential therapeutic targets. KSHV encodes genes that stimulate cellular proliferation and migration, prevent apoptosis, and counter the host immune response. The combined effect of these genes is thought to drive the proliferation and survival of infected spindle cells and influence the lesional microenvironment. Large-scale gene expression analyses have revealed that KSHV infection also induces dramatic reprogramming of the EC transcriptome. These changes in cellular gene expression likely contribute to the development of the KS lesion. In addition to KS, KSHV is also present in B cell neoplasias including primary effusion lymphoma and multicentric Castleman disease. A combination of virus and virus-induced host factors are similarly thought to contribute to establishment and progression of these malignancies. A number of lymphocyte- and EC-based systems have been developed that afford some insight into the means by which KSHV contributes to malignant transformation of host cells. Whereas KSHV is well maintained in PEL cells cultured in vitro, explanted spindle cells rapidly lose the viral episome. Thus, endothelial cell-based systems for studying KSHV gene expression and function, as well as the effect of infection on host cell physiology, have required in vitro infection of primary or life-extended EC. This chapter includes a review of these in vitro cell culture systems, acknowledging their strengths and weaknesses and putting into perspective how each has contributed to our understanding of the complex KS lesional environment. In addition, we present a model of KS lesion progression based on findings culled from these models as well as recent clinical advances in KS chemotherapy. Thus this unifying model describes our current understanding of KS pathogenesis by drawing together multiple theories of KS progression that by themselves cannot account for the complexities of tumor development.

1 Description and Classification of Kaposi Sarcoma

Moritz Kaposi, a dermatologist in Vienna, described the case histories of five middle-aged and elderly men who presented with "multiple idiopathic pigmented sarcomas of the skin" in 1872 (Kaposi 1872); this disease was named Kaposi sarcoma (KS) a few years later, and remained a medical rarity in industrialized countries for over a hundred years. In 1981, an aggressive variant of KS became one of the sentinel diseases of the AIDS pandemic (Friedman-Kien et al. 1982). The poor prognosis of AIDS-associated KS led one playwright to describe the disease as "the wine-dark kiss of the angel of death" (Kushner 1995). Four distinct epidemiological forms of KS are now recognized (Dourmishev et al. 2003) and are distinguished by relative severity, HIV serostatus, and geographic distribution: classic, the variant first described in elderly Mediterranean men; AIDS-KS, severe and often fatal, the most common AIDS-associated neoplasm; endemic, an aggressive but non-HIV-associated variant common in sub-Saharan Africa; and iatrogenic, a form occasionally complicating organ transplantation. In all forms, KS usually presents as multiple reddish-purple dermal or mucosal lesions that may be flat or raised. Lesions may coalesce as disease progresses, becoming nodular and even ulcerated. Dissemination to visceral organs, including the lungs, liver, lymph nodes, and gut may also occur and is associated with poor prognosis. The four epidemiological forms of KS are histologically indistinguishable, characterized by disorganized networks of abnormal microvasculature composed of spindle-shaped cells (Flaitz et al. 1996); these cells do not maintain the integrity of microvascular channels, accounting for lesional edema and abundant extravasation of inflammatory cells and erythrocytes. Spindle cells, thought to be of endothelial lineage, are present in early patch-stage lesions but become the predominant cell type in later plaque- and nodular-stage lesions (Aluigi et al. 1996; Beckstead et al. 1985; Regezi et al. 1993, 1993b; Roth et al. 1992, 1988; Rutgers et al. 1986; Scully et al. 1988). Patch-stage lesions are typically polyclonal, but oligoclonal and monoclonal later-stage lesions have been described (Gill et al. 1998; Judde et al. 2000; Rabkin et al. 1997). Thus, KS exhibits characteristics of both a reactive hyperproliferative lesion and a truly transformed sarcoma.

2 Kaposi Sarcoma-Associated Herpesvirus

Representational difference analysis (RDA) is a technique used to identify minor differences between complex genomic samples. DNA from KS tissue and normal skin was compared by RDA, leading to the discovery of Kaposi sarcoma-associated herpesvirus (KSHV) in 1994 (Chang et al., 1994). Sequences from this new human herpesvirus were detected in nearly all KS tumors but not in most other pathologic samples examined. Importantly, viral sequences were found in lymph node biopsies from gay men with AIDS, which suggested a link not only to patients with KS but also to groups known to be at increased risk of developing KS. Seroprevalence studies have since shown that KSHV infection is more common in populations known to be at greater risk for developing KS (Herndier and Ganem 2001) and that KSHV viral load increases before the onset of disease [although the utility of viral load as a predictor of KS development is debated (Engels et al. 2003; Polstra et al. 2004; Serraino et al. 2001)]. KSHV is now considered necessary for all clinical forms of KS (Boshoff et al. 1995; Chang et al. 1994; Cheung 2004a) as well as two rare lymphoproliferative disorders, primary effusion lymphoma (PEL) (Cesarman et al. 1995, 1996; Cheung, 2004b) and multicentric Castleman disease (MCD) (Soulier et al. 1995). However, as the seroprevalence of KSHV far exceeds the incidence of these diseases (even in populations known to be at higher risk), we can conclude that KSHV is necessary but not sufficient for the development of disease (Blauvelt 1999).

Complete sequencing of the KSHV genome revealed relatedness to members of the *Gammaherpesvirinae* subfamily (Neipel et al. 1997; Russo et al. 1996; Zhong et al. 1996). KSHV has been placed in the *Rhadinovirus* or gamma-2 group within this subfamily, becoming the first human herpesvirus thus categorized. The KSHV genome is ~165 kb with a central unique coding region of ~140 kb flanked on either end by variable numbers of noncoding, GC-rich repeat sequences (Lagunoff and Ganem 1997; Renne et al. 1996a).

The coding region of KSHV consists of seven blocks of conserved herpesvirus genes that align closely with those of the prototypical gamma-2 virus, herpesvirus saimiri (HVS), a virus that causes fatal T cell lymphomas in New World primates (1997 IARC Working Group; Fickenscher and Fleckenstein, 2001). These conserved genes are labeled open reading frames (ORF) 1 through 75, and most encode proteins required for production of progeny virions (transcription factors, DNA synthesis-related enzymes, as well as capsid, tegument, and envelope proteins) or proteins involved in the establishment and maintenance of latency. Interspersed between the conserved blocks are genes that are unique to gammaherpesviruses or to KSHV, labeled K1 through 15. Many of the K genes have homologs to cellular genes involved in intra- and intercellular signaling including cytokines, chemokines, chemokine receptors and interferon response factors (Bubman and Cesarman 2003).

Both in vitro and in vivo, KSHV is maintained as a viral episome in a primarily latent state characterized by expression of few viral gene products (Ascherl et al. 1999; Brousset et al. 2001; Davis et al. 1997; Dupin et al. 1999; Katano et al. 2000; Parravicini et al. 2000; Polstra et al. 2003; Sturzl et al. 1997). At least three viral gene products, vFLIP, vCYC, and LANA-1, are believed to comprise the minimal latency expression program of KSHV and are consistently expressed in all virally infected cells in KS, PEL, and MCD (Davis et al., 1997; Dupin et al., 1999; Sadler et al., 1999; Sturzl et al., 1997). The gene product of ORF 71, vFLIP (viral FLICE-inhibitory protein; K13) inhibits apoptosis (Djerbi et al. 1999; Thome et al. 1997) and activates NF-κB (Liu et al., 2002). ORF 72, the viral cyclin (v-Cyclin), is a homologue of cellular D-type cyclin that can drive quiescent cells into S phase in part by inhibiting retinoblastoma protein (Rb) activity (Chang et al. 1996; Child and Mann 2001; Godden-Kent et al. 1997; Verschuren et al. 2002). In addition, v-Cyclin inactivates the cyclin-dependent kinase inhibitor p27 (Swanton et al. 1997) and promotes cytoplasmic export of human Orc1, a component of the host origin recognition complex (Laman et al. 2001). The latency-associated nuclear antigen (LANA-1; ORF 73) targets the tumor suppressor proteins p53 and Rb (Friborg et al. 1999; Radkov et al. 2000), providing both antiapoptotic and cell cycle-regulatory functions. LANA-1 is also a transcriptional modulator of various cellular and viral promoters (Renne et al. 2001) and tethers multiple copies of the viral episome to host cell chromosomes, a function indispensable for maintenance of the viral genome during cell division (Ballestas and Kaye 2001).

Two additional viral genes, K10.5 and K12, are transcribed during latency but have additional features deserving of mention. K10.5 is expressed during latency in the B cell disorders MCD and PEL, but not KS (Rivas et al. 2001). The gene product of K10.5, LANA-2, also known as viral interferon regulatory factor 3 (vIRF3), has partial homology to members of the cellular IRF family (particularly IRF4) and two other viral IRFs, K9 (vIRF1) and K11.1 (vIRF2) (Lubyova and Pitha 2000; Rivas et al. 2001). A role for LANA-2 in inhibition of p53-induced apoptosis has been proposed, both through direct interaction with p53 (Rivas et al. 2001) and via inhibition of NF-κB activity (Seo et al. 2004); thus LANA-2 could contribute to proliferative or neoplastic expansion of KSHV-infected B cells.

Kaposin (K12) was initially identified as an abundant latent transcript in KS tumor samples (Zhong et al. 1996); unlike other KSHV latency genes, however, K12 transcript levels increase after lytic cycle induction, suggesting a more complicated transcriptional regulation pattern. Questions awaiting elucidation include whether Kaposin utilizes different transcription units during latency and lytic replication, as well as how cell cycle and tumor type influence transcriptional regulation. The Kaposin locus encodes at least three proteins (Kaposin A, B, and C) via a translational program that is similarly complex and incompletely understood (Li et al. 2002; Muralidhar et al. 1998, 2000; Sadler et al. 1999). Transforming functions have been ascribed to the Kaposin family, particularly to Kaposin A (Muralidhar et al. 1998), and recently a role for Kaposin B in stabilization of cytokine transcripts via activation of the cellular kinase MK2 was proposed (McCormick and Ganem 2005). Overall, the latency program of KSHV enables expansion of a population of latently infected spindle cells by inducing proliferation and preventing apoptosis of infected cells as well as maintaining the viral genome in daughter cells.

In vivo, the majority of KSHV-infected spindle cells and neoplastic B cells maintain the virus as a latent infection, with only a small percentage of cells expressing lytic cycle genes (Cannon et al. 1999; Guo et al. 1997; Staskus et al. 1999, 1997; Zhong et al. 1996). The KSHV gene ORF 50, a homologue of Epstein-Barr virus (EBV) Rta, encodes a replication and transcriptional activator that is necessary and sufficient to trigger lytic replication and production of viral progeny (reviewed by West and Wood 2003). ORF 50 is one of the earliest immediate-early genes induced upon lytic reactivation in B cells and endothelial cells (EC) (Krishnan et al. 2004; Lukac et al. 1999; Sun et al. 1999) and is considered the master regulator of the switch from latency to lytic replication. In vitro, KSHV-infected cells can be induced to enter the lytic cycle by ectopic expression of ORF 50 (Lukac et al. 1999; Sun et al. 1999; Vieira and O'Hearn 2004) or by treatment with phorbol esters (Renne et al. 1996b) or sodium butyrate (Sun et al. 1996; Yu et al. 1999). The biological signals that initiate the lytic cascade are not fully understood, although hypoxia (Davis et al. 2001; Haque et al. 2003), inflammatory cytokines (IC) (Chang et al. 2000; Mercader et al. 2000; Monini et al. 1999), HIV-1 *tat* (Chang et al. 2000; Harrington et al. 1997), and CMV coinfection (Vieira et al. 2001) have all been implicated. A role for LANA-1 in controlling lytic induction via regulation of ORF 50 expression and function has also been proposed (Lan et al. 2004).

Upon reactivation, KSHV expresses a wide repertoire of gene products in a coordinated cascade, culminating in release of progeny virions from infected cells. Chemical induction significantly increases the percentage of cells that express lytic gene products and was used by two groups for initial classification of KSHV lytic genes into kinetic classes with microarray analysis (Jenner et al. 2001; Paulose-Murphy et al. 2001). Lytic replication is essential for viral dissemination, and the observation that treatment of KSHV-seropositive AIDS patients with ganciclovir decreases the incidence of KS development suggests that lytic replication plays an important role in viral pathogenesis (Martin et al. 1999). A number of KSHV lytic genes code for regulatory proteins that could conceivably drive proliferation or transformation as well as genes that modulate apoptotic signals and recognition by the host immune system (see Moore and Chang 2003 and Direkze and Laman 2004 for recent comprehensive reviews). These genes include homologues of cellular cytokines (vIL-6/K2), chemokines (vMIP-I/K6, vMIP-II/K4, vMIP-III/K4.1), chemokine receptors (vGPCR/ORF74), interferon regulatory factors (vIRF-1/K9), antiapoptotic factors (vIAP/K7), and modulators of immune recognition (K3/MIR1 and K5/MIR2). Thus, although lytic replication is incompatible with host cell survival, reservoirs of lytic infection are thought to exert essential paracrine mechanisms that sustain tumorigenesis (see Direkze and Laman 2004 and Viejo-Borbolla et al. 2003 for recent reviews). Additionally, some speculate that a limited subset of lytic genes could be expressed under certain conditions in the absence of full lytic replication (Hayward 2003; Krishnan et al. 2004; Moses et al. 1999; Naranatt et al. 2004). Such lytic gene expression could contribute to pathogenesis without leading to host cell death resulting from production and release of progeny virus. Of particular importance here may be expression of the immediate-early gene K5, which can be expressed independent of ORF50 expression in PEL cells (Okuno et al. 2002) and experimentally infected 293A cells (Jae Jung, personal communication). K5, along with another IE gene, K3, perform a crucial immune evasion function via downregulation of MHC-1 on infected cells (Ishido et al. 2000).

3
KSHV-Infected Cell Lines Derived from Primary Effusion Lymphomas

PEL (also body cavity-based lymphomas) are KSHV-related neoplasms that can present independently of HIV infection (Cesarman et al. 1996) but are more frequent and severe in AIDS patients (Cesarman et al. 1995). Because most PEL tumors are CD20 negative, PEL patients cannot be treated with B cell-targeted therapies; the prognosis of PEL, therefore, is poor. KSHV was first recovered by cultivation of cell lines derived from PEL (Renne et al. 1996b). Numerous clonal PEL cell lines have now been established (Arvanitakis et al. 1996; Boshoff et al. 1998; Cannon et al. 2000; Gaidano et al. 1996; Herndier et al. 1994; Katano et al. 1999; Komanduri et al. 1996; Picchio et al. 1997), about half of which are coinfected with EBV (Boshoff et al. 1998; Carbone et al. 1996; Cesarman et al. 1996; Matsushima et al. 1999; Renne et al. 1996b; Strauchen et al. 1996). Cell lines can be readily established from PEL tumors and, unlike explanted KS spindle cells, maintain the KSHV genome in 100% of cells even after extensive tissue culture passage (Dupin et al. 1999; Fakhari and Dittmer 2002; Jenner et al. 2001; Kedes et al. 1997; Kellam et al. 1997, 1999; Paulose-Murphy et al. 2001; Rainbow et al. 1997; Rivas et al. 2001). In PEL lines, the KSHV genome is maintained as 50–150 latent episomes per cell, which is approximately 10-fold higher than in KS (Boshoff et al. 1995; Lallemand et al. 2000). Although PEL infection is predominantly latent, lytic replication can be induced by chemical induction or ectopic expression of ORF50 (Lukac et al. 1999; Renne et al. 1996a). Collectively, these characteristics have made PEL lines an extremely consistent and convenient model in which to study KSHV biology. PEL were used for the initial in vitro studies of KSHV biology, including those that yielded the viral genomic sequences, and have since been widely used to determine the kinetics of viral gene expression and mechanisms of viral latency and reactivation (Jenner et al. 2001; Lu et al. 2004; Lukac et al. 1999; Nakamura et al. 2003; Neipel et al. 1997; Nicholas et al. 1998; Paulose-Murphy et al. 2001; Russo et al. 1996; Sun et al. 1996, 1999; Zhu et al. 2004). PEL systems have also been used to develop novel KSHV-specific therapeutic strategies that may find clinical application. Klass et al. (Klass et al. 2005), demonstrated that inducing replication in PEL cells with valproate (an antiseizure medication with histone deacetylase inhibitor action) while simultaneously blocking herpesviral DNA replication with ganciclovir and phosphonoformic acid led to apoptosis in the tumor cells without increasing viral load. Godfrey et al. (Godfrey et al. 2005) tested RNA interference as a therapeutic strategy for PEL. With the use of lentivirus-delivered short hairpin RNA (shRNA) to target latency genes, inhibition of v-Cyclin and v-FLIP led to apoptosis in all four PEL lines (BC-3, BCP-1, JSC-1, and HBL-6) tested. Non-KSHV cell lines including the Burkitt lymphoma line RAMOS were unaffected by this treatment. Interestingly, LANA-1 was not as reliable a target; stable cell lines maintaining the sh-LANA vector, and consequent lower KSHV copy numbers, could be generated. Whether this result reveals a capacity for KSHV to integrate into the host genome, thus becoming LANA-independent for segregation during cell division, remains to be established. Another interesting finding emerging from this study was the fact that LANA knockdown increased expression of ORF50, a result consistent with a recent report that LANA inhibits ORF50 expression and function (Lan et al. 2004). In a murine model, injection of the sh-v-cyclin vector prevented development of PEL-driven ascites and reduced established ascites. Other researchers have also used PEL lines to establish animal models for the study of KSHV biology and PEL tumorigenesis in vivo (Boshoff et al. 1998; Picchio et al. 1997; Salahuddin et al. 1988; Staudt et al. 2004; Zenger et al. 2002). In the future, such models should greatly facilitate in vivo screening and validation of anti-KSHV drugs that may have efficacy for KS as well as PEL.

Despite their considerable contribution to our current understanding of KSHV biology, PEL cell lines are not amenable to studying KSHV induction of transformation because the transformation event precedes cell line establishment in vitro. Interestingly, primary B cells and some established B cell lines can be infected with KSHV in vitro, but infection does not lead to prolonged maintenance of the viral genome or to cellular transformation (Bechtel et al. 2003; Blackbourn et al. 2000; Naranatt et al. 2004). This is in marked contrast to B cell infection by two closely related gammaherpesviruses, EBV and RRV, both of which readily transform B cells and establish persistent infections. The reasons for this disparity are unknown, and further studies are required to clarify these issues. Additionally, PEL are of limited utility in studying the cell-specific role of KSHV in KS, where the target cell is of endothelial lineage. Therefore, development of EC-based systems that support de novo KSHV-infection has been essential for studying KSHV infection and gene expression patterns in non-B-lineage cells, for identifying KSHV-induced changes in EC physiology that contribute to KS pathogenesis, and for developing effective KS-targeted therapies.

4
Experimental Infection of Endothelial Cells with KSHV

Spindle cells harbor the KSHV genome in vivo, but all described cell lines derived from KS lesions appear to rapidly lose the KSHV genome upon serial in vitro passage (Aluigi et al. 1996; Ambroziak et al. 1995; Benelli et al. 1996; Boshoff et al. 1995; Herndier et al. 1994; Li et al. 1996; Staskus et al. 1997). Genome loss could be explained by an insufficiency of the episomal maintenance mechanism in endothelial-lineage cells. Others have argued that this phenomenon is artifactual, reflecting the different selective pressures inherent in tissue culture conditions versus the KS lesional microenvironment. Because explanted KS cells lose the KSHV genome after serial passage, the development of in vitro models for KS using experimental de novo infection has proven invaluable for investigation of KSHV-EC interaction.

EC are the presumptive precursors of the characteristic spindle cells in KS lesions (Aluigi et al. 1996; Beckstead et al. 1985; Roth et al. 1988; Rutgers et al. 1986; Scully et al. 1988) although, as discussed in more detail below, whether KSHV preferentially infects lymphatic, vascular, or precursor EC is a matter of debate. Spindle cells harbor KSHV DNA in vivo (Boshoff et al. 1995; Li et al. 1996; Staskus et al. 1997), but the loss of the genome from explanted tumor cells, as well as the need to evaluate the consequences of de novo infection, stimulated efforts to develop in vitro EC-based culture systems that supported KSHV infection (Ciufo et al. 2001; Flore et al. 1998; Krug et al. 2004; Lagunoff et al. 2002; Moses et al. 1999). All EC-based KS models aim to generate age- and passage-matched infected and uninfected cultures, illuminating an important advantage over PEL cell lines that lack uninfected counterparts for use as controls. The EC-based in vitro systems described to date differ to varying degrees with respect to the protocol used and the observed consequences of infection, but key similarities have also been noted. Collectively, when evaluated in the context of the differences in experimental design, these models have yielded valuable information about KSHV biology and pathogenesis in EC. The degree to which any of these models reproduces the spindle component within the complex tumor microenvironment will be discussed further toward the end of this chapter.

Flore et al. were the first to describe infection of primary EC in vitro, using adult bone marrow-derived EC as initial targets and HUVEC as secondary recipients to demonstrate both productive infection and paracrine influences (Flore et al. 1998). Virus inoculum was prepared from concentrated TPA-induced supernatants from the BC-3 PEL line. The KSHV-exposed cultures acquired telomerase activity, exhibited extended life spans, and formed colonies in soft agar, suggesting a KSHV-mediated transformation event. However, the frequency of initial infection was low and KSHV was maintained in only a minority of cells (<10%) after serial passage. Thus, the long-term survival of these cultures was attributed to indirect influences that could also be reproduced by transferring supernatants from infected cultures to naive, uninfected HUVEC. Specifically, the type 2 VEGF receptor, KDR, was upregulated on all cells, regardless of infection status, via a KSHV-initiated paracrine effect, rendering the KDR+ cells responsive to VEGF in the culture medium. Although more recent models have allowed a higher percentage of infected cells, the paradigm whereby a subset of infected cells influences the larger microenvironment remains extremely relevant to KS pathogenesis.

Ciufo et al. describe infection of primary adult and neonatal dermal microvascular EC (DMVEC) using infectious KSHV prepared from the supernatants of three different TPA-induced PEL lines, BCP-1, BC-3 and JSC1 (Ciufo et al. 2001). Virus was concentrated to generate high-titer inoculae allowing for a high initial infection rate. Interestingly, inoculum prepared from the JSC1 line allowed the most efficient infection, due to the production of a higher number of progeny virions and perhaps some qualitative differences between PEL lines as well. Upon infection, EC morphology changed from a cobblestone shape with a flat aspect to aggregates of cells with elongated spindloid morphology. Infection of the primary DMVEC appeared to be predominantly latent; almost all converted spindle cells were LANA-1-positive, with approximately 5%–10% of cells expressing immediately early or early lytic proteins (e.g., K8, K5, ORF50, and ORF59) and 1%–2% expressing the late lytic glycoprotein K8.1A. K8.1A-positive cells were typically a subfraction of rounded up cells exhibiting CPE associated with the final stages of productive replication. In some of the K8.1A-positive or ORF59-positive cells, intranuclear inclusions resembling herpesviral replication compartments were visible. In summary, infection of primary DMVEC with JSC1-derived KSHV at high MOI led to the establishment of a predominantly latently infected cell pool that showed marked morphological changes reminiscent of KS spindle cells, while a subset of cells underwent spontaneous lytic replication with release of infectious progeny. These findings are reminiscent of KS tumor cells in vivo, with respect to both morphology and viral gene expression patterns. The KSHV-infected cultures could not be maintained indefinitely, however, both because of a chronic loss of cells from lytic infection and failure of the latently infected cells to be maintained for more than a few cycles as a population of dividing cells that maintained the viral episome. Addition of uninfected cells to the cultures at a 1-to-10 ratio provided fresh targets for infection, providing a means to generate additional infected cultures without establishing new PEL-initiated infections. Thus, multiple rounds of de novo infection, as opposed to long-term maintenance of latently infected cells, define this culture model.

Similar protocols have since been used to generate KSHV-infected EC cultures to examine patterns of viral (Krishnan et al. 2004) or host (Hong et al. 2004; Naranatt et al. 2004; Poole et al. 2002; Wang et al. 2004) gene expression by microarray analysis, RT-PCR, and/or immunostaining after de novo infection with PEL-derived virus. Krishnan and colleagues (Krishnan et al. 2004) were the first to comprehensively examine viral gene expression patterns in primary EC after de novo infection. In this report, adult DMVEC were infected with concentrated BCBL-1-derived KSHV and the kinetics of latent and lytic gene expression were examined by whole genome array (Celonex HHV8 viruChip), RT-PCR, and immunostaining at time points from 30 min to 5 days postinfection (PI). This report revealed a number of interesting findings. Expression of a limited set of immediate-early (IE) and early (E) lytic genes was initiated concurrently with latent genes in the majority of cells immediately after infection, with a sharp decline thereafter. Compared to TPA-induced BCBL cells, only a limited number of ORF50-activated genes were expressed, and the majority of genes involved in DNA replication and viral assembly were not expressed. Of particular interest, the IE gene K5/MIR2 was expressed at early time points and continued to be well expressed after a decline in other IE and E genes. Another interesting finding was confirmation of the absence of LANA 2 (ORFK10.5) expression in primary EC. Expression of a limited set of IE and E genes with immune evasion and antiapoptotic function, with a subsequent decline thereafter, may play a crucial role in vivo in allowing establishment of infection and tumor initiation/progression. Signals that determine which cells progress through a full lytic cycle and produce infectious progeny, versus those that establish latency, remain to be determined; indeed their elucidation may require the multifactorial tumor environment to be more fully represented in in vitro KS models.

Host EC gene expression has been evaluated after only 2 and 4 h PI (Naranatt et al. 2004), after several days (Hong et al. 2004; Wang et al. 2004), or after 3 weeks (Poole et al. 2002). Poolc et al. (Poole et al. 2002) also performed microarray analysis after reseeding of infected cultures with uninfected cells (Poole et al. 2002). Naranatt and colleagues (Naranatt et al. 2004) examined gene expression in KSHV-infected primary DMVEC with Affymetrix HG-U133A gene arrays as early as 2 and 4 h PI and discovered a significant early reprogramming of the host transcriptome that included alterations in genes involved in signaling, apoptosis, transcription, host defense, cell cycle, metabolism, inflammation, angiogenesis, and tumorigenesis. The authors also evaluated infected fibroblasts and the B cell line BJAB, to create a database that illuminated both cell type specific and common responses to infection. The study by Poole and colleagues (Poole et al. 2002) differed in that the authors' aim was to interrogate host reprogramming after establishment of a latent infection and after exposure of latently infected cells to TPA; hence, cells were harvested 3 weeks PI as well as 2 weeks after reseeding infected cells with new uninfected targets. Clontech Human Atlas or Incyte Human UniGemV2.0 cDNA arrays were used, with RT-PCR analysis confirming changes in selected individual genes. This study revealed that, even after establishment of latency, KSHV-infected cells exhibited a profound alteration in gene expression patterns; between 1.4% and 2.5% of genes represented were significantly upregulated or downregulated. Of particular interest was the induction of interferon-induced genes, genes involved in cell signaling and angiogenesis, and genes involved in cell cycle progression and apoptosis. The studies by Hong and colleagues and Wang and colleagues (Hong et al. 2004; Wang et al. 2004) were specifically designed to evaluate lymphatic reprogramming and are discussed below, along with a similar study using immortalized EC (Carroll et al. 2004).

Primary infection of EC has also been used to study KSHV binding and entry and the signaling pathways induced in the earliest phases of infection (Naranatt et al. 2003). This study identified a role for integrin $\alpha_3\beta_1$-FAK-dependent phosphatidylinositol 3- (PI 3-) kinase activation in KSHV entry and a role for the PKC-ζ-MEK-ERK signaling cascade during the earliest stages of KSHV infection. Activation of a mitogenic cascade during KSHV entry may have important implications for establishment of latency and the division of latently infected cells, as well as creation of a host microenvironment conducive to the expression of immediate early lytic cycle proteins.

A recent report by Krug et al. is worthy of mention because infection was performed at low MOI to specifically explore the ability of latently infected EC to replicate as well as the ability of infected EC to produce infectious virus for dissemination via de novo infection (Krug et al. 2004). Pooled neonatal DMVEC were infected with KSHV derived from TPA-induced BCBL-1 cells. Although less than 5% of cells were initially infected, infection spread primarily because of proliferation of latently infected cell with episome maintenance. A limited contribution by de novo infection from virions produced from the small percentage that entered the lytic cycle was also noted. Spread of infection could not be blocked by treating cultures with compounds that block herpesviral lytic replication; therefore, proliferation of latently infected cells characterizes this culture system. A growth advantage for LANA-1-positive cells as compared to uninfected cells was noted, and cell spindling with loss of contact inhibition after postconfluent growth was observed. Thus, of all primary infection protocols, this is the most similar to the system described by Moses et al. that is discussed in Sect. 5. Of note, infection of the primary EC induced expression of cellular antiviral genes, specifically dsRNA-activated protein kinase (PKR) and 2′,5′-oligoadenylate synthetase (2′5′-OAS), a phenotype that could be reproduced by IFN-γ treatment. IFN-γ treatment was also effective in preventing lytic activation and viral replication. However, because infection was initiated at low MOI and was maintained primarily by proliferation of latently infected cells, induction of antiviral genes did not eliminate viral infection in this system. Other protocols for infection of primary cells that utilize high-titer inoculae may induce higher levels of these host-defense genes, which may in turn interfere with establishment and maintenance of infection.

The life span of primary EC in vitro is limited because of replicative senescence. To overcome this limitation, some investigators have employed life-extended EC prepared by ectopic expression of human telomerase or genes from other transforming viruses (Lagunoff et al. 2002; Moses et al. 1999). Expression of these genes does not induce laboratory evidence of transformation other than extending the length of time cells can be serially propagated in culture. A system described by Moses et al. was the first to use life-extended EC for KSHV infection and the first to describe cultures in which the majority of cells became infected and maintained the genome in a predominantly latent state (Moses et al. 1999). In this model, adult primary DMVEC are immortalized by retroviral transduction of the E6 and E7 genes of human papillomavirus (HPV) type 16. The life-extended cells exhibit no overt signs of transformation (i.e., they become contact inhibited at confluence and do not form colonies in soft agar) but with serial passage can be maintained significantly longer than their primary counterparts. HPV E6 and E7 are known to have significant effects on the cell cycle, most notably through targeting p53 and Rb. E6 participates in the ubiquitination and degradation of p53 (Werness et al. 1990), and E7 binds and sequesters Rb (Howley et al. 1989) and induces Rb phosphorylation (i.e., inactivation) (Zerfass-Thome et al. 1996). These are two pathways that KSHV latent genes also inhibit; LANA-1 disrupts the function of both p53 (Friborg et al. 1999) and Rb (Radkov et al., 2000), and v-Cyclin induces Rb phosphorylation via activation of CDK-6 (Godden-Kent et al. 1997). In addition, LANA and p53 colocalization is observed in KSHV-associated tumors in vivo where p53 is usually wild type, suggesting that LANA-mediated p53 inactivation is important for tumorigenesis (Katano et al. 2001). Therefore, immortalization of DMVEC by E6/E7 expression may augment some of the alterations in cellular physiology that are themselves induced by KSHV, thus creating a cellular microenvironment conducive to KSHV infection. It cannot be ruled out, however, that these HPV gene products complicate results obtained when using this culture system for studies of KSHV biology. Thus, key findings obtained with E6/E7-immortalized DMVEC have been verified in primary cells and KS tissue (McAllister et al. 2004; Moses et al. 1999, 2002a; Raggo et al. 2005).

For studies performed with the E6/E7-DMVEC model to date, infectious KSHV has been derived from unconcentrated TPA-induced BCBL-1 cultures, yielding a relatively low-titer inoculum and thus the expectation of a low percentage of initially infected cells. Indeed, evaluation of LANA-1 expression by IFA at early times (12 h) PI revealed that not more than 10% of EC in a treated culture were KSHV infected (Moses et al. 1999). LANA expression increased with time, such that by 14 days PI up to 80% of cells in a KSHV-exposed culture were infected. PCR for the KSHV genome at days 7 and 14 PI revealed increased intensity of the amplified product by day 14, supporting evidence of virus spread. A similar trend was seen with expression of the early lytic protein ORF 59: At 1 week PI <1% of cells were ORF 59 positive, but by 8 weeks PI up to 5% of cells expressed this lytic marker. Expression of the late lytic glycoprotein K8.1A/B followed similar expression kinetics, though to a consistently 5- to 10-fold lesser degree than ORF 59. Lytic replication could be induced in infected cells by treatment with TPA, sodium butyrate, or corticosteroids, but never in more than 40% of LANA-positive cells.

Before exposure to KSHV, E6/E7-immortalized DMVEC retained a classic cobblestone appearance. KSHV-infection, however, induced marked changes in cellular morphology reminiscent of the spindle cells observed in KS lesions including elongated cells with oval cell bodies, uniformly narrow elongated cells, and extremely narrow light-refractile cells displaying scattering. A low percentage of rounded up cells that tended to detach from the monolayer were also observed; the nuclei of these cells as well as the nuclei of extremely spindled cells displayed intranuclear inclusions resembling typical herpesviral CPE. The extent of morphologic change within virus-exposed cultures increased with time PI and correlated strongly with the percentage of KSHV-infected cells, suggesting a direct effect of the virus on cell morphology. Evaluation of gene expression in concert with morphology revealed the following: LANA-1 expression was sufficient for spindling; ORF59-positive cells were spindle-shaped, with a proportion displaying intranuclear inclusions, and all rounded cells were strongly ORF59 positive; rare K8.1A/B positive cells displayed severe spindling but were more frequently rounded. The presence of infectious virions in conditioned supernatants that could transfer infection to naive cultures, albeit at a low MOI, confirmed that DMVEC could support a fully permissive replication cycle. The relative ratios of ORF73- to ORF59- to K8.1A/B-positive cells at any one time suggested that the majority of cells harbored latent infection and that completion of the lytic cycle occurred in only a fraction of the lytic cell population. These observations are very similar to those reported by Ciufo and colleagues on infection of primary DMVEC (Ciufo et al. 2001).

Evaluation of KSHV gene expression and cell morphology suggests that the attainment of completely infected cultures in the E6/E7-DMVEC model reflects a combination of the proliferation of latently infected cells and de novo infection by EC-generated virus. Once all cells are infected and naive viral targets are absent, latently infected cells continue to proliferate with only a minor fraction of cells lost due to productive lytic infection. Infected cultures can be expanded by passaging, but the genome is best maintained when low split ratios are used; presumably a delicate balance exists between cell division and episome replication and segregation to daughter cells. A key difference between the immortalized DMVEC and primary EC is the increased length of time for which latently infected E6/E7-DMVEC can be passaged with genome maintenance. It is possible that the presence of E6/E7 allows for a degree of episome maintenance more akin to what is seen in PEL cells that, unlike KS tumor cells, are able to maintain the KSHV episome in vitro as well as in vivo.

The latent gene ORFK10.5/LANA-2 that inhibits p53 function is well expressed in PEL cell but not in primary DMVEC (Krishnan et al. 2004; Parravicini et al. 2000); possessing multiple ways to disable this tumor suppressor protein may be a reason for efficient episome maintenance in PEL cells. Distinct from their role as oncoproteins, E6 and E7 are also known to be required for stable maintenance of HPV episomes in undifferentiated human keratinocytes (Thomas et al. 1999). The HPV oncogenes in the immortalized DMVEC may assist in episome maintenance by duplicating the function of LANA-2 or by acting as surrogates for other currently unappreciated mechanisms occurring in the lesional microenvironment.

Interestingly, after several weeks in culture, there is a decrease in both spontaneous ORF 59-expression and inducibility of lytic cycle proteins by chemical induction in KSHV-infected E6/E7-DMVEC. Thus, a predominantly latent infection is established, possibly by selection of clones that are resistant to lytic induction but that can propagate as continually dividing, latently infected cells. In such cultures, CD31 is strongly expressed on the LANA-positive cells, suggesting that K5 is not expressed and thus a true latency exists (Mandana Mansouri and Klaus Früh, unpublished observations).

As described above, the E6/E7-DMVEC model appears to represent two distinct stages: one in which the viral infection and genome spread occurs via both lytic and latent means and a second where latently infected cells survive for multiple passages and maintain the genome in a latent state. When such cells are not passaged before achieving tissue culture confluence, a third state is generated, characterized by the continued postconfluent growth of the cells into three-dimensional foci. Importantly, uninfected cells grow to confluence and enter a quiescent state under similar conditions. Postconfluent growth reflects loss of contact inhibition, one of the hallmarks of cellular transformation. Infected cells are also able to form colonies in soft agar, a measure of anchorage-independent growth, and form tumors in mice when injected into the tail base in a matrigel solution (Shane McAllister and Ashlee Moses, unpublished observations). Krug and colleagues (Krug et al. 2004) similarly report that when primary DMVEC are infected at low MOI with subsequent spread of infection primarily through division of latently infected cells, loss of contact inhibition is observed. The piled-up aggregates of infected cells reported by Ciufo and colleagues (Ciufo et al. 2001) are also reminiscent of three-dimensional focus formation. Thus acquisition of a transformed phenotype is not unique to the E6/E7 DMVEC system, although the potential for additional contributions from the HPV proteins should be considered.

Latently infected E6/E7-DMVEC have been used in gene expression profiling studies to examine KSHV-induced cellular gene reprogramming and identify potential therapeutic targets for KS. These studies have employed cDNA arrays (Moses et al. 2002b) as well as Affymetrix U95A and U133A and B GeneChips (Moses et al. 2002a; Raggo et al. 2005) and have interrogated several different KSHV-infected cultures relative to age- and passage-matched uninfected controls. For all of these studies, KSHV-infected DMVEC were infected with BCBL-1-derived KSHV at low MOI and passaged when confluent at low split ratios. Cells were harvested for microarray analysis when immunofluorescent staining of parallel cultures revealed that >90% of KSHV-infected cultures were LANA positive. This typically took 3–4 weeks and encompassed approximately five to seven tissue culture passages. A minimum of two biological replicates were used for each comparison. The complete data sets from these different microarray experiments are available online (ohsu.edu/vgti/fruh.htm), and further details can be found in the specific papers referenced. To date, this database has been used to identify a handful of potential targets for KS chemotherapy (McAllister et al. 2004; Moses et al. 2002b; Raggo et al. 2005). Briefly, the receptor tyrosine kinase c-Kit was identified as a gene induced by KSHV with cDNA arrays (Moses et al. 2002b) and confirmed with Affymetrix arrays, RT-PCR analysis, and immunofluorescence, the latter being on primary DMVEC. An independent analysis of gene expression in primary DMVEC with U133A GeneChips also reported KSHV induction of c-Kit (Hong et al. 2004). A role for c-Kit in proliferation and post-confluent growth of KSHV-infected DMVEC was then demonstrated with both a pharmacological inhibitor of c-Kit [Imatinib Mesylate, Gleevec; formerly STI 571 (Moses et al. 2002b)] and gene knockdown approaches (Moses et al. 2002a). These data, combined with evidence of c-Kit expression in KS tumors (Pantanowitz et al. 2004), contributed to a recent clinical trial that demonstrated the efficacy of Imatinib Mesylate as a therapeutic regimen for KS (Koon et al. 2004). Because both c-Kit and PDGF-R are expressed in KS, the relative importance of these two targets remains unclear. However, the study is notable because it was the first to attempt to identify novel drug targets for KSHV with microarrays and to test the efficacy of a pharmacologic agent in a disease model (Jenner and Young 2005). The same conceptual approach was used to identify two KSHV-induced cellular proteins, RDC-1 and Neuritin, with novel oncogenic properties (Raggo et al. 2005). RDC-1 was also identified as one of the most highly induced genes in the microarray studies performed on primary DMVEC by Poole and colleagues (Poole et al. 2002). E6/E7-DMVEC were recently used in a proteomics-based screen to identify KSHV induction of the enzyme heme oxygenase-1 (HO-1) (McAllister et al. 2004). HO-1 expression in KS tissue was confirmed by the authors and by a recent SAGE study (Cornelissen et al. 2003). Inhibition of HO-1-induced proliferation of infected cells by treatment with mesoporphyrin compounds may, on further study, offer an additional treatment option for KS (McAllister et al. 2004; Yarchoan 2004).

A second efficient culture system for studying KSHV biology based on the use of immortalized EC was described by Lagunoff and colleagues (Lagunoff et al. 2002). In this model, human neonatal DMVEC were immortalized by retroviral transduction of the telomerase reverse transcriptase subunit (hTERT) (Venetsanakos et al. 2002). The telomerase-immortalized microvascular endothelial cells (TIME cells) retained a normal karyotype as well as many of the properties of the primary cells from which they were derived, including expression of CD31 and $\alpha_V\beta_3$-integrin, LDL uptake, and tubule formation in matrigel. TIME cells grow well when serially passaged and become contact inhibited at confluence, with the only morphologic change being a mild cell spindling.

TIME cells are infected at a high MOI with KSHV concentrated from TPA-induced BCBL-1 supernatants. Nearly all TIME cells are LANA positive at 48 h after infection, and latency is the predominant outcome; only about 1% of infected cells express the early lytic protein ORF 59, and a smaller subset express the late structural glycoprotein K8.1. This system thus accurately reflects the state of the viral genome in vivo (Staskus et al. 1997) and is reminiscent of what has been observed after de novo infection of primary and E6/E7-DMVEC (Ciufo et al. 2001; Moses et al. 1999). In addition, because of the high frequency of initial infection, TIME cells provide a valuable system for studying early events in de novo infection. With increasing time PI, however, there is a rapid reduction in the infected TIME cell population, such that by tissue culture passage 7, less than 0.1% of cells are infected. Such loss of latently infected cells suggests inefficient maintenance of the viral episome and resembles what is seen with explanted KS spindle cells. However, because a percentage of TIME cells are lytically infected, infectious virus can be serially transferred or cultures can be maintained by addition of uninfected cells. An advantage of the TIME system is the ease with which TIME cells can be cultivated and infected with high efficiency. In addition, potential effects from oncoproteins such as E6 and E7 are not a concern in this model.

While TIME cells establish a predominantly latent infection, they can be induced to lytically reactivate by infection with an ORF50-expressing adenovirus vector (Bechtel et al. 2003). Using this system, Glausinger and Ganem (Glaunsinger and Ganem 2004) demonstrated that lytic KSHV infection strongly inhibits host gene expression by accelerating global mRNA turnover. Shut-off is mediated by the viral SOX (shutoff exonuclease) protein, the product of the HSV alkaline exonuclease homologue ORF37. The TIME cells were subsequently used to demonstrate that a subset of host transcripts, including IL-6 and the IL-1 type 1 receptor, escape host shut-off (Glaunsinger and Ganem 2004).

TIME cells express very little or no Prox-1 protein before KSHV infection, suggesting that the original immortalized clone was derived from blood vascular, as opposed to lymphatic, endothelium. Prox-1 is required for expression of two key markers that differentiate lymphatic from vascular endothelium, VEGFR3 and podoplanin (Wigle and Oliver 1999). TIME cells have thus also proved useful for investigating the hypothesis that KSHV infection drives EC to a more lymphatic phenotype, a hypothesis supported by the robust expression of lymphatic EC markers on KS spindle cells (Jussila et al. 1998; Skobe et al. 1999). Carroll and colleagues (Carroll et al. 2004) used RT-PCR and cDNA microarray techniques to investigate expression of genes specific to lymphatic EC after KSHV infection of TIME cells and reported significant induction of such genes including Prox-1, VEGFR3, podoplanin, and LYVE-1. The microarray studies were performed after 24, 48, and 96 h of infection; 147 genes (about 1% of genes on the array) were significantly induced (>1.8 fold; $P < 0.001$) at all time points and 61 genes significantly repressed. This list comprises another valuable data set with which to examine KSHV reprogramming of the host transcriptome, particularly at early times PI and when considering the blood vascular phenotype of the uninfected controls. Array analysis performed with primary cells may reflect a mixed population of both blood and lymphatic EC, because both types can be present in primary cultures (Makinen et al. 2001). In addition to the differential expression of lymphatic markers observed in KSHV-infected TIME cells, other interesting findings included significant upregulation of host IL-6 and significant downregulation of IL-8. Induction of IL-6 expression is consistent with the finding that IL-6 escapes host shut-off; loss of IL-8 protein expression in KSHV-infected E6/E7-DMVEC stimulated with IL-1 has also been observed (Ashlee Moses and Michael Jarvis, unpublished observations).

Telomerase-immortalized KSHV-infected DMVEC have also been described by Tomescu and colleagues (Tomescu et al. 2003). The cells used in this study, TIME-T4 cells, were derived from a different parental cell than that of Lagunoff et al. but were transduced with the same retroviral hTERT expression vector. TIME-T4 cells were infected with concentrated BCBL-1-derived virus or via coculture with BCBL cells. The TIME-T4 cells respond differently to KSHV infection than the original TIME cells in that a dramatic spindling was observed as soon as 24 h PI. This may be due to an inherent difference in the immortalized cell clone or to quantitative or qualitative differences in the inoculae used. The T4 cells demonstrated downregulation of major histocompatibility complex (MHC) class I and the adhesion molecules ICAM-1 and CD31, within 48 h of infection, indicating that this system is extremely useful for the study of KSHV immune evasion mechanisms in EC. The effective downregulation of MHC1 suggests that the

T4 cells were expressing latent and IE lytic genes concurrently, as suggested by the studies of Krishnan and colleagues in primary cells (Krishnan et al. 2004). TIME T4 cells should prove valuable for determining the kinetics of IE expression in EC, particularly for K3 and K5, as well as the window of time after their downregulation for which MHC1 and adhesion molecule surface expression remains functionally compromised.

The above text contains multiple references to gene expression profiling experiments performed on KSHV-infected EC. Taking into account the differences in experimental design, including cells and virus origin, infection and culture conditions and the microarray platforms used, it is not surprising that the correlation between these data sets is modest. Genes that are commonly induced could reveal patterns that are general responses to virus infections, as well as genes so tightly linked to KSHV pathogenesis that their dysregulation transcends differences in experimental design. On the other hand, differences may reveal valid temporal changes in gene expression or culture-specific events. Analyses performed at early times PI would no doubt include host genes deregulated by those early lytic viral genes that may be transiently expressed, whereas those performed after many tissue culture passages (with episome maintenance) would more reflect the influences of latent gene expression, as well as the delayed effects of those early events PI. Collectively, these studies should be viewed as a valuable database from which to further analyze links between KSHV pathogenesis and cellular gene expression and function. Valuable progress has already been made in this regard.

5
Parental Lineage of Spindle Cells

Identification of the cellular origin of KS lesional spindle cells has been a matter of ongoing debate. Spindle cells express EC, smooth muscle cell, macrophage, fibroblast, and dendritic cell markers (Kaaya et al. 1995; Regezi et al. 1993b) but are generally accepted to originate from an EC precursor. In addition to expressing markers of blood vascular endothelium, KS spindle cells express several markers specific for lymphatic endothelium, including VEGF-R3 and podoplanin (Jussila et al. 1998; Skobe et al. 1999; Weninger et al. 1999). These observations suggest either that KSHV preferentially infects lymphatic endothelium in vivo or that KSHV infects precursor EC, lymphatic EC (LEC), or blood vascular EC (BEC) and drives the gene expression profile to a more convergent one, where lymphatic-lineage markers are induced or retained. As discussed in detail in this chapter, KSHV infects BEC in vitro; if, however, LEC were the preferred target in vivo, this might in part explain the inability of KSHV-infected EC to maintain the genome in tissue culture for extended periods. Wang et al. recently infected both LEC and BEC with KSHV and found with quantitative PCR that KSHV genomes were maintained at higher copy number in LEC (Wang et al. 2004). This same group compared gene expression profiles of nodular KS samples and normal skin with Affymetrix U133A arrays and developed a KS expression signature of 1,482 genes by removing genes expressed at similar levels in both KS and normal dermis or epidermis. Using expression profiles similarly generated from purified LEC and BEC, the authors found that although both LEC and BEC markers were present in KS tissue, the KS expression signature was more like that of the LEC. Interestingly, infection of LEC and BEC led to a convergence of their profiles such that they were more like each other than the uninfected counterparts. Hong and colleagues (Hong et al. 2004) compared gene profiles of primary DMVEC at day 7 PI and found significant upregulation of key lymphatic lineage-specific genes after KSHV infection including Prox1, LYVE-1, reelin, follistatin, desmoplakin, and leptin receptor. In addition, retroviral transduction of BEC with LANA led to induction of Prox1, the master gene responsible for lymphatic vessel development (Wigle and Oliver 1999). This study is in agreement with a study done in TIME cells, an immortalized BEC line, reporting KSHV induction of lymphatic-lineage markers (Carroll et al. 2004).

Collectively, these studies suggest that in vitro KSHV induces a transcriptional drift in BEC and LEC toward a more convergent phenotype. Because commercial EC preparations contain both LEC and BEC, the relative ratios in cell preparations used in different laboratories may have some bearing on the observed outcome of infection. Regarding immortalized EC, the BEC lineage of TIME cells was recently established (Carroll et al. 2004). On the other hand, E6/E7-immortalized EC clones that are KSHV permissive appear to have a more LEC-like phenotype; genes for LYVE-1, podoplanin, VEGFR-3, leptin receptor, oncostatin M receptor, c-MAF, and reelin are all expressed (Patrick Rose and Ashlee Moses; unpublished observations). Because KSHV genomes appear to be better maintained in LEC than BEC (Wang et al. 2004), these findings may contribute to the ability of E6/E7-DMVEC to maintain the KSHV episome for long periods of time. The relevance of the above studies to KS tumors is clearly the appreciation that lymphangiogenic molecules are involved in KS pathogenesis; if KSHV infection of EC results in a cell type with characteristics of both vascular and lymphatic EC, understanding this unique tumor phenotype will be important for understanding the disease and developing effective clinical approaches for KS therapy.

6
A Model of KS Lesion Progression

The observed frequency of immunoreactivity for KSHV markers in EC infected in vitro with a limited panel of antibodies is LANA ≫ ORF 59 ≫ K8.1 (i.e., latent ≫ early lytic ≫ late lytic). As described elsewhere in this chapter, these ratios have been observed in studies using both primary and immortalized EC and different viral infection protocols. These ratios also accurately reflect gene expression patterns found in KS lesions by immunohistochemistry and in situ hybridization (Parravicini et al. 2000). Two theoretical outcomes of lytic reactivation could account for the relative excess of early lytic versus late lytic gene expression. First, if all cells activated to the lytic cycle consistently support viral DNA replication with assembly and release of infectious progeny, then all early lytic-positive cells would, in time, progress to cell lysis and death. Alternately, if only a fraction of lytically activated cells produce infectious progeny, then a pool of cells must exist that express several IE and early genes without supporting productive viral replication. Evidence that antiherpes agents such as ganciclovir and foscarnet, which inhibit lytic but not latent herpesviral infection, can improve the clinical outcome of KS strengthens the assertion that ongoing lytic gene expression is involved in KS pathogenesis (Glesby et al. 1996; Martin et al. 1999; Mocroft et al. 1996). If every lytic event leads to cell death, then a continual pool of new targets for de novo infection must be recruited; on the other hand, limited expression of IE and E genes would allow for immune evasion and antiapoptotic mechanisms, as well for as the angiogenic and chemoattractant properties of lytic genes, to contribute to tumor formation without a net loss of cells. The clinical responsiveness of KS to drugs that block proliferation of latently infected cells (Koon et al. 2004) indicates that lytic reactivation is not the only mechanism driving KS lesion progression. Accumulation of latently infected cells in KS lesions is due to proliferation and tissue invasion of latently infected cells. The rapid loss of the KSHV genome on culture of explanted spindle cells may reflect an insufficiency of the episomal maintenance machinery in endothelial lineage cells. It is unknown whether this occurs in vivo when virally infected spindle cells are stimulated to divide. Loss of the cytoprotective functions of some KSHV genes could lead to apoptotic cell death of spindle cells after cell division without episomal maintenance. Another possibility is that spindle cells that have lost the viral genome could become targets for reinfection; multiple rounds of reinfection could contribute to the accumulation of mutations leading eventually to the outgrowth of truly transformed clonal populations of spindle cells.

The lack of a well-characterized animal model currently precludes the in vivo examination of KSHV deletion mutants, so the specific gene expression

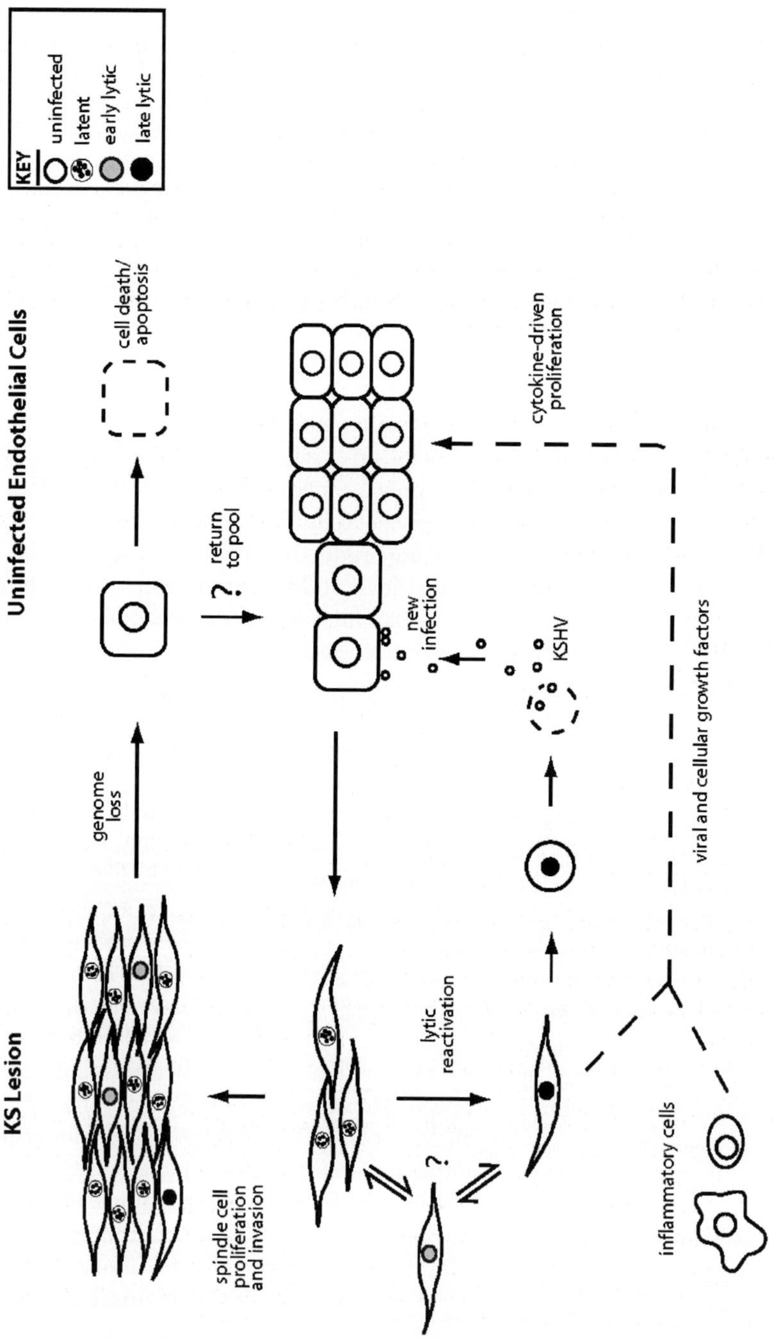
KS Lesion
Uninfected Endothelial Cells
KEY
uninfected
latent
early lytic
late lytic
genome loss
cell death/ apoptosis
? return to pool
new infection
KSHV
cytokine-driven proliferation
viral and cellular growth factors
spindle cell proliferation and invasion
lytic reactivation
?
inflammatory cells

Fig. 1 The model of KS lesion progression presented here is based on recent clinical advances in KS chemotherapy and draws together multiple theories of KS that by themselves cannot account for the complexities of KS lesions: The accumulation of spindle cells as KS lesions progress is driven by both continuous de novo infection and proliferation of latently infected cells. Lytic reactivation results in both the production of infectious virus but also paracrine signaling that contributes to lesion progression by inducing proliferation of uninfected neighboring cells. Loss of the KSHV genome on multiple rounds of cell division may be followed by repeated rounds of infection, contributing ultimately to the outgrowth of truly transformed clones that have accumulated multiple mutations. It is currently unknown whether proliferating spindle cells lose the KSHV genome in vivo

requirements for KS disease states must be inferred indirectly. The clinical responsiveness of KS to drugs that block herpesvirus lytic cycle, and to drugs that block proliferation of latently infected cells, indicates that both of these mechanisms can drive KS lesion establishment and progression, perhaps cooperatively (Fig. 1).

References

(1997) Proceedings of the IARC Working Group on the Evaluation of Carcinogenic Risks to Humans. Epstein-Barr Virus and Kaposi's Sarcoma Herpesvirus/Human Herpesvirus 8. Lyon, France, 17–24 June 1997. IARC Monogr Eval Carcinog Risks Hum 70:1–492

Aluigi MG, Albini A, Carlone S, Repetto L, De Marchi R, Icardi A, Moro M, Noonan D, Benelli R (1996) KSHV sequences in biopsies and cultured spindle cells of epidemic, iatrogenic and Mediterranean forms of Kaposi's sarcoma. Res Virol 147(5):267–75

Ambroziak JA, Blackbourn DJ, Herndier BG, Glogau RG, Gullett JH, McDonald AR, Lennette ET, Levy JA (1995) Herpes-like sequences in HIV-infected and uninfected Kaposi's sarcoma patients. Science 268(5210):582–3

Arvanitakis L, Mesri EA, Nador RG, Said JW, Asch AS, Knowles DM, Cesarman E (1996) Establishment and characterization of a primary effusion (body cavity-based) lymphoma cell line (BC-3) harboring Kaposi's sarcoma-associated herpesvirus (KSHV/HHV-8) in the absence of Epstein-Barr virus. Blood 88(7):2648–54

Ascherl G, Hohenadl C, Monini P, Zietz C, Browning PJ, Ensoli B, Sturzl M (1999) Expression of human herpesvirus-8 (HHV-8) encoded pathogenic genes in Kaposi's sarcoma (KS) primary lesions. Adv Enzyme Regul 39:331–9

Ballestas ME, Kaye KM (2001) Kaposi's sarcoma-associated herpesvirus latency-associated nuclear antigen 1 mediates episome persistence through *cis* -acting terminal repeat (TR) sequence and specifically binds TR DNA. J Virol 75(7):3250–8

Bechtel JT, Liang Y, Hvidding J, Ganem D (2003) Host range of Kaposi's sarcoma-associated herpesvirus in cultured cells. J Virol 77(11):6474–81

Beckstead JH, Wood GS, Fletcher V (1985) Evidence for the origin of Kaposi's sarcoma from lymphatic endothelium. Am J Pathol 119(2):294–300

Benelli R, Albini A, Parravicini C, Carlone S, Repetto L, Tambussi G, Lazzarin A (1996) Isolation of spindle-shaped cell populations from primary cultures of Kaposi's sarcoma of different stage. Cancer Lett 100(1–2):125–32

Blackbourn DJ, Lennette E, Klencke B, Moses A, Chandran B, Weinstein M, Glogau RG, Witte MH, Way DL, Kutzkey T, Herndier B, Levy JA (2000) The restricted cellular host range of human herpesvirus 8. AIDS 14(9):1123–33

Blauvelt A (1999) The role of human herpesvirus 8 in the pathogenesis of Kaposi's sarcoma. Adv Dermatol 14:167–206; discussion 207

Boshoff C, Gao S-J, Healy LE, Matthews S, Thomas AJ, Coignet L, Warnke RA, Strauchen JA, Matutes E, Karnel OW, Moore PS, Weiss RA, Chang Y (1998) Establishing a KSHV+ cell line (BCP-1) from peripheral blood and characterizing its growth in nod/SCID mice. Blood 91(5):1671–1679

Boshoff C, Schulz TF, Kennedy MM, Graham AK, Fisher C, Thomas A, McGee JO, Weiss RA, O'Leary JJ (1995) Kaposi's sarcoma-associated herpesvirus infects endothelial and spindle cells. Nat Med 1(12):1274–8

Brousset P, Cesarman E, Meggetto F, Lamant L, Delsol G (2001) Colocalization of the viral interleukin-6 with latent nuclear antigen-1 of human herpesvirus-8 in endothelial spindle cells of Kaposi's sarcoma and lymphoid cells of multicentric Castleman's disease. Hum Pathol 32(1):95–100

Bubman D, Cesarman E (2003) Pathogenesis of Kaposi's sarcoma. Hematol Oncol Clin North Am 17(3):717–45

Cannon J, Hamzeh F, Moore S, Nicholas J, Ambinder R (1999) Human herpesvirus 8-encoded thymidine kinase and phosphotransferase homologues confer sensitivity to ganciclovir. J Virol 73:4786–4793

Cannon JS, Ciufo D, Hawkins AL, Griffin CA, Borowitz MJ, Hayward GS, Ambinder RF (2000) A new primary effusion lymphoma-derived cell line yields a highly infectious Kaposi's sarcoma herpesvirus-containing supernatant. J Virol 74(21):10187–93

Carbone A, Gloghini A, Vaccher E, Zagonel V, Pastore C, Dalla Palma P, Branz F, Saglio G, Volpe R, Tirelli U, Gaidano G (1996) Kaposi's sarcoma-associated herpesvirus DNA sequences in AIDS-related and AIDS-unrelated lymphomatous effusions. Br J Haematol 94(3):533–43

Carroll PA, Brazeau E, Lagunoff M (2004) Kaposi's sarcoma-associated herpesvirus infection of blood endothelial cells induces lymphatic differentiation. Virology 328(1):7–18

Cesarman E, Chang Y, Moore PS, Said JW, Knowles DM (1995) Kaposi's sarcoma-associated herpesvirus-like DNA sequences in AIDS-related body-cavity-based lymphomas [see comments]. N Engl J Med 332(18):1186–91

Cesarman E, Nador RG, Aozasa K, Delsol G, Said JW, Knowles DM (1996) Kaposi's sarcoma-associated herpesvirus in non-AIDS related lymphomas occurring in body cavities. Am J Pathol 149(1):53–7

Chang J, Renne R, Dittmer D, Ganem D (2000) Inflammatory cytokines and the reactivation of Kaposi's sarcoma-associated herpesvirus lytic replication. Virology 266(1):17–25

Chang Y, Cesarman E, Pessin MS, Lee F, Culpepper J, Knowles DM, Moore PS (1994) Identification of herpesvirus-like DNA sequences in AIDS-associated Kaposi's sarcoma [see comments]. Science 266(5192):1865–9

Chang Y, Moore PS, Talbot SJ, Boshoff CH, Zarkowska T, Godden K, Paterson H, Weiss RA, Mittnacht S (1996) Cyclin encoded by KS herpesvirus [letter]. Nature 382(6590):410

Cheung TW (2004a) AIDS-related cancer in the era of highly active antiretroviral therapy (HAART): a model of the interplay of the immune system, virus, and cancer. "On the offensive – the Trojan Horse is being destroyed" – Part A: Kaposi's sarcoma. Cancer Invest 22(5):774–86

Cheung TW (2004b) AIDS-related cancer in the era of highly active antiretroviral therapy (HAART): a model of the interplay of the immune system, virus, and cancer. "On the offensive – the Trojan Horse is being destroyed" – Part B: Malignant lymphoma. Cancer Invest 22(5):787–98

Child ES, Mann DJ (2001) Novel properties of the cyclin encoded by human herpesvirus 8 that facilitate exit from quiescence. Oncogene 20(26):3311–22

Ciufo DM, Cannon JS, Poole LJ, Wu FY, Murray P, Ambinder RF, Hayward GS (2001) Spindle cell conversion by Kaposi's sarcoma-associated herpesvirus: formation of colonies and plaques with mixed lytic and latent gene expression in infected primary dermal microvascular endothelial cell cultures. J Virol 75(12):5614–26

Cornelissen M, Van Der Kuyl AC, Van Den Burg R, Zorgdrager F, Van Noesel CJ, Goudsmit J (2003) Gene expression profile of AIDS-related Kaposi's sarcoma. BMC Cancer 3(1):7

Davis DA, Rinderknecht AS, Zoeteweij JP, Aoki Y, Read-Connole EL, Tosato G, Blauvelt A, Yarchoan R (2001) Hypoxia induces lytic replication of Kaposi sarcoma-associated herpesvirus. Blood 97(10):3244–50

Davis MA, Sturzl MA, Blasig C, Schreier A, Guo HG, Reitz M, Opalenik SR, Browning PJ (1997) Expression of human herpesvirus 8-encoded cyclin D in Kaposi's sarcoma spindle cells. J Natl Cancer Inst 89(24):1868–74

Direkze S, Laman H (2004) Regulation of growth signalling and cell cycle by Kaposi's sarcoma-associated herpesvirus genes. Int J Exp Pathol 85(6):305–19

Djerbi M, Screpanti V, Catrina AI, Bogen B, Biberfeld P, Grandien A (1999) The inhibitor of death receptor signaling, FLICE-inhibitory protein defines a new class of tumor progression factors. J Exp Med 190(7):1025–32

Dourmishev LA, Dourmishev AL, Palmeri D, Schwartz RA, Lukac DM (2003) Molecular genetics of Kaposi's sarcoma-associated herpesvirus (human herpesvirus-8) epidemiology and pathogenesis. Microbiol Mol Biol Rev 67(2):175–212, table of contents

Dupin N, Fisher C, Kellam P, Ariad S, Tulliez M, Franck N, Van Marck E, Salmon D, Gorin I, Escande J-P, Weiss RA, Alitalo K, Boshoff C (1999) Distribution of human herpesvirus-8 latently infected cells in Kaposi's sarcoma, multicentric Castleman's disease, and primary effusion lymphoma. Proc Natl Acad Sci USA 96(8):4546–4551

Engels EA, Biggar RJ, Marshall VA, Walters MA, Gamache CJ, Whitby D, Goedert JJ (2003) Detection and quantification of Kaposi's sarcoma-associated herpesvirus to predict AIDS-associated Kaposi's sarcoma. AIDS 17(12):1847–51

Fakhari FD, Dittmer DP (2002) Charting latency transcripts in Kaposi's sarcoma-associated herpesvirus by whole-genome real-time quantitative PCR. J Virol 76(12):6213–23

Fickenscher H, Fleckenstein B (2001) Herpesvirus saimiri. Philos Trans R Soc Lond B Biol Sci 356(1408):545–67

Flaitz CM, Nichols CM, Hicks MJ (1996) Herpesviridae-associated persistent mucocutaneous ulcers in acquired immunodeficiency syndrome. A clinicopathologic study. Oral Surg Oral Med Oral Pathol Oral Radiol Endod 81(4):433–41

Flore O, Rafii S, Ely S, O'Leary JJ, Hyjek EM, Cesarman E (1998) Transformation of primary human endothelial cells by Kaposi's sarcoma-associated herpesvirus. Nature 394(6693):588–92

Friborg J, Jr., Kong W, Hottiger MO, Nabel GJ (1999) p53 inhibition by the LANA protein of KSHV protects against cell death. Nature 402(6764):889–94

Friedman-Kien AE, Laubenstein LJ, Rubinstein P, Buimovici-Klein E, Marmor M, Stahl R, Spigland I, Kim KS, Zolla-Pazner S (1982) Disseminated Kaposi's sarcoma in homosexual men. Ann Intern Med 96(6 Pt 1):693–700

Gaidano G, Cechova K, Chang Y, Moore PS, Knowles DM, Dalla-Favera R (1996) Establishment of AIDS-related lymphoma cell lines from lymphomatous effusions. Leukemia 10(7):1237–40

Gill PS, Tsai YC, Rao AP, Spruck CH, 3rd, Zheng T, Harrington WA, Jr., Cheung T, Nathwani B, Jones PA (1998) Evidence for multiclonality in multicentric Kaposi's sarcoma. Proc Natl Acad Sci USA 95(14):8257–61

Glaunsinger B, Ganem D (2004) Highly selective escape from KSHV-mediated host mRNA shutoff and its implications for viral pathogenesis. J Exp Med 200(3):391–8

Glesby MJ, Hoover DR, Weng S, Graham NM, Phair JP, Detels R, Ho M, Saah AJ (1996) Use of antiherpes drugs and the risk of Kaposi's sarcoma: data from the Multicenter AIDS Cohort Study. J Infect Dis 173(6):1477–80

Godden-Kent D, Talbot SJ, Boshoff C, Chang Y, Moore P, Weiss RA, Mittnacht S (1997) The cyclin encoded by Kaposi's sarcoma-associated herpesvirus stimulates cdk6 to phosphorylate the retinoblastoma protein and histone H1. J Virol 71(6):4193–8

Godfrey A, Anderson J, Papanastasiou A, Takeuchi Y, Boshoff C (2005) Inhibiting primary effusion lymphoma by lentiviral vectors encoding short hairpin RNA. Blood 105(6):2510–8

Guo HG, Browning P, Nicholas J, Hayward GS, Tschachler E, Jiang YW, Sadowska M, Raffeld M, Colombini S, Gallo RC, Reitz MS, Jr. (1997) Characterization of a chemokine receptor-related gene in human herpesvirus 8 and its expression in Kaposi's sarcoma. Virology 228(2):371–8

Haque M, Davis DA, Wang V, Widmer I, Yarchoan R (2003) Kaposi's sarcoma-associated herpesvirus (human herpesvirus 8) contains hypoxia response elements: relevance to lytic induction by hypoxia. J Virol 77(12):6761–8

Harrington W, Jr., Sieczkowski L, Sosa C, Chan-a-Sue S, Cai JP, Cabral L, Wood C (1997) Activation of HHV-8 by HIV-1 tat. Lancet 349(9054):774–5

Hayward GS (2003) Initiation of angiogenic Kaposi's sarcoma lesions. Cancer Cell 3(1):1–3

Herndier B, Ganem D (2001) The biology of Kaposi's sarcoma. Cancer Treat Res 104:89–126

Herndier BG, Werner A, Arnstein P, Abbey NW, Demartis F, Cohen RL, Shuman MA, Levy JA (1994) Characterization of a human Kaposi's sarcoma cell line that induces angiogenic tumors in animals. AIDS 8(5):575–81

Hong YK, Foreman K, Shin JW, Hirakawa S, Curry CL, Sage DR, Libermann T, Dezube BJ, Fingeroth JD, Detmar M (2004) Lymphatic reprogramming of blood vascular endothelium by Kaposi sarcoma-associated herpesvirus. Nat Genet 36(7):683–5

Howley PM, Munger K, Werness BA, Phelps WC, Schlegel R (1989) Molecular mechanisms of transformation by the human papillomaviruses. Princess Takamatsu Symp 20:199–206

Ishido S, Choi JK, Lee BS, Wang C, DeMaria M, Johnson RP, Cohen GB, Jung JU (2000) Inhibition of natural killer cell-mediated cytotoxicity by Kaposi's sarcoma-associated herpesvirus K5 protein. Immunity 13(3):365–74

Jenner RG, Alba MM, Boshoff C, Kellam P (2001) Kaposi's sarcoma-associated herpesvirus latent and lytic gene expression as revealed by DNA arrays. J Virol 75(2):891–902

Jenner RG, Young RA (2005) Insights into host responses against pathogens from transcriptional profiling. Nat Rev Microbiol 3(4):281–94

Judde JG, Lacoste V, Briere J, Kassa-Kelembho E, Clyti E, Couppie P, Buchrieser C, Tulliez M, Morvan J, Gessain A (2000) Monoclonality or oligoclonality of human herpesvirus 8 terminal repeat sequences in Kaposi's sarcoma and other diseases. J Natl Cancer Inst 92(9):729–36

Jussila L, Valtola R, Partanen TA, Salven P, Heikkila P, Matikainen MT, Renkonen R, Kaipainen A, Detmar M, Tschachler E, Alitalo R, Alitalo K (1998) Lymphatic endothelium and Kaposi's sarcoma spindle cells detected by antibodies against the vascular endothelial growth factor receptor-3. Cancer Res 58(8):1599–604

Kaaya EE, Parravicini C, Ordonez C, Gendelman R, Berti E, Gallo RC, Biberfeld P (1995) Heterogeneity of spindle cells in Kaposi's sarcoma: comparison of cells in lesions and in culture. J Acquir Immune Defic Syndr Hum Retrovirol 10(3):295–305

Kaposi M (1872) Idiopathisches multiples Pigmentsarkom der Haut. Archiv fur Dermatologie und Syphilis 3:265–73

Katano H, Hoshino Y, Morishita Y, Nakamura T, Satoh H, Iwamoto A, Herndier B, Mori S (1999) Establishing and characterizing a CD30-positive cell line harboring HHV-8 from a primary effusion lymphoma. J Med Virol 58(4):394–401

Katano H, Ogawa-Goto K, Hasegawa H, Kurata T, Sata T (2001) Human-herpesvirus-8-encoded K8 protein colocalizes with the promyelocytic leukemia protein (PML) bodies and recruits p53 to the PML bodies. Virology 286(2):446–55

Katano H, Sato Y, Kurata T, Mori S, Sata T (2000) Expression and localization of human herpesvirus 8-encoded proteins in primary effusion lymphoma, Kaposi's sarcoma, and multicentric Castleman's disease. Virology 269(2):335–44

Kedes DH, Lagunoff M, Renne R, Ganem D (1997) Identification of the gene encoding the major latency-associated nuclear antigen of the Kaposi's sarcoma-associated herpesvirus. J Clin Invest 100(10):2606–10

Kellam P, Boshoff C, Whitby D, Matthews S, Weiss RA, Talbot SJ (1997) Identification of a major latent nuclear antigen, LNA-1, in the human herpesvirus 8 genome. J Hum Virol 1(1):19–29

Kellam P, Bourboulia D, Dupin N, Shotton C, Fisher C, Talbot S, Boshoff C, Weiss RA (1999) Characterization of monoclonal antibodies raised against the latent nuclear antigen of human herpesvirus 8. J Virol 73(6):5149–55

Klass CM, Krug LT, Pozharskaya VP, Offermann MK (2005) The targeting of primary effusion lymphoma cells for apoptosis by inducing lytic replication of human herpesvirus 8 while blocking virus production. Blood 105(10):4028–34

Komanduri KV, Luce JA, McGrath MS, Herndier BG, Ng VL (1996) The natural history and molecular heterogeneity of HIV-associated primary malignant lymphomatous effusions. J Acquir Immune Defic Syndr Hum Retrovirol 13(3):215–26

Koon HB, Bubley GJ, Pantanowitz L, Masiello D, Smith B, Crosby K, Proper J, Weeden W, Miller TE, Chatis P, Egorin MJ, Tahan SR, Dezube BJ (2004) Imatinib-induced regression of AIDS-related Kaposi's sarcoma. J Clin Oncol

Krishnan HH, Naranatt PP, Smith MS, Zeng L, Bloomer C, Chandran B (2004) Concurrent expression of latent and a limited number of lytic genes with immune modulation and antiapoptotic function by Kaposi's sarcoma-associated herpesvirus early during infection of primary endothelial and fibroblast cells and subsequent decline of lytic gene expression. J Virol 78(7):3601–20

Krug LT, Pozharskaya VP, Yu Y, Inoue N, Offermann MK (2004) Inhibition of infection and replication of human herpesvirus 8 in microvascular endothelial cells by alpha interferon and phosphonoformic acid. J Virol 78(15):8359–71

Kushner T (1995) *Angels in America.* Theatre Communications Group, Inc., 520 8th Ave., New York, NY 10018-4156

Lagunoff M, Bechtel J, Venetsanakos E, Roy AM, Abbey N, Herndier B, McMahon M, Ganem D (2002) De novo infection and serial transmission of Kaposi's sarcoma-associated herpesvirus in cultured endothelial cells. J Virol 76(5):2440–8

Lagunoff M, Ganem D (1997) The structure and coding organization of the genomic termini of Kaposi's sarcoma-associated herpesvirus. Virology 236(1):147–54

Lallemand F, Desire N, Rozenbaum W, Nicolas JC, Marechal V (2000) Quantitative analysis of human herpesvirus 8 viral load using a real-time PCR assay. J Clin Microbiol 38(4):1404–8

Laman H, Peters G, Jones N (2001) Cyclin-mediated export of human Orc1. Exp Cell Res 271(2):230–7

Lan K, Kuppers DA, Verma SC, Robertson ES (2004) Kaposi's sarcoma-associated herpesvirus-encoded latency-associated nuclear antigen inhibits lytic replication by targeting Rta: a potential mechanism for virus-mediated control of latency. J Virol 78(12):6585–94

Li H, Komatsu T, Dezube BJ, Kaye KM (2002) The Kaposi's sarcoma-associated herpesvirus K12 transcript from a primary effusion lymphoma contains complex repeat elements, is spliced, and initiates from a novel promoter. J Virol 76(23):11880–8

Li JJ, Huang YQ, Cockerell CJ, Friedman-Kien AE (1996) Localization of human herpes-like virus type 8 in vascular endothelial cells and perivascular spindle-shaped cells of Kaposi's sarcoma lesions by in situ hybridization. Am J Pathol 148(6):1741–8

Liu L, Eby MT, Rathore N, Sinha SK, Kumar A, Chaudhary PM (2002) The human herpes virus 8-encoded viral FLICE inhibitory protein physically associates with and persistently activates the IκB kinase complex. J Biol Chem 277(16):13745–51

Lu M, Suen J, Frias C, Pfeiffer R, Tsai MH, Chuang E, Zeichner SL (2004) Dissection of the Kaposi's sarcoma-associated herpesvirus gene expression program by using the viral DNA replication inhibitor cidofovir. J Virol 78(24):13637–52

Lubyova B, Pitha PM (2000) Characterization of a novel human herpesvirus 8-encoded protein, vIRF-3, that shows homology to viral and cellular interferon regulatory factors. J Virol 74(17):8194–201

Lukac DM, Kirshner JR, Ganem D (1999) Transcriptional activation by the product of open reading frame 50 of Kaposi's sarcoma-associated herpesvirus is required for lytic viral reactivation in B cells. J Virol 73(11):9348–61

Makinen T, Veikkola T, Mustjoki S, Karpanen T, Catimel B, Nice EC, Wise L, Mercer A, Kowalski H, Kerjaschki D, Stacker SA, Achen MG, Alitalo K (2001) Isolated lymphatic endothelial cells transduce growth, survival and migratory signals via the VEGF-C/D receptor VEGFR-3. EMBO J 20(17):4762–73

Martin DF, Kuppermann BD, Wolitz RA, Palestine AG, Li H, Robinson CA (1999) Oral ganciclovir for patients with cytomegalovirus retinitis treated with a ganciclovir implant. Roche Ganciclovir Study Group. N Engl J Med 340(14):1063–70

Matsushima AY, Strauchen JA, Lee G, Scigliano E, Hale EE, Weisse MT, Burstein D, Kamel O, Moore PS, Chang Y (1999) Posttransplantation plasmacytic proliferations related to Kaposi's sarcoma-associated herpesvirus. Am J Surg Pathol 23(11):1393–400

McAllister SC, Hansen SG, Ruhl RA, Raggo CM, DeFilippis VR, Greenspan D, Fruh K, Moses AV (2004) Kaposi sarcoma-associated herpesvirus (KSHV) induces heme oxygenase-1 expression and activity in KSHV-infected endothelial cells. Blood 103(9):3465–73

McCormick C, Ganem D (2005) The kaposin B protein of KSHV activates the p38/MK2 pathway and stabilizes cytokine mRNAs. Science 307(5710):739–41

Mercader M, Taddeo B, Panella JR, Chandran B, Nickoloff BJ, Foreman KE (2000) Induction of HHV-8 lytic cycle replication by inflammatory cytokines produced by HIV-1-infected T cells. Am J Pathol 156(6):1961–71

Mocroft A, Youle M, Gazzard B, Morcinek J, Halai R, Phillips AN (1996) Anti-herpesvirus treatment and risk of Kaposi's sarcoma in HIV infection. Royal Free/Chelsea and Westminster Hospitals Collaborative Group. AIDS 10(10):1101–5

Monini P, Colombini S, Sturzl M, Goletti D, Cafaro A, Sgadari C, Butto S, Franco M, Leone P, Fais S, Leone P, Melucci-Vigo G, Chiozzini C, Carlini F, Ascherl G, Cornali E, Zietz C, Ramazzotti E, Ensoli F, Andreoni M, Pezzotti P, Rezza G, Yarchoan R, Gallo RC, Ensoli B (1999) Reactivation and persistence of human herpesvirus-8 infection in B cells and monocytes by Th-1 cytokines increased in Kaposi's sarcoma. Blood 93(12):4044–58

Moore PS, Chang Y (2003) Kaposi's sarcoma-associated herpesvirus immunoevasion and tumorigenesis: two sides of the same coin? Annu Rev Microbiol 57:609–39

Moses AV, Fish KN, Ruhl R, Smith PP, Strussenberg JG, Zhu L, Chandran B, Nelson JA (1999) Long-term infection and transformation of dermal microvascular endothelial cells by human herpesvirus 8. J. Virol. 73(8):6892–6902

Moses AV, Jarvis MA, Raggo C, Bell YC, Ruhl R, Luukkonen BG, Griffith DJ, Wait CL, Druker BJ, Heinrich MC, Nelson JA, Fruh K (2002a) A functional genomics approach to Kaposi's sarcoma. Ann NY Acad Sci 975:180–91

Moses AV, Jarvis MA, Raggo C, Bell YC, Ruhl R, Luukkonen BGM, Griffith DJ, Wait CL, Druker BJ, Heinrich MC, Nelson JA, Früh K (2002b) KSHV-induced upregulation of the c-Kit proto-oncogene, as identified by gene expression profiling, is essential for the transformation of endothelial cells. J Virol (in press)

Muralidhar S, Pumfery AM, Hassani M, Sadaie MR, Kishishita M, Brady JN, Doniger J, Medveczky P, Rosenthal LJ (1998) Identification of kaposin (open reading frame K12) as a human herpesvirus 8 (Kaposi's sarcoma-associated herpesvirus) transforming gene. J Virol 72(6):4980–8

Muralidhar S, Veytsmann G, Chandran B, Ablashi D, Doniger J, Rosenthal LJ (2000) Characterization of the human herpesvirus 8 (Kaposi's sarcoma-associated herpesvirus) oncogene, kaposin (ORF K12). J Clin Virol 16(3):203–13

Nakamura H, Lu M, Gwack Y, Souvlis J, Zeichner SL, Jung JU (2003) Global changes in Kaposi's sarcoma-associated virus gene expression patterns following expression of a tetracycline-inducible Rta transactivator. J Virol 77(7):4205–20

Naranatt PP, Akula SM, Zien CA, Krishnan HH, Chandran B (2003) Kaposi's sarcoma-associated herpesvirus induces the phosphatidylinositol 3-kinase-PKC-ζ-MEK-ERK signaling pathway in target cells early during infection: implications for infectivity. J Virol 77(2):1524–39

Naranatt PP, Krishnan HH, Svojanovsky SR, Bloomer C, Mathur S, Chandran B (2004) Host gene induction and transcriptional reprogramming in Kaposi's sarcoma-associated herpesvirus (KSHV/HHV-8)-infected endothelial, fibroblast, and B cells: insights into modulation events early during infection. Cancer Res 64(1):72–84

Neipel F, Albrecht JC, Fleckenstein B (1997) Cell-homologous genes in the Kaposi's sarcoma-associated rhadinovirus human herpesvirus 8: determinants of its pathogenicity? J Virol 71(6):4187–92

Nicholas J, Zong JC, Alcendor DJ, Ciufo DM, Poole LJ, Sarisky RT, Chiou CJ, Zhang X, Wan X, Guo HG, Reitz MS, Hayward GS (1998) Novel organizational features, captured cellular genes, and strain variability within the genome of KSHV/HHV8. J Natl Cancer Inst Monogr 23:79–88

Okuno T, Jiang YB, Ueda K, Nishimura K, Tamura T, Yamanishi K (2002) Activation of human herpesvirus 8 open reading frame K5 independent of ORF50 expression. Virus Res 90(1–2):77–89

Pantanowitz L, Dezube BJ, Pinkus GS, Tahan SR (2004) Histological characterization of regression in acquired immunodeficiency syndrome-related Kaposi's sarcoma. J Cutan Pathol 31(1):26–34

Parravicini C, Chandran B, Corbellino M, Berti E, Paulli M, Moore PS, Chang Y (2000) Differential viral protein expression in Kaposi's sarcoma-associated herpesvirus-infected diseases: Kaposi's sarcoma, primary effusion lymphoma, and multicentric Castleman's disease. Am J Pathol 156(3):743–9

Paulose-Murphy M, Ha NK, Xiang C, Chen Y, Gillim L, Yarchoan R, Meltzer P, Bittner M, Trent J, Zeichner S (2001) Transcription program of human herpesvirus 8 (Kaposi's sarcoma-associated herpesvirus). J Virol 75(10):4843–53

Picchio GR, Sabbe RE, Gulizia RJ, McGrath M, Herndier BG, Mosier DE (1997) The KSHV/HHV8-infected BCBL-1 lymphoma line causes tumors in SCID mice but fails to transmit virus to a human peripheral blood mononuclear cell graft. Virology 238(1):22–9

Polstra AM, Cornelissen M, Goudsmit J, van der Kuyl AC (2004) Retrospective, longitudinal analysis of serum human herpesvirus-8 viral DNA load in AIDS-related Kaposi's sarcoma patients before and after diagnosis. J Med Virol 74(3):390–6

Polstra AM, Goudsmit J, Cornelissen M (2003) Latent and lytic HHV-8 mRNA expression in PBMCs and Kaposi's sarcoma skin biopsies of AIDS Kaposi's sarcoma patients. J Med Virol 70(4):624–7

Poole LJ, Yu Y, Kim PS, Zheng QZ, Pevsner J, Hayward GS (2002) Altered patterns of cellular gene expression in dermal microvascular endothelial cells infected with Kaposi's sarcoma-associated herpesvirus. J Virol 76(7):3395–420

Rabkin CS, Janz S, Lash A, Coleman AE, Musaba E, Liotta L, Biggar RJ, Zhuang Z (1997) Monoclonal origin of multicentric Kaposi's sarcoma lesions. N Engl J Med 336(14):988–93

Radkov SA, Kellam P, Boshoff C (2000) The latent nuclear antigen of Kaposi sarcoma-associated herpesvirus targets the retinoblastoma-E2F pathway and with the oncogene Hras transforms primary rat cells. Nat Med 6(10):1121–7

Raggo C, Ruhl R, McAllister S, Koon H, Dezube BJ, Früh K, Moses A (2005) Novel cellular genes essential for transformaton of endothelial cells by Kaposi1s sarcoma-associated herpesvirus. Cancer Res

Rainbow L, Platt GM, Simpson GR, Sarid R, Gao SJ, Stoiber H, Herrington CS, Moore PS, Schulz TF (1997) The 222- to 234-kilodalton latent nuclear protein (LNA) of Kaposi's sarcoma-associated herpesvirus (human herpesvirus 8) is encoded by orf73 and is a component of the latency-associated nuclear antigen. J Virol 71(8):5915–21

Regezi JA, MacPhail LA, Daniels TE, DeSouza YG, Greenspan JS, Greenspan D (1993a) Human immunodeficiency virus-associated oral Kaposi's sarcoma. A heterogeneous cell population dominated by spindle-shaped endothelial cells. Am J Pathol 143(1):240–9

Regezi JA, MacPhail LA, Daniels TE, Greenspan JS, Greenspan D, Dodd CL, Lozada-Nur F, Heinic GS, Chinn H, Silverman S, Jr., et al. (1993b) Oral Kaposi's sarcoma: a 10-year retrospective histopathologic study. J Oral Pathol Med 22(7):292–7

Renne R, Barry C, Dittmer D, Compitello N, Brown PO, Ganem D (2001) Modulation of cellular and viral gene expression by the latency-associated nuclear antigen of Kaposi's sarcoma-associated herpesvirus. J Virol 75(1):458–68

Renne R, Lagunoff M, Zhong W, Ganem D (1996a) The size and conformation of Kaposi's sarcoma-associated herpesvirus (human herpesvirus 8) DNA in infected cells and virions. J Virol 70(11):8151–4

Renne R, Zhong W, Herndier B, McGrath M, Abbey N, Kedes D, Ganem D (1996b) Lytic growth of Kaposi's sarcoma-associated herpesvirus (human herpesvirus 8) in culture. Nat Med 2(3):342–6

Rivas C, Thlick AE, Parravicini C, Moore PS, Chang Y (2001) Kaposi's sarcoma-associated herpesvirus LANA2 is a B-cell-specific latent viral protein that inhibits p53. J Virol 75(1):429–38

Roth WK, Brandstetter H, Sturzl M (1992) Cellular and molecular features of HIV-associated Kaposi's sarcoma [editorial] [published erratum appears in AIDS 1992 Nov;6(11):following 1410]. AIDS 6(9):895–913

Roth WK, Werner S, Risau W, Remberger K, Hofschneider PH (1988) Cultured, AIDS-related Kaposi's sarcoma cells express endothelial cell markers and are weakly malignant in vitro. Int J Cancer 42(5):767–73

Russo JJ, Bohenzky RA, Chien MC, Chen J, Yan M, Maddalena D, Parry JP, Peruzzi D, Edelman IS, Chang Y, Moore PS (1996) Nucleotide sequence of the Kaposi sarcoma-associated herpesvirus (HHV8). Proc Natl Acad Sci USA 93(25):14862–7

Rutgers JL, Wieczorek R, Bonetti F, Kaplan KL, Posnett DN, Friedman-Kien AE, Knowles DM, 2nd (1986) The expression of endothelial cell surface antigens by AIDS-associated Kaposi's sarcoma. Evidence for a vascular endothelial cell origin. Am J Pathol 122(3):493-9

Sadler R, Wu L, Forghani B, Renne R, Zhong W, Herndier B, Ganem D (1999) A complex translational program generates multiple novel proteins from the latently expressed kaposin (K12) locus of Kaposi's sarcoma-associated herpesvirus. J Virol 73(7):5722–30

Salahuddin SZ, Nakamura S, Biberfeld P, Kaplan MH, Markham PD, Larsson L, Gallo RC (1988) Angiogenic properties of Kaposi's sarcoma-derived cells after long-term culture in vitro. Science 242(4877):430–3

Scully PA, Steinman HK, Kennedy C, Trueblood K, Frisman DM, Voland JR (1988) AIDS-related Kaposi's sarcoma displays differential expression of endothelial surface antigens. Am J Pathol 130(2):244–51

Seo T, Park J, Lim C, Choe J (2004) Inhibition of nuclear factor kappaB activity by viral interferon regulatory factor 3 of Kaposi's sarcoma-associated herpesvirus. Oncogene 23(36):6146–55

Serraino D, Toma L, Andreoni M, Butto S, Tchangmena O, Sarmati L, Monini P, Franceschi S, Ensoli B, Rezza G (2001) A seroprevalence study of human herpesvirus type 8 (HHV8) in eastern and Central Africa and in the Mediterranean area. Eur J Epidemiol 17(9):871–6

Skobe M, Brown LF, Tognazzi K, Ganju RK, Dezube BJ, Alitalo K, Detmar M (1999) Vascular endothelial growth factor-C (VEGF-C) and its receptors KDR and flt-4 are expressed in AIDS-associated Kaposi's sarcoma. J Invest Dermatol 113(6):1047–53

Soulier J, Grollet L, Oksenhendler E, Cacoub P, Cazals-Hatem D, Babinet P, d'Agay MF, Clauvel JP, Raphael M, Degos L, et al. (1995) Kaposi's sarcoma-associated herpesvirus-like DNA sequences in multicentric Castleman's disease. Blood 86(4):1276–80

Staskus KA, Sun R, Miller G, Racz P, Jaslowski A, Metroka C, Brett-Smith H, Haase AT (1999) Cellular tropism and viral interleukin-6 expression distinguish human herpesvirus 8 involvement in Kaposi's sarcoma, primary effusion lymphoma, and multicentric Castleman's disease. J Virol 73(5):4181–7

Staskus KA, Zhong W, Gebhard K, Herndier B, Wang H, Renne R, Beneke J, Pudney J, Anderson DJ, Ganem D, Haase AT (1997) Kaposi's sarcoma-associated herpesvirus gene expression in endothelial (spindle) tumor cells. J Virol 71(1):715–9

Staudt MR, Kanan Y, Jeong JH, Papin JF, Hines-Boykin R, Dittmer DP (2004) The tumor microenvironment controls primary effusion lymphoma growth in vivo. Cancer Res 64(14):4790–9

Strauchen JA, Hauser AD, Burstein D, Jimenez R, Moore PS, Chang Y (1996) Body cavity-based malignant lymphoma containing Kaposi sarcoma-associated herpesvirus in an HIV-negative man with previous Kaposi sarcoma. Ann Intern Med 125(10):822–5

Sturzl M, Blasig C, Schreier A, Neipel F, Hohenadl C, Cornali E, Ascherl G, Esser S, Brockmeyer NH, Ekman M, Kaaya EE, Tschachler E, Biberfeld P (1997) Expression of HHV-8 latency-associated T0.7 RNA in spindle cells and endothelial cells of AIDS-associated, classical and African Kaposi's sarcoma. Int J Cancer 72(1):68–71

Sun R, Lin SF, Gradoville L, Miller G (1996) Polyadenylylated nuclear RNA encoded by Kaposi sarcoma-associated herpesvirus. Proc Natl Acad Sci USA 93(21):11883–8

Sun R, Lin SF, Staskus K, Gradoville L, Grogan E, Haase A, Miller G (1999) Kinetics of Kaposi's sarcoma-associated herpesvirus gene expression. J Virol 73(3):2232–42

Swanton C, Mann DJ, Fleckenstein B, Neipel F, Peters G, Jones N (1997) Herpes viral cyclin/Cdk6 complexes evade inhibition by CDK inhibitor proteins. Nature 390(6656):184–187

Thomas JT, Hubert WG, Ruesch MN, Laimins LA (1999) Human papillomavirus type 31 oncoproteins E6 and E7 are required for the maintenance of episomes during the viral life cycle in normal human keratinocytes. Proc Natl Acad Sci USA 96(15):8449–54

Thome M, Schneider P, Hofmann K, Fickenscher H, Meinl E, Neipel F, Mattmann C, Burns K, Bodmer JL, Schroter M, Scaffidi C, Krammer PH, Peter ME, Tschopp J (1997) Viral FLICE-inhibitory proteins (FLIPs) prevent apoptosis induced by death receptors. Nature 386(6624):517–21

Tomescu C, Law WK, Kedes DH (2003) Surface downregulation of major histocompatibility complex class I, PE-CAM, and ICAM-1 following de novo infection of endothelial cells with Kaposi's sarcoma-associated herpesvirus. J Virol 77(17):9669–84

Venetsanakos E, Mirza A, Fanton C, Romanov SR, Tlsty T, McMahon M (2002) Induction of tubulogenesis in telomerase-immortalized human microvascular endothelial cells by glioblastoma cells. Exp Cell Res 273(1):21–33

Verschuren EW, Klefstrom J, Evan GI, Jones N (2002) The oncogenic potential of Kaposi's sarcoma-associated herpesvirus cyclin is exposed by p53 loss in vitro and in vivo. Cancer Cell 2(3):229–41

Vieira J, O'Hearn P, Kimball L, Chandran B, Corey L (2001) Activation of Kaposi's sarcoma-associated herpesvirus (human herpesvirus 8) lytic replication by human cytomegalovirus. J Virol 75(3):1378–86

Vieira J, O'Hearn PM (2004) Use of the red fluorescent protein as a marker of Kaposi's sarcoma-associated herpesvirus lytic gene expression. Virology 325(2):225–40

Viejo-Borbolla A, Ottinger M, Schulz TF (2003) Human herpesvirus 8: biology and role in the pathogenesis of Kaposi's sarcoma and other AIDS-related malignancies. Curr Infect Dis Rep 5(2):169–175

Wang HW, Trotter MW, Lagos D, Bourboulia D, Henderson S, Makinen T, Elliman S, Flanagan AM, Alitalo K, Boshoff C (2004) Kaposi sarcoma herpesvirus-induced cellular reprogramming contributes to the lymphatic endothelial gene expression in Kaposi sarcoma. Nat Genet 36(7):687–93

Weninger W, Partanen TA, Breiteneder-Geleff S, Mayer C, Kowalski H, Mildner M, Pammer J, Sturzl M, Kerjaschki D, Alitalo K, Tschachler E (1999) Expression of vascular endothelial growth factor receptor-3 and podoplanin suggests a lymphatic endothelial cell origin of Kaposi's sarcoma tumor cells. Lab Invest 79(2):243–51

Werness BA, Levine AJ, Howley PM (1990) Association of human papillomavirus types 16 and 18 E6 proteins with p53. Science 248(4951):76–9

West JT, Wood C (2003) The role of Kaposi's sarcoma-associated herpesvirus/human herpesvirus-8 regulator of transcription activation (RTA) in control of gene expression. Oncogene 22(33):5150–63

Wigle JT, Oliver G (1999) Prox1 function is required for the development of the murine lymphatic system. Cell 98(6):769–78

Yarchoan R (2004) KSHV induces heme oxygenase: another trick by a wily virus. Blood 103(9):3252–3253

Yu Y, Black JB, Goldsmith CS, Browning PJ, Bhalla K, Offermann MK (1999) Induction of human herpesvirus-8 DNA replication and transcription by butyrate and TPA in BCBL-1 cells. J Gen Virol 80 (Pt 1):83–90

Zenger E, Abbey NW, Weinstein MD, Kapp L, Reis J, Gofman I, Millward C, Gascon R, Elbaggari A, Herndier BG, McGrath MS (2002) Injection of human primary effusion lymphoma cells or associated macrophages into severe combined immunodeficient mice causes murine lymphomas. Cancer Res 62(19):5536–42

Zerfass-Thome K, Zwerschke W, Mannhardt B, Tindle R, Botz JW, Jansen-Durr P (1996) Inactivation of the cdk inhibitor p27KIP1 by the human papillomavirus type 16 E7 oncoprotein. Oncogene 13(11):2323–30

Zhong W, Wang H, Herndier B, Ganem D (1996) Restricted expression of Kaposi sarcoma-associated herpesvirus (human herpesvirus 8) genes in Kaposi sarcoma. Proc. Natl. Acad. Sci. USA 93(13):6641–6646

Zhu J, Trang P, Kim K, Zhou T, Deng H, Liu F (2004) Effective inhibition of Rta expression and lytic replication of Kaposi's sarcoma-associated herpesvirus by human RNase P. Proc Natl Acad Sci USA 101(24):9073–8

CTMI (2006) 312:245–262

KSHV After an Organ Transplant: Should We Screen?

A.-G. Marcelin[1] (✉) · V. Calvez[1] · E. Dussaix[2]

[1]Department of Virology, Pitié-Salpêtrière Hospital, Pierre et Marie Curie University, UPRES EA 2387, 83 bd de l'Hôpital, 75013 Paris, France
anne-genevieve.marcelin@psl.ap-hop-paris.fr

[2]Department of Microbiology, Paul Brousse Hospital, Paris-Sud XI University, UPRES 3541, 94804 Villejuif, France

Abstract The incidence of Kaposi sarcoma (KS) related to Kaposi sarcoma-associated herpesvirus (KSHV/HHV-8) after organ transplantation is 500–1,000 times greater than in the general population, and its occurrence is associated with immunosuppressive therapy. The reported incidence of posttransplant KS ranges from 0.5% to 5%, depending on the patient's country of origin and the type of organ received, mainly after renal transplantation. Posttransplant KS is caused by two possible mechanisms: KSHV reactivation in patients who were infected before the graft and KSHV contamination from the infected organ's donor to the recipient. KSHV reactivation appears to play a greater role in the risk of KS than incident infections. However, some studies, with findings based not only on serological data but also on molecular tracing of the viral infection, have shown that organ-related transmission of KSHV could be more common than previously thought and associated in some cases with severe KSHV-related disease. Precise estimates of KSHV seroprevalence in the organ donor and recipient populations in different countries are lacking. However, studies have reported seroprevalences among donors and recipients that are similar to those among the general population of the country considered. Many studies have suggested the potential utility of screening of KSHV antibodies among organ donors and recipients. However, to date the results of these studies have argued in favor of KSHV screening, even in low-KSHV infection prevalence countries, not to exclude the graft but to have the KSHV status information in order to have the opportunity to monitor, clinically and biologically, patients at risk for KSHV-related disease development. The detection of KSHV antibodies could be done in the days after the transplantation and the results transmitted to the physicians retrospectively. In conclusion, the question of screening donors and recipients for KSHV, even in low-KSHV infection prevalence countries, is still debated, and prospective studies are needed to evaluate the benefit of pre- and posttransplantation strategies.

1 Introduction

Despite great advances in stem cell and organ transplantation, opportunistic infections are a major cause of morbidity and mortality for transplant recipients. Herpesviruses remain latent in the body after primary infection and can reactivate under immunosuppression conditions, such as after transplantation, making them particularly important pathogens in transplant recipients. As no effective antiviral drug is yet available for EBV and KSHV/HHV-8, recovery of the host immune response against these viruses is crucial in controlling virus-associated tumors in transplant recipients. Skin cancers are the most common malignant conditions in transplant recipients (Euvrard et al. 2003). Focusing on KSHV, the incidence of Kaposi sarcoma (KS) after organ transplantation is 500–1,000 times greater than in the general population (Harwood et al. 1979; Woodle et al. 2001). The occurrence of KS among transplant recipients is associated with immunosuppressive therapy, as evidenced by the remission of KS lesions after the reduction or withdrawal of immunosuppressive therapy (Nagy et al. 2000). The reported incidence of posttransplant KS ranges from 0.5% to 5%, depending on the patient's country of origin and the type of organ received (Jenkins et al. 2002; Qunibi et al. 1988) (Table 1).

Posttransplant KS develops in solid organ transplant recipients, mainly after renal transplantation, and is caused by two possible mechanisms: KSHV reactivation in patients who were infected before the graft and KSHV contamination from the infected organ's donor to the recipient. The proportion

Table 1 Incidence of KS in series of kidney transplant recipients from different countries

Series origin	Incidence of KS (%)
Germany (Behrend et al. 1997)	0.26
Spain (Gomez dos Santos et al. 1997; Munoz et al. 2002)	0.2–0.5
USA (Penn 1997)	0.5
Canada (Shepherd et al. 1997)	0.6
Italy (Montagnino et al. 1996)	0.7–3.3
France (Farge 1993; Frances 1998)	0.4–2.4
Israel (Shmueli et al. 1989)	2.4
Turkey (Ecder et al. 1998)	3
South Africa (Margolius et al. 1994)	4–5
Saudi Arabia (Qunibi et al. 1988)	4–5.3

KS, Kaposi sarcoma

of cases caused by each of these mechanisms varies significantly in relation to the prevalence of KSHV antibodies in the populations studied. In populations with a high prevalence of infection, such as in Italy, posttransplant KS was mainly caused by reactivation of KSHV, whereas in areas of lower prevalence, such as Switzerland, it was due to primary infections (Parravicini et al. 1997; Regamey et al. 1998). However, some studies, with findings based not only on serological data but also on molecular tracing of the viral infection, have shown that organ-related transmission of KSHV could be more common than previously thought. Moreover, other routes of transmission, such as blood products, could not be excluded as a possible source of contamination.

2 Mechanisms of Posttransplant KS Development in Solid Organ Transplant Recipients

2.1 Incidence of Posttransplant KS in Solid Organ Transplant Recipients

The incidence of KS in transplant patients depends mainly on the epidemiology of KSHV in the population involved and on the type of transplantation.

There is a 1%–3% risk of KS with kidney transplantation (Table 1), the median interval to diagnosis of KS being 29–31 months (Frances et al. 1999; Rabkin et al. 1999). In Saudi Arabia, where there is a higher prevalence of

KSHV, the transplantation risk of KS rises to 3%–5% (Qunibi et al. 1988). It may also reach high figures in transplant patients reported from other ethnic groups with a high incidence of KSHV infection, such as Mediterranean and Middle Eastern patients. For example, in one series, the overall incidence among transplant patients was 0.78% but up to 4% among those of Jewish or Mediterranean origin (Harwood et al. 1979). Another key factor is the type of transplantation. Incidence ranges from 0.2% to 2.8% in liver recipients (Besnard et al. 1996; Colina et al. 1996) and from 0.41% to 11% in heart transplant patients (Briz et al. 1998; Frances et al. 1991; Goldstein et al. 1995; Lanza et al. 1983; Zahger et al. 1993); it was 2% after kidney-pancreas transplantation (Kaufman et al. 1999), and just two case reports on lung recipients have been published (Sachsenberg-Studer et al. 1999; Sleiman et al. 1997). In a study from France involving 7,923 transplant recipients, the overall prevalence of KS was 0.52% but, in contrast to other studies, the occurrence of KS was more common in liver transplant recipients (1.24%) than in recipients of kidneys (0.45%) or hearts (0.41%) (Farge 1993).

2.2 KSHV Reactivation After Organ Transplantation

2.2.1 Kidney Transplantation

Most of cases of posttransplant KS apparently develop as a result of viral reactivation, because more than 80% of transplant recipients with KS are seropositive for KSHV before undergoing transplantation (Cattani et al. 2001) (Table 2). In an Italian retrospective study, 10 of 11 organ recipients (91%) who developed KS were KSHV seropositive before transplantation (Parravicini et al. 1997). These data are supported by those of another study, which showed that KS developed in three of 21 recipients (14.3%) who were KSHV seropositive before kidney transplantation (Andreoni et al. 2001). It has been shown that among patients who are seropositive for KSHV before undergoing kidney transplantation the risk of posttransplant KS is 23% to 28%, as compared with a risk of 0.7% in patients who are seronegative before receiving a kidney transplant (Cattani et al. 2001; Frances et al. 2000). A French retrospective study performed between 1990 and 1996 showed that 32 patients (8%) had antibodies to KSHV at the time of transplantation and that among these 32 patients, 3 years after transplantation, graft survival was 72% and KS prevalence was 28% (KS incidence: 8.2/yr/100 KSHV+ recipients). Multivariate analysis identified severe bacterial and/or *Pneumocystis carinii* infection (odds ratio: 8.6; P=0.019) as a factor associated with KS. This

Table 2 Solid organ transplant recipients with KSHV reactivation

Reference	Type of T.	Country	No. of recipients KSHV positive before T. (%)	No. of KS patients (%)
Cattani et al.	Kidney	Italy	26/175 (15)	6/26 (23)
Frances et al.	Kidney	France	32/400 (8)	9/32 (30)
Andreoni et al.	Kidney and liver	Italy	21/130 (16)	3/21 (14)
Marcelin et al.	Liver	France	3/122 (2.5)	0/3 (0)
Aseni et al.	Liver	Italy	4/459 (0.9)	4/4 (100)
Emond et al.	Heart	France	4/150 (2.7)	1/4 (25)
Sachsenberg et al.	Lung	Switzerland	1 recipient	1

No., number; T., transplantation; KSHV, Kaposi sarcoma-associated herpesvirus; KS, Kaposi sarcoma

suggests that the level of immunosuppression may be an important factor in the development of posttransplant KS.

2.2.2 Other Types of Organ Transplantation

Kaposi sarcoma (KS) has been reported after solid organ transplantation mostly in recipients of renal graft but also in liver, heart, lung, and bone allografts.

Concerning liver transplantation, an Italian study on 130 kidney and liver recipients reported a KSHV prevalence rate of 16.1% before transplantation (Andreoni et al. 2001). In this study, none of the liver transplant recipients developed KS after transplantation. Similarly, among 122 French liver recipients, antibodies to KSHV were detected in sera of three recipients (2.4%) before transplantation, and none of these three recipients developed a KS during the follow-up (Marcelin et al. 2004). However, another Italian study reported the clinical course and outcome of five adult patients who underwent liver transplantation and developed KS (months 9–23 after transplantation) among 459 studied patients. Four of five patients died of KS, surviving 0–6 months after pathological diagnosis, suggesting a high rate of graft involvement and mortality in this setting (Aseni et al. 2001).

In the context of heart transplantation, several cases have been reported. For example, a case report has described a primary effusion lymphoma associated with KSHV infection in a cardiac transplant recipient, who was probably infected before the graft (Jones et al. 1998). Later, a French study reported a low rate of KS in heart transplant recipients: One of 150 patients (0.67%) who was seropositive for KSHV before transplantation developed iatrogenic KS after heart transplantation. This low rate could be explained by the fact that KSHV is not endemic in France in the general population (prevalence of 2%) (Marcelin et al. 1998).

Very few cases of patients developing KS after lung transplantation have been described. A case report described a KS of the skin, but also of the lung graft. Serological tests showed KSHV seronegativity of the graft donor and KSHV seropositivity of the patient before lung transplantation, suggesting that the latter was already infected before the surgery and that immunosuppression resulted in the development of KS (Huang et al. 2003; Sachsenberg-Studer et al. 1999).

These data suggest that KS occurs predominantly among kidney allograft recipients, but that recipients of other organs are also at risk for KSHV-related disease, even those living in an area in which the virus is not very prevalent.

Finally, substantial morbidity and mortality are associated with a diagnosis of KS among transplant recipients. In a large transplant tumor registry, 143 (40.2%) of 356 patients with KS had visceral involvement, and 61 (17.1%) of 356 had KS listed as their cause of death (Penn 1997). Although the reduction or cessation of immunosuppressive treatment can lead to the complete remission of KS, among patients undergoing remission 65% had graft loss or impaired function, compared with 21% of the overall population of transplant recipients.

2.3 KSHV Contamination After Organ Transplantation

2.3.1 Kidney Transplantation

The first report suggesting some evidence that de novo infection from transplantation can occur reported the case of a kidney transplant patient who was KSHV seronegative before transplantation and who seroconverted 13 months after transplantation with onset of Castleman disease (Parravicini et al. 1997). Other KSHV seroconversions after kidney transplantation were identified in Switzerland. In a cohort of 220 transplant recipients, KS developed within 26 months in two of 25 patients (8%) who had developed a primary KSHV infection after kidney transplantation (Regamey et al. 1998). These studies have suggested for the first time that transmission of KSHV can occur by the organ allograft (Table 3). This was further confirmed in other studies, showing rates of KSHV seroconversion after kidney transplantation of 2%

Table 3 Solid organ transplant recipients with KSHV transmission from the donor

Reference	Type of T.	Country	Cases of transmission or rate of seroconversion (%)	No. of KS patients after seroconversion (%)
Luppi et al.	Kidney	Italy	2 patients	2/2
Regamey et al.	Kidney	Switzerland	25/220 (17)	2/25 (8)
Milliancourt et al.	Kidney	France	6/287 (2)	0/6 (0)
Munoz et al.	Kidney, liver and heart	Spain	3/59 (5)	4/70 (5.7)
Emond et al.	Heart	France	1/146 (0.7)	0/1 (0)
Marcelin et al.	Liver	France	4/122 (0.3)	2/4 (50)

No., number; T., transplantation; KS, Kaposi's sarcoma

and 5% in France and Spain, respectively (Milliancourt et al. 2001; Munoz et al. 2002). Moreover, a primary KSHV infection has been described in two patients after both received kidneys from a seropositive cadaver donor. One of the patients developed disseminated KS, and the other developed an acute syndrome characterized by fever with plasmacytosis followed by bone marrow failure and ultimately died of renal and cardiac failure (Luppi et al. 2000). In a more recent study, sex and genetics markers specific to the donor were detected in neoplastic KSHV-infected cells in KS lesions isolated from renal transplant recipients. This finding suggests that not only KSHV but also KS progenitor cells may be seeded after solid organ transplantation, survive in the recipients, and undergo neoplastic transformation and progression (Barozzi et al. 2003).

2.3.2
Other Types of Organ Transplantation

Similar low rates of KSHV seroconversions have been reported after heart and liver transplantations. Among a group of 150 patients awaiting heart transplants only one of them seroconverted for KSHV antibodies after transplantation, without developing KS (Emond et al. 2002). In the context of liver transplantation, among 122 liver donors and their respective recipients, four primary KSHV infections were detected among KSHV-seronegative recipients who received a liver from a KSHV-positive donor. Among these four recipients, two particularly immunosuppressed patients developed symptomatic diseases and died a few months after transplantation, harboring disseminated KS and KSHV-positive lymphoproliferation. In these two patients, KSHV DNA genome sequences were detectable in peripheral blood mononuclear cells (PBMCs) and other tissues with high viremia levels before and at the beginning of KSHV-related diseases (Marcelin et al. 2004). This suggests that KSHV serological assay and PCR should be used in combination to monitor recipients having received an organ from a KSHV-positive donor (Pellet et al. 2002).

In conclusion, even if KSHV primary infections after organ transplantation seem to be rare events, this can lead in some cases to a fatal outcome.

2.3.3 KSHV Serology in the Context of Seroconversion

KSHV is not ubiquitous in the general healthy population, at least in United States and in Northern Europe, and therefore serology is more useful than other methods in detecting primary infections. Current serological assays are immunofluorescence assays (IFA) using KSHV-infected B cell lines and enzyme-linked immunosorbent assays (EIA) that use isolated KSHV proteins or peptides. Both types of assays can detect antibodies directed against viral proteins expressed during a lytic, replicative cycle or during latency. KSHV antigens that are targeted by the serological response come in two broad categories: latent and lytic. The major latent antigen is latent nuclear antigen 1, which is encoded by open reading frame (ORF)-73. The major lytic antigens are ORF65 (a capsid protein) and K8.1 (a membrane-associated glycoprotein). Although there is some debate regarding which assay is most accurate, the IFA using lytic cycle antigens appears to be the most sensitive assay (Spira et al. 2000; Zhu et al. 1999). Indeed, the rate of seroconversion could be underestimated with the use of only IFA latent assays. This was highlighted in a study in which no seroconversion event was observed with latent IFA assay even in two liver recipients who developed documented symptomatic primary KSHV infection (Marcelin et al. 2004). However, with lytic IFA assay, all the patients who received grafts from KSHV-infected donors seroconverted for KSHV. This can be related to the fact that antibodies targeted against lytic antigens appeared before antibodies directed against latent antigens, which is well known for EBV. Thus the use of a combination of both latent and lytic assays to evaluate seropositivity should avoid the possibility of false negative results and provide an exact evaluation of the rate of seroconversion after transplantation.

3 Risk Factors Associated with Posttransplant KS Development in Solid Organ Transplant Recipients

Almost all solid organ transplant patients with KS have been found to be seropositive against KSHV, with doubtful exceptions (Briz et al. 1998; Demirag et al. 1998; Farge et al. 1999; Qunibi et al. 1998; Regamey et al. 1998). However, several factors may influence the risk of developing the disease in patients with prior infection, including the ethnic origin of the patients, the seroprevalence rate in their geographic origin, sex, age, and degree of immunosuppression (Calabro et al. 1998; Gao et al. 1996b; Lennette et al. 1996; Penn 1997). Incidence of posttransplant KS is higher in countries or ethnic groups with high seroprevalence against KSHV (Farge et al. 1999). Seroprevalence of KSHV antibodies varies from 2% to 50%. KSHV antibodies have been found in 2% to 14% of studied subjects in Europe (Belec et al. 1998; Frances et al. 1999; Marcelin et al. 1998; Regamey et al. 1998; Whitby et al. 1998) and in 50% of patients with renal transplants in a study in Texas (Hudnall et al. 1998). KSHV antibodies exceed 50% in some African countries (Gao et al. 1996a; Mayama et al. 1998; Plancoulaine et al. 2000).

KS appears in younger patients compared with other posttransplant malignancies, with a median age around 35–40 years (Euvrard et al. 2003). Some pediatric cases have also been reported (Fournet et al. 1992; Penn 1994). Male patients predominate in most series and types of KS, with male-to-female ratios around 2.5–4.0 to 1 in solid organ transplant recipients (Ecder et al. 1998; Montagnino et al. 1994). Genetic predisposition related to the presence of some HLA antigens in KS patients is no longer sustained (Brunson et al. 1990; Qunibi et al. 1988; Sungur et al. 1995; Zong et al. 1999).

Finally, the degree of immunosuppression is a critical factor. It has been demonstrated that renal transplant recipients have a higher incidence of KS than nongrafted patients on maintenance dialysis (Montagnino et al. 1996; Qunibi et al. 1998). It is not clear, however, how the use of different immunosuppressive agents influences the development of KS. There is no definitive evidence of a higher rate of KS related to the use of a specific drug. In vitro studies have shown that immunosuppressive agents such as hydrocortisone and cyclosporine do not activate the lytic cycle of KSHV and do not modify the cell survival, thus promoting cancer progression by a direct cellular effect (Marcelin et al. 2001). However, administration of more than 10 corticosteroid pulses has been claimed to correlate with a higher incidence of the disease (Montagnino et al. 1996). In the same way, a severe KSHV-related disease was reported in two recipients who received a liver from a KSHV positive donor and who had a stronger immunosuppression compared to the other recipients, which was induced by higher tacrolimus trough levels and use of steroid boluses (Marcelin et al. 2004). The relative influence of different immunosuppressive agents is not totally clear (Cockburn and Krupp 1989; Eberhard et al. 1999; Farge et al. 1999; Hiesse et al. 1995; Rezeig et al. 1997). Poor donor-recipient matching has not been found to be a risk factor for KS (28).

4 KSHV Screening and Monitoring in the Context of Solid Organ Transplant Recipients

4.1 KSHV Screening

In the light of seroepidemiological studies showing high prevalence of KS within the studied populations of KSHV-positive transplant recipients, the possibility of transmission of the virus through the graft leading in some cases to fatal outcome, and finally the low graft survival, it is tempting to raise the question of the benefit of systematic screening of KSHV serological features before transplantation. Although studies have indicated that KSHV transmission by the organ allograft is uncommon, serological screening of donors for KSHV infection may be necessary, especially in geographic areas with high seroprevalence of KSHV infection. Although most cases of KS in transplant recipients are the result of KSHV reactivation, avoiding matches between KSHV-positive donors and KSHV-negative recipients would prevent some occurrence of KS. Although the current evidence is insufficient to recommend against performing a transplant on the basis of KSHV mismatches, cost/benefit analyses for strategies are needed that account for transplant type (renal, liver, etc.) and regional prevalence of KSHV infection. Regardless of whether donor/recipient seromatching proves to be beneficial, the use of posttransplantation KSHV diagnostics to identify recipients at high risk is likely to provide a benefit because of heightened surveillance and the possible prophylaxis or treatment for KS.

4.2 KSHV Monitoring

Knowledge of the KSHV infection status of graft recipients could help in identifying individuals at high risk for posttransplantation KS and in carefully monitoring them clinically and biologically. Moreover, the systematic screening of donors for KSHV, not to exclude the graft but rather to know the KSHV serological status and to monitor the recipient adequately, could be helpful. At the moment, routine virological assessment of organ donors includes, at least in most countries in Europe, HIV, HCV, and HTLV1 serology, where positivity excludes from organ donation. HBV testing comprises HBsAg, anti-HBs, and anti-HBc detection. Although HBsAg positivity excludes from organ donation, anti-HBc-positive donors may be used under certain life-threatening circumstances, even if their infectivity is clearly established (Roque-Afonso et al. 2002). However, HBV prophylaxis with antiviral agents or immunoglobulins may prevent HBV reactivation from the liver graft. A similar strategy could be developed for KSHV in the context of organ transplantation, to avoid the occurrence of severe KSHV-related disease, combining use of serological assays in donors and recipients and molecular detection of KSHV DNA sequences in recipients of positive donors.

This strategy could allow the use of several potential interventions suggested by our current understanding of the relationship of the virus to disease. The utility and efficacy of such interventions will need careful clinical evaluation and refinement as part of the process of developing a new standard of care for immunocompromised patients who are infected with KSHV. For example, reduction or discontinuation of immunosuppressive drugs is usually associated with regression of lesions, and this strategy could be proposed preventively in cases of high risk of development of KSHV-related disease, for example, in patients harboring a positivation and/or increase of KSHV viral load. Moreover, the value of an antiviral prophylaxis to lower the risk of KS needs to be explored. There is evidence to suggest that prophylaxis with antiherpes drugs, such as foscarnet, ganciclovir, and cidofovir, protects against the development of KS. Although no drugs are currently licensed for KSHV therapy, in a large clinical trial Martin et al. showed that, among HIV-infected patients treated for cytomegalovirus retinitis, both oral and intravenous ganciclovir led to significant decreases in the incidence of KS (Martin et al. 1999). The benefit in term of prophylaxis of some drugs, such as val-ganciclovir, which is used to prevent cytomegalovirus infection in transplant recipients, should be evaluated in KSHV-infected organ transplant recipients. Two recent studies reported remission of multicentric Castleman disease with anti-CD20 monoclonal antibody infusion correlating with reduction of KSHV DNA viral load in PBMCs (Corbellino et al. 2001; Marcelin et al. 2003). Thus the use of anti-CD20 monoclonal antibody could be evaluated in cases of documented primary KSHV infection with the aim of decreasing the risk of severe KSHV-related disease.

4.3 Limitations

However, before recommending systematic screening of KSHV in candidates for transplantation and organ donors, several questions remain to besolved.

First, there is a lack of large prospective studies evaluating the prevalence of KSHV in organ donors and recipients and the exact risk of posttransplant KS and the related rate of morbidity and mortality. Such a prospective study is ongoing in France including all the transplantation centers. Preliminary data have been presented and showed that KSHV prevalence among kidney donors and recipients was 1.05% (40/3,819 cases) and 2.87% (165/5,749 cases), respectively. To date, 21/165 (13%) of patients developed posttransplant KS among patients KSHV seropositive before transplantation (median follow-up of 608 days; range 0–1,377). The utility of the measurement of KSHV viremia as a predictor of the development of KS will be evaluated. Indeed, periodic monitoring of the KSHV load in PBMCs could help to identify the subset of patients who are at the highest risk for KS, and early detection may be important, because KS appears to be more likely to improve if it is treated at an early stage. This has been suggested by studies that found a lower KS tumor burden to be a predictor of a complete KS response to HAART in HIV-infected patients (Dupin et al. 1999) and a predictor of complete KS remission in organ transplant recipients who stop taking their immunosuppressive therapy (Shepherd et al. 1997).

Second, there is still some debate regarding which KSHV serological assay is most accurate. Because no “gold standard” for determining positivity for KSHV antibodies is yet available and there is a lack of standardization and commercial availability, it is difficult to recommend specific assays or formats for clinical use. However, the combination of a lytic and a latent nuclear antigen IFA should provide the best combination of sensitivity and specificity (Ablashi et al. 2002; Schatz et al. 2001). The use of multiple assays is particularly important in determining prevalence in asymptomatic populations (Biggar et al. 2003). The performance of such combined assays indicates that they may be used for the clinical management of individuals at risk of developing KSHV-associated tumors such as allograft recipients. Moreover, most of these assays are manual and are not adapted to emergency situations, as no EIA simultaneously had sensitivity and specificity above 90% (Engels et al. 2000). Many studies have suggested the potential utility of screening of KSHV antibodies among organ donors and recipients. However, to date the results of these studies have argued in favor of KSHV screening, even in low-KSHV infection prevalence countries, not to exclude the graft but to obtain KSHV status information in order to have the opportunity to monitor, clinically and biologically, patients at risk for KSHV-related disease development. The detection of KSHV antibodies could be done in the days after the transplantation and the results transmitted to physicians retrospectively.

In conclusion, the question of screening donors and recipients for KSHV, even in low-KSHV infection prevalence countries, is still debated, and prospective studies are needed to evaluate the benefit of pre- and posttransplantation strategies.

References

Ablashi DV, Chatlynne LG, Whitman JE Jr, Cesarman E (2002) Spectrum of Kaposi's sarcoma-associated herpesvirus, or human herpesvirus 8, diseases. *Clin Microbiol Rev* 15, 439–64

Andreoni M, Goletti D, Pezzotti P, Pozzetto A, Monini P, Sarmati L, Farchi F, Tisone G, Piazza A, Pisani F, Angelico M, Leone P, Citterio F, Ensoli B, Rezza G (2001) Prevalence, incidence and correlates of HHV-8/KSHV infection and Kaposi's sarcoma in renal and liver transplant recipients. *J Infect* 43, 195–9

Aseni P, Vertemati M, Minola E, Arcieri K, Bonacina E, Camozzi M, Osio C, Forti D (2001) Kaposi's sarcoma in liver transplant recipients: morphological and clinical description. *Liver Transpl* 7, 816–23

Barozzi P, Luppi M, Facchetti F, Mecucci C, Alu M, Sarid R, Rasini V, Ravazzini L, Rossi E, Festa S, Crescenzi B, Wolf DG, Schulz TF, Torelli G (2003) Post-transplant Kaposi sarcoma originates from the seeding of donor-derived progenitors. *Nat Med* 9, 554–61

Behrend M, Kolditz M, Kliem V, Oldhafer KJ, Brunkhorst R, Frei U, Pichlmayr R (1997) Malignancies in patients under long-term immunosuppression after kidney transplantation. *Transplant Proc* 29, 834–5

Belec L, Cancre N, Hallouin MC, Morvan J, Si Mohamed A, Gresenguet G (1998) High prevalence in Central Africa of blood donors who are potentially infectious for human herpesvirus 8. *Transfusion* 38, 771–5

Besnard V, Euvrard S, Kanitakis J, Mion F, Boillot O, Frances C, Faure M, Claudy A (1996) Kaposi's sarcoma after liver transplantation. *Dermatology* 193, 100–4

Biggar RJ, Engels EA, Whitby D, Kedes DH, Goedert JJ (2003) Antibody reactivity to latent and lytic antigens to human herpesvirus-8 in longitudinally followed homosexual men. *J Infect Dis* 187, 12–8

Briz M, Alonso-Pulpon L, Crespo-Leiro MG, Exposito C, Almagro M, Busto MJ, Fernandez MN (1998) Detection of herpesvirus-like sequences in Kaposi's sarcoma from heart transplant recipients. *J Heart Lung Transplant* 17, 288–93

Brunson ME, Balakrishnan K, Penn I (1990) HLA and Kaposi's sarcoma in solid organ transplantation. *Hum Immunol* 29, 56–63

Calabro ML, Sheldon J, Favero A, Simpson GR, Fiore JR, Gomes E, Angarano G, Chieco-Bianchi L, Schulz TF (1998) Seroprevalence of Kaposi's sarcoma-associated herpesvirus/human herpesvirus 8 in several regions of Italy. *J Hum Virol* 1, 207–13

Cattani P, Capuano M, Graffeo R, Ricci R, Cerimele F, Cerimele D, Nanni G, Fadda G (2001) Kaposi's sarcoma associated with previous human herpesvirus 8 infection in kidney transplant recipients. *J Clin Microbiol* 39, 506–8

Cockburn IT, Krupp P (1989) The risk of neoplasms in patients treated with cyclosporine A. *J Autoimmun* 2, 723–31

Colina F, Lopez-Rios F, Lumbreras C, Martinez-Laso J, Garcia IG, Moreno-Gonzalez E (1996) Kaposi's sarcoma developing in a liver graft. *Transplantation* 61, 1779–81

Corbellino M, Bestetti G, Scalamogna C, Calattini S, Galazzi M, Meroni L, Manganaro D, Fasan M, Moroni M, Galli M, Parravicini C (2001) Long-term remission of Kaposi sarcoma-associated herpesvirus-related multicentric Castleman disease with anti-CD20 monoclonal antibody therapy. *Blood* 98, 3473–5

Demirag A, Hizel N, Karakayali H, Moray G, Akkoc H, Bilgin N, Haberal M (1998) Kaposi sarcoma-associated herpesvirus/human herpesvirus type 8 and Epstein-Barr virus in immunosuppressed renal transplant recipients. *Transplant Proc* 30, 3166–7

Dupin N, Rubin De Cervens V, Gorin I, Calvez V, Pessis E, Grandadam M, Rabian C, Viard JP, Huraux JM, Escande JP (1999) The influence of highly active antiretroviral therapy on AIDS-associated Kaposi's sarcoma. *Br J Dermatol* 140, 875–81

Eberhard OK, Kliem V, Brunkhorst R (1999) Five cases of Kaposi's sarcoma in kidney graft recipients: possible influence of the immunosuppressive therapy. *Transplantation* 67, 180–4

Ecder ST, Sever MS, Yildiz A, Turkmen A, Kayacan SM, Kilicaslan I, Kocak T, Eldegez U (1998) Kaposi's sarcoma after renal transplantation in Turkey. *Clin Transplant* 12, 472–5

Emond JP, Marcelin AG, Dorent R, Milliancourt C, Dupin N, Frances C, Agut H, Gandjbakhch I, Calvez V (2002) Kaposi's sarcoma associated with previous human herpesvirus 8 infection in heart transplant recipients. *J Clin Microbiol* 40, 2217–9

Engels EA, Whitby D, Goebel PB, Stossel A, Waters D, Pintus A, Contu L, Biggar RJ, Goedert JJ (2000) Identifying human herpesvirus 8 infection: performance characteristics of serologic assays. *J Acquir Immune Defic Syndr* 23, 346–54

Euvrard S, Kanitakis J, Claudy A (2003) Skin cancers after organ transplantation. *N Engl J Med* 348, 1681–91

Farge D (1993) Kaposi's sarcoma in organ transplant recipients. The Collaborative Transplantation Research Group of Ile de France. *Eur J Med* 2, 339–43

Farge D, Lebbe C, Marjanovic Z, Tuppin P, Mouquet C, Peraldi MN, Lang P, Hiesse C, Antoine C, Legendre C, Bedrossian J, Gagnadoux MF, Loirat C, Pellet C, Sheldon J, Golmard JL, Agbalika F, Schulz TF (1999) Human herpes virus-8 and other risk factors for Kaposi's sarcoma in kidney transplant recipients. Groupe Cooperatif de Transplantation d' Ile de France (GCIF). *Transplantation* 67, 1236–42

Fournet JC, Peuchmaur M, Eckart P, Gagnadoux MF, Stephan JL, Hubert P, Niaudet P, Brousse N (1992) Multicentric Kaposi's sarcoma in a 5-year-old human immunodeficiency virus-negative renal allograft recipient. *Hum Pathol* 23, 956–60

Frances C (1998) Kaposi's sarcoma after renal transplantation. *Nephrol Dial Transplant* 13, 2768–73

Frances C, Farge D, Desruennes M, Boisnic S (1991) Kaposi's sarcoma in after heart transplantation. *Ann Dermatol Venereol* 118, 864–6

Frances C, Mouquet C, Calvez V (1999) Human herpesvirus 8 and renal transplantation. *N Engl J Med* 340, 1045; author reply 1046

Frances C, Mouquet C, Marcelin AG, Barete S, Agher R, Charron D, Benalia H, Dupin N, Piette JC, Bitker MO, Calvez V (2000) Outcome of kidney transplant recipients with previous human herpesvirus-8 infection. *Transplantation* 69, 1776–9

Gao SJ, Kingsley L, Hoover DR, Spira TJ, Rinaldo CR, Saah A, Phair J, Detels R, Parry P, Chang Y, Moore PS (1996a) Seroconversion to antibodies against Kaposi's sarcoma-associated herpesvirus-related latent nuclear antigens before the development of Kaposi's sarcoma. *N Engl J Med* 335, 233–41

Gao SJ, Kingsley L, Li M, Zheng W, Parravicini C, Ziegler J, Newton R, Rinaldo CR, Saah A, Phair J, Detels R, Chang Y, Moore PS (1996b) KSHV antibodies among Americans, Italians and Ugandans with and without Kaposi's sarcoma [see comments]. *Nat Med* 2, 925–8

Goldstein DJ, Williams DL, Oz MC, Weinberg AD, Rose EA, Michler RE (1995) De novo solid malignancies after cardiac transplantation. *Ann Thorac Surg* 60, 1783–9

Gomez dos Santos V, Burgos Revilla FJ, Pascual Santos J, Orofino Ascunce L, Fernandez-Juarez G, Crespo Martinez L, Clemente Ramos L, Carrera Puerta C, Marcen Letosa R, Escudero Barrilero A, Ortuno Mirete J (1997) [Neoplasm prevalence in renal transplantation]. *Arch Esp Urol* 50, 267–73; discussion 273–4

Harwood AR, Osoba D, Hofstader SL, Goldstein MB, Cardella CJ, Holecek MJ, Kunynetz R, Giammarco RA (1979) Kaposi's sarcoma in recipients of renal transplants. *Am J Med* 67, 759–65

Hiesse C, Larue JR, Kriaa F, Blanchet P, Bellamy J, Benoit G, Charpentier B (1995) Incidence and type of malignancies occurring after renal transplantation in conventionally and in cyclosporine-treated recipients: single-center analysis of a 20-year period in 1600 patients. *Transplant Proc* 27, 2450–1

Huang PM, Chang YL, Chen JS, Hsu HH, Ko WJ, Kuo SH, Lee YC (2003) Human herpesvirus-8 associated Kaposi's sarcoma after lung transplantation: a case report. *Transplant Proc* 35, 447–9

Hudnall SD, Rady PL, Tyring SK, Fish JC (1998) Serologic and molecular evidence of human herpesvirus 8 activation in renal transplant recipients. *J Infect Dis* 178, 1791–4

Jenkins FJ, Hoffman LJ, Liegey-Dougall A (2002) Reactivation of and primary infection with human herpesvirus 8 among solid-organ transplant recipients. *J Infect Dis* 185, 1238–43

Jones D, Ballestas ME, Kaye KM, Gulizia JM, Winters GL, Fletcher J, Scadden DT, Aster JC (1998) Primary-effusion lymphoma and Kaposi's sarcoma in a cardiac-transplant recipient. *N Engl J Med* 339, 444–9

Kaufman DB, Leventhal JR, Stuart J, Abecassis MM, Fryer JP, Stuart FP (1999) Mycophenolate mofetil and tacrolimus as primary maintenance immunosuppression in simultaneous pancreas-kidney transplantation: initial experience in 50 consecutive cases. *Transplantation* 67, 586–93

Lanza RP, Cooper DK, Cassidy MJ, Barnard CN (1983) Malignant neoplasms occurring after cardiac transplantation. *JAMA* 249, 1746–8

Lennette ET, Blackbourn DJ, Levy JA (1996) Antibodies to human herpesvirus type 8 in the general population and in Kaposi's sarcoma patients. *Lancet* 348, 858–61

Luppi M, Barozzi P, Schulz TF, Setti G, Staskus K, Trovato R, Narni F, Donelli A, Maiorana A, Marasca R, Sandrini S, Torelli G (2000) Bone marrow failure associated with human herpesvirus 8 infection after transplantation. *N Engl J Med* 343, 1378–85

Marcelin AG, Aaron L, Mateus C, Gyan E, Gorin I, Viard JP, Calvez V, Dupin N (2003) Rituximab therapy for HIV-associated Castleman disease. *Blood* 102, 2786–8

Marcelin AG, Dupin N, Bossi P, Calvez V (1998) Seroprevalence of human herpesvirus-8 in healthy subjects and patients with AIDS-associated and classical Kaposi's sarcoma in France. *AIDS* 12, 539–40

Marcelin AG, Milliancourt C, Dupin N, Wirden M, Huraux JM, Agut H, Calvez V (2001) Effects of cyclosporine and hydrocortisone on Kaposi's sarcoma-associated herpesvirus genome replication and cell apoptosis induction. *Transplantation* 72, 1700–3

Marcelin AG, Roque-Afonso AM, Hurtova M, Dupin N, Tulliez M, Sebagh M, Arkoub ZA, Guettier C, Samuel D, Calvez V, Dussaix E (2004) Fatal disseminated Kaposi's sarcoma following human herpesvirus 8 primary infections in liver-transplant recipients. *Liver Transpl* 10, 295–300

Margolius L, Stein M, Spencer D, Bezwoda WR (1994) Kaposi's sarcoma in renal transplant recipients. Experience at Johannesburg Hospital, 1966–1989. *S Afr Med J* 84, 16–7

Martin DF, Kuppermann BD, Wolitz RA, Palestine AG, Li H, Robinson CA (1999) Oral ganciclovir for patients with cytomegalovirus retinitis treated with a ganciclovir implant. Roche Ganciclovir Study Group [see comments]. *N Engl J Med* 340, 1063–70

Mayama S, Cuevas LE, Sheldon J, Omar OH, Smith DH, Okong P, Silvel B, Hart CA, Schulz TF (1998) Prevalence and transmission of Kaposi's sarcoma-associated herpesvirus (human herpesvirus 8) in Ugandan children and adolescents. *Int J Cancer* 77, 817–20

Milliancourt C, Barete S, Marcelin AG, Mouquet C, Dupin N, Frances C, Agut H, Bitker MO, Calvez V (2001) Human herpesvirus-8 seroconversions after renal transplantation. *Transplantation* 72, 1319–20

Montagnino G, Bencini PL, Tarantino A, Caputo R, Ponticelli C (1994) Clinical features and course of Kaposi's sarcoma in kidney transplant patients: report of 13 cases. *Am J Nephrol* 14, 121–6

Montagnino G, Lorca E, Tarantino A, Bencini P, Aroldi A, Cesana B, Braga M, Lonati F, Ponticelli C (1996) Cancer incidence in 854 kidney transplant recipients from a single institution: comparison with normal population and with patients under dialytic treatment. *Clin Transplant* 10, 461–9

Munoz P, Alvarez P, de Ory F, Pozo F, Rivera M, Bouza E (2002) Incidence and clinical characteristics of Kaposi sarcoma after solid organ transplantation in Spain: importance of seroconversion against HHV-8. *Medicine (Baltimore)* 81, 293–304

Nagy S, Gyulai R, Kemeny L, Szenohradszky P, Dobozy A (2000) Iatrogenic Kaposi's sarcoma: HHV8 positivity persists but the tumors regress almost completely without immunosuppressive therapy. *Transplantation* 69, 2230–1

Parravicini C, Olsen SJ, Capra M, Poli F, Sirchia G, Gao SJ, Berti E, Nocera A, Rossi E, Bestetti G, Pizzuto M, Galli M, Moroni M, Moore PS, Corbellino M (1997) Risk of Kaposi's sarcoma-associated herpes virus transmission from donor allografts among Italian posttransplant Kaposi's sarcoma patients. *Blood* 90, 2826–9

Pellet C, Chevret S, Frances C, Euvrard S, Hurault M, Legendre C, Dalac S, Farge D, Antoine C, Hiesse C, Peraldi MN, Lang P, Samuel D, Calmus Y, Agbalika F, Morel P, Calvo F, Lebbe C (2002) Prognostic value of quantitative Kaposi sarcoma-associated herpesvirus load in posttransplantation Kaposi sarcoma. *J Infect Dis* 186, 110–3

Penn I (1994) De novo tumors in pediatric organ transplant recipients. *Transplant Proc* 26, 1–2

Penn I (1997) Kaposi's sarcoma in transplant recipients. *Transplantation* 64, 669–73

Plancoulaine S, Abel L, van Beveren M, Tregouet DA, Joubert M, Tortevoye P, de The G, Gessain A (2000) Human herpesvirus 8 transmission from mother to child and between siblings in an endemic population. *Lancet* 356, 1062–5

Qunibi W, Akhtar M, Sheth K, Ginn HE, Al-Furayh O, DeVol EB, Taher S (1988) Kaposi's sarcoma: the most common tumor after renal transplantation in Saudi Arabia. *Am J Med* 84, 225–32

Qunibi W, Al-Furayh O, Almeshari K, Lin SF, Sun R, Heston L, Ross D, Rigsby M, Miller G (1998) Serologic association of human herpesvirus eight with posttransplant Kaposi's sarcoma in Saudi Arabia. *Transplantation* 65, 583–5

Rabkin CS, Shepherd FA, Wade JA (1999) Human herpesvirus 8 and renal transplantation. *N Engl J Med* 340, 1045–6

Regamey N, Tamm M, Wernli M, Witschi A, Thiel G, Cathomas G, Erb P (1998) Transmission of human herpesvirus 8 infection from renal-transplant donors to recipients. *N Engl J Med* 339, 1358–63

Rezeig MA, Fashir BM, Hainau B, Al Ashgar HI (1997) Kaposi's sarcoma in liver transplant recipients on FK506: two case reports. *Transplantation* 63, 1520–1

Roque-Afonso AM, Feray C, Samuel D, Simoneau D, Roche B, Emile JF, Gigou M, Shouval D, Dussaix E (2002) Antibodies to hepatitis B surface antigen prevent viral reactivation in recipients of liver grafts from anti-HBC positive donors. *Gut* 50, 95–9

Sachsenberg-Studer EM, Dobrynski N, Sheldon J, Schulz TF, Pechere M, Nador RG, Spiliopoulos A, Nicod L, Saurat JH (1999) Human herpes-virus 8 seropositive patient with skin and graft Kaposi's sarcoma after lung transplantation. *J Am Acad Dermatol* 40, 308–11

Schatz O, Monini P, Bugarini R, Neipel F, Schulz TF, Andreoni M, Erb P, Eggers M, Haas J, Butto S, Lukwiya M, Bogner JR, Yaguboglu S, Sheldon J, Sarmati L, Goebel FD, Hintermaier R, Enders G, Regamey N, Wernli M, Sturzl M, Rezza G, Ensoli B (2001) Kaposi's sarcoma-associated herpesvirus serology in Europe and Uganda: multicentre study with multiple and novel assays. *J Med Virol* 65, 123–32

Shepherd FA, Maher E, Cardella C, Cole E, Greig P, Wade JA, Levy G (1997) Treatment of Kaposi's sarcoma after solid organ transplantation. *J Clin Oncol* 15, 2371–7

Shmueli D, Shapira Z, Yussim A, Nakache R, Ram Z, Shaharabani E (1989) The incidence of Kaposi sarcoma in renal transplant patients and its relation to immunosuppression. *Transplant Proc* 21, 3209–10

Sleiman C, Mal H, Roue C, Groussard O, Baldeyrou P, Olivier P, Fournier M, Pariente R (1997) Bronchial Kaposi's sarcoma after single lung transplantation. *Eur Respir J* 10, 1181–3

Spira TJ, Lam L, Dollard SC, Meng YX, Pau CP, Black JB, Burns D, Cooper B, Hamid M, Huong J, Kite-Powell K, Pellett PE (2000) Comparison of serologic assays and PCR for diagnosis of human herpesvirus 8 infection. *J Clin Microbiol* 38, 2174–80

Sungur C, Bozdogan O, Sungur A, Oymak O, Yasavul U, Turgan C, Caglar S (1995) Lymphadenopathic aggressive Kaposi's sarcoma in a renal transplant recipient with HLA-DR2 and HLA-DR5 antigens. *Nephron* 69, 122–3

Whitby D, Luppi M, Barozzi P, Boshoff C, Weiss RA, Torelli G (1998) Human herpesvirus 8 seroprevalence in blood donors and lymphoma patients from different regions of Italy. *J Natl Cancer Inst* 90, 395–7

Woodle ES, Hanaway M, Buell J, Gross T, First MR, Trofe J, Beebe T (2001) Kaposi sarcoma: an analysis of the US and international experiences from the Israel Penn International Transplant Tumor Registry. *Transplant Proc* 33, 3660–1

Zahger D, Lotan C, Admon D, Klapholz L, Kaufman B, Shimon D, Woolfson N, Gotsman MS (1993) Very early appearance of Kaposi's sarcoma after cardiac transplantation in Sephardic Jews. *Am Heart J* 126, 999–1000

Zhu L, Wang R, Sweat A, Goldstein E, Horvat R, Chandran B (1999) Comparison of human sera reactivities in immunoblots with recombinant human herpesvirus (HHV)-8 proteins associated with the latent (ORF73) and lytic (ORFs 65, K8.1A, and K8.1B) replicative cycles and in immunofluorescence assays with HHV-8-infected BCBL-1 cells. *Virology* 256, 381–92

Zong JC, Ciufo DM, Alcendor DJ, Wan X, Nicholas J, Browning PJ, Rady PL, Tyring SK, Orenstein JM, Rabkin CS, Su IJ, Powell KF, Croxson M, Foreman KE, Nickoloff BJ, Alkan S, Hayward GS (1999) High-level variability in the ORF-K1 membrane protein gene at the left end of the Kaposi's sarcoma-associated herpesvirus genome defines four major virus subtypes and multiple variants or clades in different human populations. *J Virol* 73, 4156–70

CTMI (2006) 312:263–287

Kaposi Sarcoma-Associated Herpesvirus and Other Viruses in Human Lymphomagenesis

E. Cesarman[1] (✉) · E. A. Mesri[2]

[1]Department of Pathology and Laboratory Medicine, Weill Medical College of Cornell University, New York, NY 10021, USA
ecesarm@med.cornell.edu

[2]Viral Oncology Program and Department of Microbiology and Immunology, Sylvester Comprehensive Cancer Center and University of Miami Miller School of Medicine, Miami, FL 33136, USA

Abstract Kaposi sarcoma-associated herpesvirus (KSHV), also called human herpesvirus 8 (HHV-8), is associated with a specific subset of lymphoproliferative disorders. These include two main categories. The first is primary effusion lymphomas and related solid variants. The second is multicentric Castleman disease, from which KSHV-positive plasmablastic lymphomas can arise. KSHV contributes to lymphomagenesis by subverting the host cell molecular signaling machinery to deregulate cell growth and survival. KSHV expresses a selected set of genes in the lymphoma cells, encoding viral proteins that play important roles in KSHV lymphomagenesis. Deregulation of the NF-κB pathway is an important strategy used by KSHV to promote lymphoma cell survival, and the viral protein vFLIP is essential for this process. Two other viruses that are well documented to be causally associated with lymphoid neoplasia in humans are Epstein-Barr virus (EBV/HHV-4) and human T-cell lymphotropic virus (HTLV-1). Both of these are similar to KSHV in their use of viral proteins to promote cell survival by deregulating the NF-κB pathway. Here we review the basic information and recent developments that have contributed to our knowledge of lymphomas caused by KSHV and other viruses. The understanding of the mechanisms of viral lymphomagenesis should lead to the identification of novel therapeutic targets and to the development of rationally designed therapies.

1 Introduction

In the early twentieth century Ellerman and Bang made the first observation that a viral infection could lead to the development of leukemia (Ellerman and Bang 1908). Since that time, many associations between viruses and leukemia/lymphoma have been shown in a variety of animals, but only a few firmly have been demonstrated in humans. Nowadays, in order to consider that a viral entity is causally associated with a human cancer, it must fulfill the criteria of four so-called Koch-like postulates (Zur Hauzen 1999): (1) epidemiologic evidence that the virus represents a risk factor for the development of the tumor, (2) consistent presence or persistence of the viral genome in the tumor cells, (3) documented oncogenic effects of viral infection or introduction of viral genes to cells, and (4) demonstration that the malignant phenotype of tumor cells depends on functions conferred by the viral genomes. Based on these rigorous criteria only three human viruses are clearly directly associated with lymphoid neoplasia: Epstein-Barr virus [EBV; formally called human herpesvirus 4 (HHV-4)], Kaposi sarcoma-associated herpesvirus (KSHV; human herpesvirus 8 (HHV-8)], and human T cell lymphotropic virus 1 (HTLV-1).

Although the prevalence of lymphomagenic viruses in the normal population is quite high (>95% of adults for EBV), only a small number of infected immunocompetent people ever develop the virus-induced malignancy and do so only after a long latency period. This observation reflects the multistep nature of oncogenesis, with the viral infection representing only one of these steps. In the following sections, the current understanding and recent developments concerning the pathobiology of KSHV and its role in lymphomagenesis are reviewed. We also briefly review the similarities between KSHV and two other human lymphomagenic viruses, EBV and HTLV-1.

2
KSHV

2.1
Association of KSHV and Malignant Lymphomas

KSHV is found invariably in Kaposi sarcoma, and compelling evidence suggests that it is an etiologic agent for this disease (Chang et al. 1994; Moore and Chang 1995). KSHV is also present in several lymphoproliferative disorders. Shortly after its first discovery, the presence of the KSHV genome was recognized in a subset of lymphomas in HIV-positive patients, all of which occurred as lymphomatous effusions in body cavities and were first called body cavity-based lymphomas and subsequently primary effusion lymphomas (PELs) (Cesarman et al. 1995a; Nador et al. 1996). Although they are more common in HIV-positive males, PELs also occur in HIV-negative men and women (Nador et al. 1995; Said et al. 1996). PELs contain many KSHV genomes, ranging between 40 and 80 copies per cell, which contrasts with an average of one copy per cell in KS lesions (Fig. 1A). The presence of KSHV in this subset of lymphomas allowed the development of cell lines that have been used as a tool for its propagation and for serologic assays (Arvanitakis et al. 1996; Boshoff et al. 1998; Cesarman et al. 1995b; Renne et al. 1996). Although these cell lines contain mostly latent virus, expressing a restricted number of viral genes, they can be induced to undergo lytic reactivation with phorbol esters or butyrate, thus producing infectious virus (Fig. 1A). Purified virus from PEL cell lines has been used to demonstrate its ability to infect B cells (Mesri et al. 1996), endothelial cells (Cannon et al. 2000; Flore et al. 1998; Moses et al. 1999; Panyutich et al. 1998), and cell lines from a variety of derivations (Bechtel et al. 2003).

Because some KSHV-negative lymphomas, such as Burkitt lymphoma (BL), can involve body cavities as lymphomatous effusions, and KSHV-positive lymphomas can present as solid-tissue masses, diagnostic criteria for PEL have been proposed (Cesarman and Knowles 1999; Nador et al. 1996). These criteria include immunoblastic-anaplastic large-cell morphology (Fig. 2), null-cell phenotype (including the lack of B cell-associated antigen and immunoglobulin expression in most cases), and B-cell genotype. The diagnosis of a KSHV-associated malignancy now can be easily confirmed by immunohistochemistry, as antibodies to viral proteins are commercially available (Fig. 2). The expression of CD138/syndecan-1 (Gaidano et al. 1997) and hypermutation of the immunoglobulin genes (Fais et al. 1999; Matolcsy et al. 1998) suggest that PELs are at a postgerminal center stage of B-cell differentiation and close to the plasma cell stage. This assumption was more recently confirmed by gene expression profiling of PEL (Jenner et al. 2003;

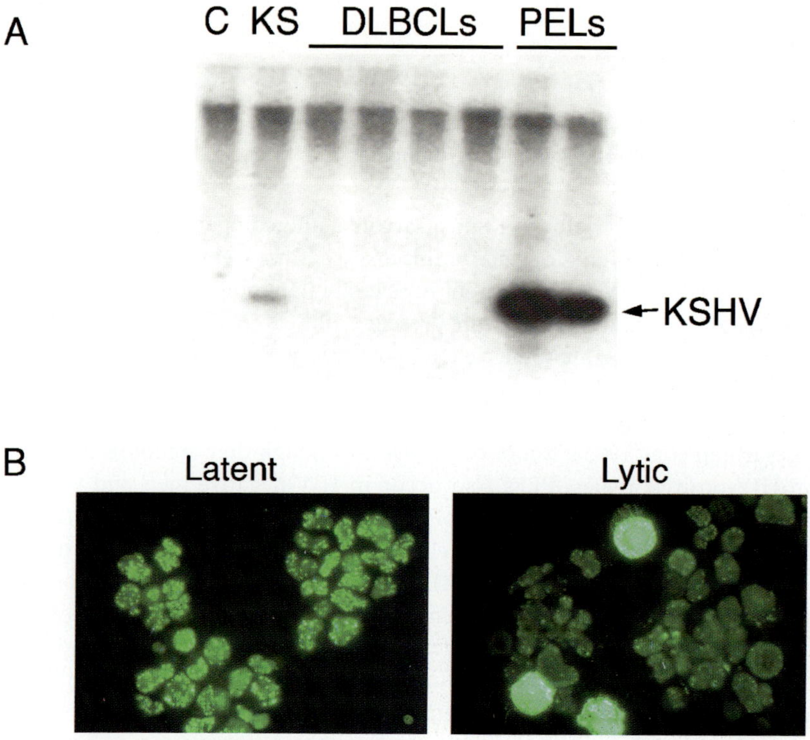

Fig. 1 Detection of the KSHV genome and latent and lytic antigens in PEL cell. Lines. **A** Southern blot hybridization using a probe from the KSHV genome detected a strong band in *BamH* 1-digested DNA extracted from two cases of PEL and a weaker band in Kaposi sarcoma, but not in a panel of diffuse large B cell lymphoma (*DLBCLs*) or in the HL60 cell lines used as the negative control (*C*). **B** Immunofluorescence assays were performed on the BC-3 PEL cell line, which was either untreated (*left*) or treated with tetradecanoyl phorbol acetate (*TPA*; *right*), using serum from a patient with KS as primary antibody. Positivity in latent cells is seen by a speckled nuclear pattern due to recognition of LANA-1 by the antiserum. TPA induces lytic reactivation in three cells in this field, which is visualized as the recognition of numerous nuclear and cellular proteins

Klein et al. 2003), where PEL resembled plasma cells and had a profile between multiple myeloma and EBV-associated immunoblastic lymphoma.

Lymphomas containing KSHV can also occur at presentation as solid tissue masses, similar to other AIDS-related non-Hodgkin lymphomas. Although some of these lymphomas subsequently develop an effusion, others apparently do not. They usually present as solid extranodal lymphomas and are diagnosed as diffuse large-cell, immunoblastic, or anaplastic large-cell lym-

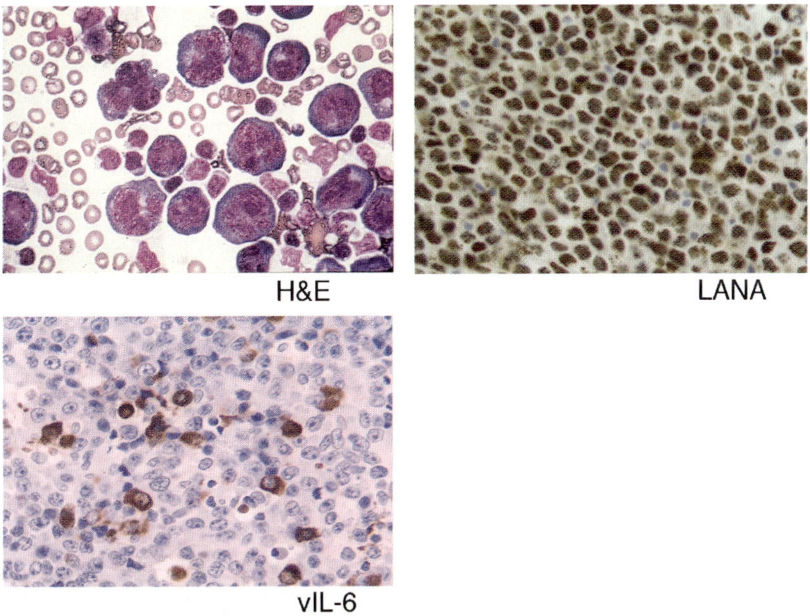

Fig. 2 Morphology and immunohistochemistry for KSHV of primary effusion lymphoma. An air-dried cytocentrifuge preparation of a KSHV-positive primary effusion lymphoma was stained with Wright-Giemsa (*left panel*). The tumor cells in this image are considerably larger than normal benign lymphocytes and monocytes. The cells display significant polymorphism and possess moderately abundant basophilic cytoplasm. A prominent, clear perinuclear Golgi zone can be appreciated in several cells. The nuclei vary from large and round to highly irregular, multilobated, and pleomorphic and often contain one or more prominent nucleoli (original magnification ×1,000). Immunohistochemical staining with the following antibodies is shown: LANA (ORF 73), a monoclonal rat antibody was used, showing brown speckled nuclear positivity in all the neoplastic cells; vIL-6, immunohistochemical staining with a polyclonal rabbit antiserum shows abundant cytoplasmic expression (*brown*) in many lymphoma cells (original magnification ×400)

phomas, in which the presence of KSHV in practically all the lymphoma cells can be demonstrated by immunohistochemistry or molecular techniques (Carbone et al. 2005; Chadburn et al. 2004; Deloose et al. 2005; Engels et al. 2003). Most of these are immunoblastic in appearance and have a high mitotic rate and variable amounts of apoptotic debris. These lymphomas appear to fall in the spectrum of PEL, as they frequently lack expression of B-cell antigens and immunoglobulin, they have a similar morphology, and the majority are coinfected with EBV.

Plasmablastic lymphomas, associated with multicentric Castleman disease, have also been described in HIV-positive patients (Dupin et al. 2000). Although these plasmablastic lymphomas are KSHV positive, they differ from PEL in a number of ways. Plasmablastic lymphomas are EBV negative, do not contain mutations in the Ig genes, and are thought to arise from naive IgM lambda-expressing B cells rather than terminally differentiated B cells (Du et al. 2001). In addition, KSHV has been documented in germinotropic lymphoproliferative disorders in HIV-negative patients (Du et al. 2002), suggesting that this virus is present in a heterogeneous but distinct group of lymphoproliferative diseases and may be more common than previously thought.

2.2
Role of KSHV in PEL

The almost invariable presence of KSHV in PELs suggests that it is necessary for their development. However, PELs are rare tumors, even in populations with high KSHV seroprevalence, accounting for about 3% of AIDS-related lymphomas and 0.4% of all AIDS-unrelated diffuse large-cell NHLs (Carbone et al. 1996). Therefore, it is evident that KSHV infection represents only one of several events involved in the development of PEL. Another cofactor appears to be EBV, because most PELs contain both viral genomes. The majority of PELs in vivo, as well as in culture, are latently infected with both viruses, but there is a small proportion of cells in which EBV and/or KSHV productive infection takes place. Analysis of the pattern of EBV gene expression in PELs revealed that only EBNA1 was expressed, corresponding to Latency I (Horenstein et al. 1997; Szekely et al. 1998). In addition, PELs lack structural alterations in most cell-transforming genes frequently involved in lymphomagenesis, with the possible exception of mutation in the regulatory region of Bcl-6 (Carbone et al. 1998; Nador et al. 1996). Although there are many cytogenetic abnormalities, reflecting genomic instability, none of these is consistent in all PELs. KSHV expresses a small number of viral proteins in PEL cells, but practically all of these have been documented to have oncogenic potential. These observations suggest that KSHV plays a transforming role in PELs.

2.3
Molecular Mechanisms of KSHV-Induced Lymphomagenesis

The majority of PELs are latently infected by KSHV. Latency allows the virus to remain in the infected cell, ensuring that the cell survives and is not recognized as infected by the host immune system. On initial infection, KSHV produces viral proteins that inhibit innate antiviral responses, and subsequently during latency it produces a protein that ensures maintenance of viral

DNA in the form of extrachromosomal circles, called episomes, in dividing cells. It also produces proteins during latency that promote survival of the infected cells. Because promotion of cell survival is a main feature of cancer, it is not surprising that KSHV is associated with malignancies. In fact, given the number of proteins produced by KSHV with oncogenic potential, the surprise comes from the relative rarity of KSHV-associated cancers.

2.3.1 Latent Viral Proteins

Five major KSHV proteins have been confirmed to be produced in latently infected lymphoma cells. They are called LANA-1 (also LANA or LNA), vCyclin (vCYC), vFLIP, vIL-6, and LANA-2 (or vIRF-3). Of these, LANA-1, vCYC, and vFLIP are transcribed from the same region into a single tricistronic mRNA, which gets alternatively spliced into a transcript containing LANA-1 and another encoding for vCYC and vFLIP, the expression of which comes from an internal ribosomal entry site (IRES) (Bieleski and Talbot 2001; Low et al. 2001; Talbot et al. 1999). Other proteins have also been reported to be expressed during latency (like K15 and the Kaposins), but these data are either controversial or not confirmed at the protein level. All of these genes bear potential to participate in lymphomagenesis and the maintenance of the malignant phenotype by affecting cellular survival and/or proliferation. Their function is illustrated in Fig. 3 and can be summarized as follows:

1. LANA-1: This protein, essential for episome maintenance (Ballestas et al. 1999), has been shown to have the ability to transform cells (Radkov et al. 2000) and induce lymphoid tumors in transgenic mice (Fakhari et al. 2006). LANA-1 has the ability to affect several pathways that are involved in tumorigenesis: It can bind and inactivate the retinoblastoma (Rb) protein (Radkov et al. 2000) and similarly bind and inactivate p53 (Friborg et al. 1999), protecting cells from apoptosis and inducing genetic instability (Si and Robertson 2006). LANA-1 can also bind and inactivate GSK-3β, leading to activation of the β-catenin pathway involved in solid tumors (Fujimuro and Hayward 2004). It has also been shown to have transcriptional effects, both positive and negative, on a variety of cellular and viral genes.
2. vCYC: This protein is a functional cyclin that can associate with CDK6 and induce phosphorylation of Rb protein and overcome Rb-mediated cell cycle arrest (Godden-Kent et al. 1997; Li et al. 1997). vCYC differs from the cellular cyclin D in its ability to induce degradation of the CDK inhibitor p27Kip when complexed with CDK6 (Ellis et al. 1999; Mann

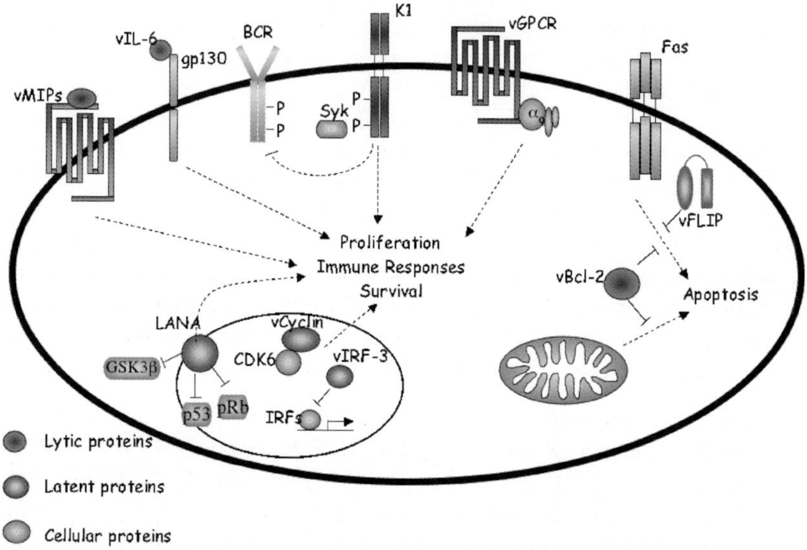

Fig. 3 Representative KSHV-encoded proteins that can affect cellular proliferation, survival, differentiation, and/or angiogenesis. KSHV contains several open reading frames encoding homologues to cellular proteins involved in vital proliferative and survival functions. Among these are (1) secreted autocrine/paracrine factors (vIL-6 and vMIPs); (2) transmembrane signaling molecules (vGPCR and K1); (3) transcriptional regulators (vIRFs and LANA); (4) a cell cycle regulator (vCYC); (5) tumor suppressor gene-binding activity (LANA) and (6) apoptosis inhibitors (vBCL-2 and vFLIP). Those genes clearly demonstrated to be expressed in latently infected cells of Kaposi's sarcoma are illustrated in *ovals*. (Reprinted from Cesarman and Mesri 2006)

et al. 1999). Therefore, vCYC is likely to modulate the cell cycle in PEL cells, avoiding normal regulatory checkpoints. Transgenic mice with vCYC develop lymphomas, but only in the absence of p53, and a model has been proposed whereby vCYC induces genome instability and loss of p53 subsequently allows expansion of tumorigenic clones (Verschuren et al. 2004, 2002).

3. vFLIP (K13): This protein is homologous to cFLIP, which is an inhibitor of the proapoptotic molecule FLICE/Caspase-8 (Thome et al. 1997). vFLIP can bind to TRAFs and localize to the IKK complex, activating NF-κB though both the classic and alternative pathways (Chaudhary et al. 1999; Field et al. 2003; Guasparri et al. 2006; Matta and Chaudhary 2004), and through this mechanism induce expression of antiapoptotic genes and protect cells from Fas-induced cell death in vitro and in vivo (Djerbi et al.

1999; Guasparri et al. 2004). vFLIP protects PEL cells from spontaneous apoptosis and is the first viral protein shown to be essential for tumor cell survival in vitro and in vivo (Godfrey et al. 2005; Guasparri et al. 2004).

4. LANA-2 (vIRF-3): Although this protein is expressed by only rare cells in KS lesions, indicating a lytic expression pattern in endothelial cells, it is expressed by most PEL cells, being a latent protein in lymphomas (Rivas et al. 2001). This protein was shown to potently inhibit p53 in reporter assays (Rivas et al. 2001) and thus may be involved in deregulation of apoptosis. Although LANA-2 is not a DNA-binding protein, it is recruited to the interferon promoters via its interaction with cellular IRF-3 and IRF-7 and stimulates their transcriptional activity (Lubyova et al. 2004).

5. vIL-6: This is sometimes considered to be a lytic gene because its expression is increased on lytic reactivation. However, it is expressed by a variable but significant proportion of latently infected PEL cells. Considering the fact that this protein can be secreted, it may affect other tumor cells that do not express it and play a role in their proliferation. vIL-6 differs from cellular IL-6 in that it is selectively glycosylated and can bind the gp130 receptor in the absence of the high-affinity IL-6 receptor to activate IL-6-responsive genes and promote B-cell survival (Molden et al. 1997; Moore et al. 1996; Nicholas et al. 1997). vIL-6 has been shown to induce expression of vascular endothelial growth factor (VEGF) (Aoki et al. 1999), which in turn can induce effusion development (Aoki and Tosato 1999; Liu et al. 2001), therefore being involved in the pathogenesis of PEL.

2.3.2 Lytic Viral Proteins

Although mostly latent genes are expressed in PELs, some lytic genes are expressed to some extent in a minority of cells, and several of these have oncogenic potential. In addition, patterns of KSHV gene expression in de novo infection and during initial steps of lymphomagenesis are not known. These lytic genes encode for some functionally interesting proteins. The viral interferon regulatory factor 1 (K9, vIRF-1) (Li et al. 1998), the viral G protein-coupled receptor (ORF 74, vGPCR) (Bais et al. 1998), and ORF K1 (Lee et al. 1998b) can transform rodent cells and/or cause tumors in animal models. Three transmembrane proteins are potentially important in PEL lymphomagenesis because they can trigger signaling cascades relevant for B- and T-cell growth. These are:

1. vGPCR: This is a constitutively active G protein-coupled receptor able to trigger the mitogen-activated protein (MAP) kinase signaling cascades and induce secretion of VEGF (Bais et al. 1998). MAP kinase cascades such as those triggered by vGPCR are activated by inflammatory cytokines and mitogens. VEGF is an angiogenic and vascular permeability factor that could contribute to the effusion phenotype (Aoki and Tosato 1999). In PEL cells, vGPCR can activate a variety of signaling cascades activating the transcription factors AP-1, NF-κB, CREB, and NFAT (Cannon et al. 2003; Cannon and Cesarman 2004). vGPCR can induce cell cycle arrest that is mediated by p21 (Cannon et al. 2006). Cell cycle arrest can be induced by other KSHV lytic genes (Wu et al. 2002) and is common during the lytic phase of other viruses including EBV (Rodriguez et al. 2001a, 2001b). However, in the case of vGPCR, this induction of p21 results in inhibition of Cdk2 activity that can delay, and perhaps abort, lytic replication. This ability of vGPCR to inhibit viral replication may explain how this lytic gene product may still have oncogenic effects by inducing the expression of multiple inflammatory and angiogenic factors.

2. K1: This protein has an ITAM motif that can activate cytoplasmic tyrosine kinases and mimic signaling by the B-cell antigen receptor (Lee et al. 1998a). K1 can activate the MAP kinase and NF-κB signaling pathways and has potential antiapoptotic functions (Brinkmann et al. 2003; Sharp et al. 2002). This viral protein has also been shown to activate the Akt and Src kinases (Prakash et al. 2005; Tomlinson and Damania 2004). Therefore, like vGPCR, it appears to have very broad signaling capabilities.

3. K15: This open reading frame encodes a protein with complex splicing and multiple transmembrane motifs, structurally resembling EBV LMP-2A. It has SH2 and SH3 motifs, and also a TRAF binding motif, and can also activate signaling cascades that include NF-κB and MAP kinase (Brinkmann et al. 2003). Also like EBV LMP-2A, K15 can inhibit B cell receptor signaling (Choi et al. 2000). K15 protein has been found to be present in most latently infected cells by one study (Sharp et al. 2002), but the RNA expression data are controversial, and most studies argue for lytic expression. K15 has been proposed to have antiapoptotic functions (Sharp et al., 2002).

Other interesting molecules that are expressed during lytic replication and have a potential role in PEL pathogenesis are vIRF1, which can inhibit interferon-induced transcriptional activation (Li et al. 1998; Zimring et al. 1998), and vBCL-2 (ORF16), a homologue of cellular BCL-2, which is a cellular oncogene involved in the development of follicular lymphomas. vBCL-2 can block apoptosis as efficiently as cellular Bcl-2, Bcl-xL, or the EBV Bcl-2 homologue, BHRF1 (Cheng et al. 1997). Interestingly, the KSHV Bcl-2 cannot homodimerize or heterodimerize with other Bcl-2 family members, suggesting that it may have evolved to escape any negative regulatory effects of the cellular Bax and Bak proteins.

2.4 Animal Models for KSHV-Associated Lymphomas

Animal models are useful tools to reproduce human disease under controlled experimental conditions in order to study the disease pathobiology and be able to test the effectiveness of drugs. It was first found that PELs injected into SCID/NOD mice developed highly angiogenic immunoblastic lymphomas (Picchio et al. 1997). In efforts to more accurately reproduce PEL disease presentation, PEL cells were injected intravenously (i.v.) and intraperitoneally (i.p.) into NOD/SCID mice (Boshoff et al. 1998; Said et al. 1997). Intravenous injections of PEL cell led to organ infiltration but did not lead to formation of tumors. On the other hand, i.p. injection of PEL led to the development of ascites and diffuse infiltration of organs, without solid lymphoma formation, resembling the diffuse nature of human PEL (Boshoff 1998). This model has been very useful for the preclinical testing of anti-PEL therapeutic approaches (Keller et al. 2006; Wu et al., 2005) (see Sect. 2.5) and for studies on PEL pathobiology. PEL cells are able to secrete VEGF, which is not only angiogenic but also a powerful vascular permeability factor. Thus secreted VEGF could play a role in the effusion phenotype of PELs. Indeed, it was found that anti-VEGF blocking antibodies were able to prevent the formation of PEL ascites, indicating that the VEGF–VEGF receptor axis plays a major role in effusion formation (Aoki and Tosato 1999). Thus it appears that observations made in PELs growing in vivo are able to recapitulate PEL cell-host interactions that may affect viral and host gene expression. This was recently demonstrated by Staudt et al. (Staudt et al. 2004), who found that PEL cells injected in matrigel generated well-organized angiogenic tumors formed by KSHV-infected PEL cells, displaying a pattern of lytic and latent KSHV gene expression that was quite different from the pattern found in PEL cells growing in culture.

2.5 Current and Future Therapeutic Approaches for KSHV-Associated Lymphomas

Survival of PEL patients with conventional chemotherapy is dismal, so it is important to develop novel therapeutic approaches. A promising cellular target is the transcription factor NF-κB, which is constitutively active in PEL cells, and pharmacologic inhibition with Bay 11-7082 leads to tumor cell apoptosis (Keller et al. 2000). This compound also resulted in improved survival and cures in a mouse model of PEL as well as a model of EBV-lymphoma (Keller et al. 2006), indicating that NF-κB inhibition is a viable approach for the treatment of lymphomas associated with herpesviral infections. Although this particular inhibitor is not being pursued with therapeutic intent, other compounds such as the proteosomal inhibitor Bortezomib (Velcade) affect NF-κB and may be useful for the treatment of PEL. A promising new approach that also targets the NF-kB pathway is the combination of AZT, a nucleoside analog that also functions as an NF-kB inhibitor, and IFN-α, an inducer of apoptosis. This combination leads to apoptosis in HHV-8+/EBV-PEL cells in culture, by induction of a tumor necrosis factor-related apoptosis-inducing ligand (TRAIL)-mediated suicide program This approach has been proven effective in one reported patient with PEL and has the advantage of using available and clinically approved agents (Ghosh et al. 2003). Recent experiments in the PEL ascites animal model show that this treatment was effective in increasing the mean survival time in mice (Wu et al. 2005). This suggests that induction of apoptosis in PEL cells may be exploited as an effective, relatively nontoxic therapy targeting KSHV-infected PELs.

Although cellular targets such as NF-κB have the advantage that they are being developed at a rapid pace for other diseases, in particular other cancers and inflammation, viral targets would have the added advantage of specificity and low toxicity. KSHV encodes several genes that can activate NF-κB, but only vFLIP is largely responsible for the constitutive NF-κB activity in PEL cells. Elimination of vFLIP results in marked reduction of NF-κB, decreased expression of NF-κB-regulated genes and cellular apoptosis (Guasparri et al. 2004). Therefore, vFLIP functions as a viral oncogene in KSHV-associated lymphomas and is a promising therapeutic target.

3 Other Lymphomagenic Viruses

3.1 Epstein-Barr Virus

The association of EBV with several specific lymphoid malignancies is quite consistent, indicating an etiopathogenic role in their development. EBV infection is practically ubiquitous in healthy adults, so it has been difficult to establish the exact role of this virus in lymphomagenesis. Nevertheless, extensive epidemiologic and experimental data support the notion that EBV is an oncogenic virus. EBV can infect and transform normal human B-cells in vitro, resulting in their "immortalization" and leading to continuously growing lymphoblastoid cell lines (LCLs) (Rickinson and Kieff 1996). In addition, loss of EBV episomes from transformed lymphoma cell lines results in loss of tumorigenicity.

The pattern of expression of EBV latent genes is associated with specific lymphoma subtypes. In some cases EBV gene expression can be linked specific molecular and oncogenic phenotypes of the lymphoid malignancies. Lymphomas such as Burkitt lymphoma (BL) only express EBNA1. Although EBNA1 transgenic mice have an increased incidence of lymphomas (Wilson et al. 1996), the role of EBNA1 in tumorigenesis is unclear. There is evidence that infection of B cells with EBV lacking EBNA1 can lead to the development of LCLs that are tumorigenic in mice, but this process is much less efficient than with wild-type EBV (Humme et al. 2003). In general it is thought that EBNA1 is not a major oncogene of EBV. BLs bear specific chromosomal translocations involving the c-myc oncogene, and these are probably essential because EBNA1 is insufficient to drive cellular proliferation. EBV-associated lymphomas frequently present in immunodeficient individuals, including those infected with HIV and organ transplant recipients. In these, tumor cells usually express a larger number of EBV-encoded genes, which tend to be more oncogenic as well as more immunogenic. These proteins include EBNA2, LMP1, and LM2A, all essential for transformation by EBV in vitro. EBNA2 is thought to represent a constitutively active member of the Notch signaling pathway (Grossman et al. 1994; Hsieh and Hayward 1995). The LMP2A protein is structurally reminiscent of two KSHV transmembrane proteins, K1 and K15. Like K1, it contains an ITAM motif, therefore mimicking antigen receptor signaling, and like both, it can signal through specific tyrosine kinases, including fyn and lyn. The LMP1 protein is transforming and tumorigenic in vitro (Wang et al. 1985) and in vivo. Transgenic mice expressing LMP1 under the control of immunoglobulin gene regulatory elements develop B-cell lymphomas (Kulwichit et al. 1998). LMP1 functions as a constitutively active CD40 receptor, a member of the TNF (tumor necrosis factor)-receptor family. LMP1 aggregates in the membrane as its cytoplasmic tail interacts with TNF receptor-associated factors (TRAFs) and TNFR-1-associated death domain protein (TRADD), leading to activation of nuclear factor (NF)-κB and the c-Jun amino-terminal kinase (JNK) (Eliopoulos et al. 1999; Kieser et al. 1997; Kilger et al. 1998), a kinase cascade activated by inflammatory cytokines and involved in bcr-abl leukemogenesis (Dickens et al. 1997). Thus signaling by EBV LMP-1 is similar to that of KSHV vFLIP; although these two proteins are not structural homologues, they have subverted the very same signaling pathway to promote the survival of infected B cells.

3.2 Human T-Cell Lymphotropic Virus-1 and Adult T-Cell Leukemia/Lymphoma

Human T-cell lymphotropic virus-1, the first described human lymphotropic retrovirus (Poiesz et al. 1980), is the etiologic agent for adult T-cell leukemia/lymphoma (ATLL). This is an often aggressive malignancy of mature CD4+ lymphocytes. Cumulative evidence points to HTLV-1 as the causal agent of ATLL (Cann and Chen 1996; Dalgleish 1998): (1) Seroepidemiologic studies show a close association between ATLL and HTLV-1 infection; (2) throughout malignant progression ATLL cell clonality or oligoclonality correlates with the clonality or oligoclonality of retroviral insertion; (3) HTLV-1 can transform T cells in vitro, resulting in immortalized cell lines; and (4) HLTV-1 can induce cancers in animal models.

Human T-cell lymphotropic virus-1 seems to play a role in the early steps of ATLL malignant progression and not necessarily in the maintenance of the transformed phenotype (Cann and Chen 1996), because ATLL tumor cells lack expression of HTLV-1 genes (Reitz et al. 1983) and carry viral genomes that tend to be heavily deleted (Yoshida et al. 1985, 1984). ATLL tumor cells do have oncogenic alterations in p53 (Cesarman et al. 1992; Pise-Masison et al. 1998) and CDKN2/p16 tumor suppressor genes (Uchida et al. 1998) and chromosomal aberrations (Fukuhara et al. 1983; Maruyama et al. 1990), the complexity of which correlates with tumor aggressiveness (Sanada et al. 1985). Therefore, oncogenesis of ATLL seems to involve an HTLV-1-dependent step in which transformation of T cells results in a polyclonal population of proliferating immortalized T cells, with subsequent acquisition of new oncogenic genetic alterations, clonal expansion, and progression to full malignancy and HTVL-1 independence (Cann and Chen 1996) (Fig. 4). In contrast to other animal retroviruses that transform cells by transduction of oncogenes or activation of oncogenes adjacent to insertion sites, HTLV-1 transforms T cells by a rather complex mechanism that is mediated by its transactivator protein, Tax (Cann and Chen 1996; Felber et al. 1985; Flint and Shenk 1997; Hiscott et al. 1995). Evidence supporting an oncogenic role for Tax stems from its ability to induce T-cell immortalization (Grassmann et al. 1992, 1989) and mesenchymal tumors in transgenic mice (Hinrichs et al. 1987; Nerenberg et al. 1987). Transgenic mice in which the expression of Tax was specifically targeted to T cells developed a large granular T-cell lymphocytic leukemia (Grossman et al. 1995).

Although HTLV-1 Tax is not generally able to directly interact with DNA (Cann and Chen 1996; Lenzmeier et al. 1998), it is able to exert powerful biological responses by increasing the activity of transcription factors that are able to transactivate both the HTLV-1 LTR and genes involved in T-cell proliferation (Felber et al. 1985; Flint and Shenk 1997; Hiscott et al. 1995).

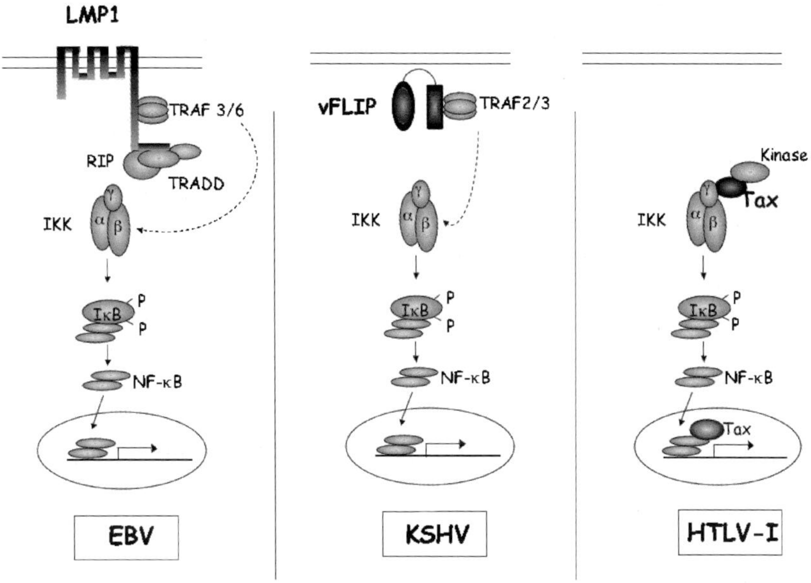

Fig. 4 Viral proteins can activate NF-κB. Transforming proteins of EBV, KSHV, and HTLV-1 include LMP-1, vFLIP and Tax, respectively. These three proteins share the ability to activate the NF-κB pathway. Although all three have been shown to activate this pathway through canonical and noncannonical mechanisms, only the canonical pathway is shown. This pathway involves activation of the IKK complex by the corresponding viral protein with subsequent phosphorylation of IκBα that leads to its degradation. This results in the release of Rel proteins, which are translocated to the nucleus and dimerize to form different NF-κB complexes that bind to specific sequences to activate the transcription of multiple genes. (Reprinted from Cesarman and Mesri 2006)

Tax can interact with members of the NF-κB family of transcription factors (Cann and Chen 1996; Flint and Shenk 1997; Leung and Nabel 1988; Ruben et al. 1988), its physiological IκB inhibitors (Nicot et al. 1998; Petropoulos and Hiscott 1998), and its activating kinases (Geleziunas et al. 1998), leading to increased transcription from NF-κB-responsive promoters. NF-κB can activate transcription of many genes related to T-cell proliferation and growth such as c-rel, c-myc, and both IL-2 and IL-2 receptor. Simultaneous activation of IL-2 and its receptor can lead to IL-2-independent proliferation of T cells expressing Tax (Iwanaga et al. 1999). Although not all HTLV-1-transformed cells express IL-2, the creation of an IL-2 autocrine loop by Tax may be important in the early stages of malignant progression (Cann and Chen 1996).

As in the case of KSHV- and EBV-mediated lymphomagenesis, NF-κB activation by Tax plays a central role in HTLV-I-mediated oncogenesis. It appears that HTLV-1 has evolved multiple strategies to subvert the NF-κB pathway, a major regulator of T-cell survival, activation, differentiation, and proliferation (Yoshida 2001). HTLV-1 Tax can activate NF-κB through both the canonical or classic and the noncanonical or alternative pathways. HTLV-I Tax can physically interact with IKKγ and activate phosphorylation and degradation of IKKα and IKKβ, thus activating NF-κB through the canonical pathway (Harhaj and Sun 1999; Sun and Ballard 1999). In addition, HTLV-I-transformed cells consistently express p52, indicating that the noncanonical NF-κB pathway is constitutively activated. Indeed, it was found that HTLV-1 Tax is able to stimulate p100 processing into p52 by bridging the interaction between p100 and IKKα (Xiao et al. 2001). The multiple roles that NF-κB could play in T-cell tumorigenesis, in addition to the multiple strategies devised by HTLV-I to subvert it, indicate that this pathway is critical for HTLV-1-mediated lymphomagenesis, and therefore it could be an attractive therapeutic target. Bay 11-7082, an NF-κB inhibitor that acts by inhibiting IκBα degradation, was shown to induce apoptosis of HTLV-1-infected T-cells but not uninfected T-cells or PMBC (Mori et al. 2002). Moreover, Bay 11-7082 efficiently suppressed ATLL tumor growth in a new model in immunodeficient NOD/SCID gamma-null (NOG) mice (Dewan et al. 2003). These data demonstrated the involvement of the NF-κB pathway in HTLV-1 tumorigenesis and point to NF-κB as an attractive target for therapy that can be inhibited to achieve selective cytotoxicity in HTLV-1-associated ATLL tumors

4 Conclusions

KSHV, EBV, and HTLV-1 are associated with specific subsets of malignant lymphomas, and accumulated evidence indicates that they play an etiopathogenic role in their development. Although these viruses differ in many respects, they all carry genes that subvert the host molecular machinery to deregulate cell growth and immortalize infected lymphoid cells, setting the stage for further genetic alterations to take place so that the cells will progress to full malignancy. However, their ability to transform cells and to lead to lymphomagenesis is highly dependent on host factors. Understanding the oncogenic and immunologic mechanisms involved in the pathobiology of virus-associated lymphomas is important for the development of targeted therapeutic and preventive approaches.

Acknowledgements Supported in part by National Institutes of Health Grants CA-68939 and CA-103646 and a Leukemia and Lymphoma Society Translational Research grant to EC and CA-75918 to EAM.

References

Aoki Y, Jaffe ES, Chang Y, Jones K, Teruya-Feldstein J, Moore PS, Tosato G (1999) Angiogenesis and hematopoiesis induced by Kaposi's sarcoma-associated herpesvirus-encoded interleukin-6. Blood 93:4034–4043.

Aoki Y, Tosato G (1999) Role of vascular endothelial growth factor/vascular permeability factor in the pathogenesis of Kaposi's sarcoma-associated herpesvirus-infected primary effusion lymphomas. Blood 94:4247–4254

Arvanitakis L, Mesri EA, Nador R, Said JW, Asch AS, Knowles DM, Cesarman E (1996) Establishment and characterization of a primary effusion (body cavity-based) lymphoma cell line (BC-3) harboring Kaposi's sarcoma-associated herpesvirus (KSHV/HHV-8) in the absence of Epstein-Barr virus. Blood 88:2648–2654

Bais C, Santomasso B, Coso O, Arvanitakis L, Geras Raaka E, Gutkind JS, Asch AS, Cesarman E, Gershengorn MC, Mesri E (1998) G-protein-coupled receptor of Kaposi's sarcoma-associated herpesvirus is a viral oncogene and angiogenesis activator. Nature 391:86–89

Ballestas ME, Chatis PA, Kaye KM (1999) Efficient persistence of extrachromosomal KSHV DNA mediated by latency-associated nuclear antigen. Science 284:641–644

Bechtel JT, Liang Y, Hvidding J, Ganem D (2003) Host range of Kaposi's sarcoma-associated herpesvirus in cultured cells. J Virol 77:6474–6481

Bieleski L, Talbot SJ (2001) Kaposi's sarcoma-associated herpesvirus vCyclin open reading frame contains an internal ribosome entry site. J Virol 75:1864–1869

Boshoff C, Gao SJ, Healy LE, Matthews S, Thomas AJ, Coignet L, Warnke RA, Strauchen JA, Matutes E, Kamel OW, Moore PS, Weiss RA, Chang Y (1998) Establishing a KSHV+ cell line (BCP-1) from peripheral blood and characterizing its growth in Nod/SCID mice. Blood 91:1671–1679

Brinkmann MM, Glenn M, Rainbow L, Kieser A, Henke-Gendo C, Schulz TF (2003) Activation of mitogen-activated protein kinase and NF-κB pathways by a Kaposi's sarcoma-associated herpesvirus K15 membrane protein. J Virol 77:9346–9358

Cann AJ, Chen, I. S. Y. (1996). Human T-cell leukemia virus types I and II. In "Fields Virology" (Fields, B.N., Knipe , D.M., and Howley, P.M., eds.), pp. 1849–1880. Lippincott-Raven, Philadelphia.

Cannon JS, Ciufo D, Hawkins AL, Griffin CA, Borowitz MJ, Hayward GS, Ambinder RF (2000) A new primary effusion lymphoma-derived cell line yields a highly infectious Kaposi's sarcoma herpesvirus-containing supernatant. J Virol 74:10187–10193.

Cannon M, Cesarman E, Boshoff C (2006) KSHV G protein-coupled receptor inhibits lytic gene transcription in primary-effusion lymphoma cells via p21-mediated inhibition of Cdk2. Blood 107:277–284

Cannon M, Philpott NJ, Cesarman E (2003) The Kaposi's sarcoma-associated herpesvirus G protein-coupled receptor has broad signaling effects in primary effusion lymphoma cells. J Virol 77:57–67.

Cannon ML, Cesarman E (2004) The KSHV G protein-coupled receptor signals via multiple pathways to induce transcription factor activation in primary effusion lymphoma cells. Oncogene 23:514–523

Carbone A, Gaidano G, Gloghini A, Larocca LM, Capello D, Canzonieri V, Antinori A, Tirelli U, Falini B, Dalla-Favera R (1998) Differential expression of BCL-6, CD138/syndecan-1, and Epstein-Barr virus-encoded latent membrane protein-1 identifies distinct histogenetic subsets of acquired immunodeficiency syndrome-related non-Hodgkin's lymphomas. Blood 91:747–755

Carbone A, Gloghini A, Vaccher E, Cerri M, Gaidano G, Dalla-Favera R, Tirelli U (2005) Kaposi's sarcoma-associated herpesvirus/human herpesvirus type 8-positive solid lymphomas: a tissue-based variant of primary effusion lymphoma. J Mol Diagn 7:17–27

Carbone A, Gloghini A, Vaccher E, Zagonel V, Pastore C, Dalla Palma P, Branz F, Saglio G, Volpe R, Tirelli U, Gaidano G (1996) Kaposi's sarcoma-associated herpesvirus DNA sequences in AIDS-related and AIDS-unrelated lymphomatous effusions. Br J Haematol 94:533–543

Cesarman E, Chadburn A, Inghirami G, Gaidano G, Knowles DM (1992) Structural and functional analysis of oncogenes and tumor suppressor genes in adult T-cell leukemia/lymphoma shows frequent p53 mutations. Blood 80:3205–3216

Cesarman E, Chang Y, Moore PS, Said JW, Knowles DM (1995a) Kaposi's sarcoma-associated herpesvirus-like DNA sequences in AIDS-related body cavity-based lymphomas. N Engl J Med 332:1186–1191

Cesarman E, Knowles DM (1999) The role of Kaposi's sarcoma-associated herpesvirus (KSHV/HHV-8) in lymphoproliferative diseases. Semon Cancer Biol 9:165–174

Cesarman E, Mesri, E. A. (2006). Pathogenesis of viral lymphomas. In "Hodgkin's and Non-Hodgkin's Lymphoma" (Leonard , J.P., and Coleman, M., eds.), pp. 49–88. Springer, New York.

Cesarman E, Moore PS, Rao P, Inghirami G, Knowles DM, Chang Y (1995b) In vitro establishment and characterization of two acquired immunodeficiency syndrome-related lymphoma cell lines (BC-1 and BC-2) containing Kaposi's sarcoma-associated herpesvirus-like (KSHV) DNA sequences. Blood 86:2708–2714

Chadburn A, Hyjek E, Mathew S, Cesarman E, Said J, Knowles DM (2004) KSHV-positive solid lymphomas represent an extra-cavitary variant of primary effusion lymphoma. Am J Surg Pathol 28:1401–1416

Chang Y, Cesarman E, Pessin MS, Lee F, Culpepper J, Knowles DM, Moore PS (1994) Identification of herpesvirus-like DNA sequences in AIDS-associated Kaposi's sarcoma. Science 266:1865–1869

Chaudhary PM, Jasmin A, Eby MT, Hood L (1999) Modulation of the NF-κB pathway by virally encoded death effector domains-containing proteins. Oncogene 14:5738–5746

Cheng, E. H. Y., Nicholas J, Bellows DS, Hayward GS, Guo HG, Reitz MS, Hardwick JM (1997) A Bcl-2 homolog encoded by Kaposi sarcoma-associated virus, human herpesvirus 8, inhibits apoptosis but does not heterodimerize with Bax or Bak. Proc Natl Acad Sci USA 94:690–694

Choi JK, Lee BS, Shim SN, Li M, Jung JU (2000) Identification of the novel K15 gene at the rightmost end of the Kaposi's sarcoma-associated herpesvirus genome. J Virol 74:436–446.

Dalgleish AG (1998) Human T-cell lymphotropic virus type 1—Infections and pathogenesis. Curr Opin Infect Dis 11:195–199

Deloose ST, Smit LA, Pals FT, Kersten MJ, van Noesel CJ, Pals ST (2005) High incidence of Kaposi sarcoma-associated herpesvirus infection in HIV-related solid immunoblastic/plasmablastic diffuse large B-cell lymphoma. Leukemia 19:851–855

Dewan MZ, Terashima K, Taruishi M, Hasegawa H, Ito M, Tanaka Y, Mori N, Sata T, Koyanagi Y, Maeda M, Kubuki Y, Okayama A, Fujii M, Yamamoto N (2003) Rapid tumor formation of human T-cell leukemia virus type 1-infected cell lines in novel NOD-SCID/γc(null) mice: suppression by an inhibitor against NF-κB. J Virol 77:5286–5294

Dickens M, Rogers JS, Cavanagh J, Raitano A, Xia Z, Halpern JR, Greenberg ME, Sawyers CL, Davis RJ (1997) A cytoplasmic inhibitor of the JNK signal transduction pathway. Science 277:693–696

Djerbi M, Screpanti V, Catrina AI, Bogen B, Biberfeld P, Grandien A (1999) The inhibitor of death receptor signaling, FLICE-inhibitory protein defines a new class of tumor progression factors [see comments]. J Exp Med 190:1025–1032

Du MQ, Diss TC, Liu H, Ye H, Hamoudi RA, Cabecadas J, Dong HY, Harris NL, Chan JK, Rees JW, Dogan A, Isaacson PG (2002) KSHV- and EBV-associated germinotropic lymphoproliferative disorder. Blood 100:3415–3418

Du MQ, Liu H, Diss TC, Ye H, Hamoudi RA, Dupin N, Meignin V, Oksenhendler E, Boshoff C, Isaacson PG (2001) Kaposi sarcoma-associated herpesvirus infects monotypic (IgM lambda) but polyclonal naive B cells in Castleman disease and associated lymphoproliferative disorders. Blood 97:2130–2136

Dupin N, Diss TL, Kellam P, Tulliez M, Du MQ, Sicard D, Weiss RA, Isaacson PG, Boshoff C (2000) HHV-8 is associated with a plasmablastic variant of Castleman disease that is linked to HHV-8-positive plasmablastic lymphoma. Blood 95:1406–1412.

Eliopoulos AG, Blake SM, Floettmann JE, Rowe M, Young LS (1999) Epstein-Barr virus-encoded latent membrane protein 1 activates the JNK pathway through its extreme C terminus via a mechanism involving TRADD and TRAF2. J Virol. 73:1023–1035

Ellerman V, Bang O (1908) Experimentelle Leukamie bei Huhnern. Zentalbl Bakteriol Alet I 46:595–597

Ellis M, Chew YP, Fallis L, Freddersdorf S, Boshoff C, Weiss RA, Lu X, Mittnacht S (1999) Degradation of p27(Kip) cdk inhibitor triggered by Kaposi's sarcoma virus cyclin-cdk6 complex. EMBO J 18:644–653

Engels EA, Pittaluga S, Whitby D, Rabkin C, Aoki Y, Jaffe ES, Goedert JJ (2003) Immunoblastic lymphoma in persons with AIDS-associated Kaposi's sarcoma: a role for Kaposi's sarcoma-associated herpesvirus. Mod Pathol 16:424–429

Fais F, Gaidano G, Capello D, Gloghini A, Ghiotto F, Roncella S, Carbone A, Chiorazzi N, Ferrarini M (1999) Immunoglobulin V region gene use and structure suggest antigen selection in AIDS-related primary effusion lymphomas. Leukemia 13:1093–1099

Fakhari FD, Jeong JH, Kanan Y, Dittmer DP (2006) The latency-associated nuclear antigen of Kaposi sarcoma-associated herpesvirus induces B cell hyperplasia and lymphoma. J Clin Invest 116:735–742

Felber BK, Paskalis H, Kleinman-Ewing C, Wong-Staal F, Pavlakis GN (1985) The pX protein of HTLV-I is a transcriptional activator of its long terminal repeats. Science 229:675–679

Field N, Low W, Daniels M, Howell S, Daviet L, Boshoff C, Collins M (2003) KSHV vFLIP binds to IKK-γ to activate IKK. J Cell Sci 116:3721–3728
Flint J, Shenk T (1997) Viral transactivating proteins. Annu Rev Genet 31:177–212
Flore O, Rafii S, Ely S, O'Leary JJ, Hyjek EM, Cesarman E (1998) Transformation of primary human endothelial cells by Kaposi's sarcoma-associated herpesvirus. Nature 394:588–592
Friborg J, Jr., Kong W, Hottiger MO, Nabel GJ (1999) p53 inhibition by the LANA protein of KSHV protects against cell death. Nature 402:889–894
Fujimuro M, Hayward SD (2004) Manipulation of glycogen-synthase kinase-3 activity in KSHV-associated cancers. J Mol Med
Fukuhara S, Hinuma Y, Gotoh YI, Uchino H (1983) Chromosome aberrations in T lymphocytes carrying adult T-cell leukemia-associated antigens (ATLA) from healthy adults. Blood 61:205–207
Gaidano G, Gloghini A, Gattei V, Rossi MF, Cilia AM, Godeas C, Degan M, Perin T, Canzonieri V, Aldinucci D, Saglio G, Carbone A, Pinto A (1997) Association of Kaposi's sarcoma-associated herpesvirus-positive primary effusion lymphoma with expression of the CD138/syndecan-1 antigen. Blood 90:4894–4900
Geleziunas R, Ferrell S, Lin X, Mu Y, Cunningham ET, Jr., Grant M, Connelly MA, Hambor JE, Marcu KB, Greene WC (1998) Human T-cell leukemia virus type 1 Tax induction of NF-κB involves activation of the IκB kinase α(IKKα) and IKKβ cellular kinases. Mol Cell Biol 18:5157–5165
Ghosh SK, Wood C, Boise LH, Mian AM, Deyev VV, Feuer G, Toomey NL, Shank NC, Cabral L, Barber GN, Harrington WJJ, and 15;101(6). B. M. 2003. Potentiation of TRAIL-induced apoptosis in primary effusion lymphoma through azidothymidine-mediated inhibition of NF-κB. Blood 101:2321–2327
Godden-Kent D, Talbot SJ, Boshoff C, Chang Y, Moore P, Weiss RA, Mittnacht S (1997) The cyclin encoded by Kaposi's sarcoma-associated herpesvirus stimulates cdk6 to phosphorylate the retinoblastoma protein and histone H1. J Virol 71:4193–4198
Godfrey A, Anderson J, Papanastasiou A, Takeuchi Y, Boshoff C (2005) Inhibiting primary effusion lymphoma by lentiviral vectors encoding short hairpin RNA. Blood 105:2510–2518
Grassmann R, Berchtold S, Radant I, Alt M, Fleckenstein B, Sodroski JG, Haseltine WA, Ramstedt U (1992) Role of human T-cell leukemia virus type 1 X region proteins in immortalization of primary human lymphocytes in culture. J Virol 66:4570–4575
Grassmann R, Dengler C, Muller-Fleckenstein I, Fleckenstein B, McGuire K, Dokhelar MC, Sodroski JG, Haseltine WA (1989) Transformation to continuous growth of primary human T lymphocytes by human T-cell leukemia virus type I X-region genes transduced by a Herpesvirus saimiri vector. Proc Natl Acad Sci USA 86:3351–3355
Grossman SR, Johannsen E, Tong X, Yalamanchili R, Kieff E (1994) The Epstein-Barr virus nuclear antigen 2 transactivator is directed to response elements by the J κ recombination signal binding protein. Proc Natl Acad Sci USA 91:7568–7572
Grossman WJ, Kimata JT, Wong FH, Zutter M, Ley TJ, Ratner L (1995) Development of leukemia in mice transgenic for the tax gene of human T-cell leukemia virus type I. Proc Natl Acad Sci USA 92:1057–1061
Guasparri I, Keller SA, Cesarman E (2004) KSHV vFLIP Is essential for the survival of infected lymphoma cells. J Exp Med 199:993–1003

Guasparri I, Wu H, Cesarman E (2006) The KSHV oncoprotein vFLIP contains a TRAF-interacting motif and requires TRAF2 and TRAF3 for signalling. EMBO Rep 7:114–119

Harhaj EW, Sun SC (1999) IKKγ serves as a docking subunit of the IκB kinase (IKK) and mediates interaction of IKK with the human T-cell leukemia virus Tax protein. J Biol Chem 274:22911–22914

Hinrichs SH, Nerenberg M, Reynolds RK, Khoury G, Jay G (1987) A transgenic mouse model for human neurofibromatosis. Science 237:1340–1343

Hiscott J, Petropoulos L, Lacoste J (1995) Molecular interactions between HTLV-1 Tax protein and the NF-κB/κB transcription complex. Virology 214:3–11

Horenstein MG, Nador RG, Chadburn A, Hyjek EM, Inghirami G, Knowles DM, Cesarman E (1997) Epstein-Barr virus latent gene expression in primary effusion lymphomas containing Kaposi's sarcoma-associated herpesvirus human herpesvirus-8. Blood 90:1186–1191

Hsieh JJ, Hayward SD (1995) Masking of the CBF1/RBPJκ transcriptional repression domain by Epstein-Barr virus EBNA2. Science 268:560–563

Humme S, Reisbach G, Feederle R, Delecluse HJ, Bousset K, Hammerschmidt W, Schepers A (2003) The EBV nuclear antigen 1 (EBNA1) enhances B cell immortalization several thousandfold. Proc Natl Acad Sci USA 100:10989–10994

Iwanaga Y, Tsukahara T, Ohashi T, Tanaka Y, Arai M, Nakamura M, Ohtani K, Koya Y, Kannagi M, Yamamoto N, Fujii M (1999) Human T-cell leukemia virus type 1 tax protein abrogates interleukin-2 dependence in a mouse T-cell line. J Virol 73:1271–1277

Jenner RG, Maillard K, Cattini N, Weiss RA, Boshoff C, Wooster R, Kellam P (2003) Kaposi's sarcoma-associated herpesvirus-infected primary effusion lymphoma has a plasma cell gene expression profile. Proc Natl Acad Sci USA 100:10399–10404

Keller SA, Dunne CE, Schattner EJ, Cesarman E (2000) Inhibition of NF-kB induces apoptosis of KSHV-infected primary effusion lymphoma cells. Blood 96:830a

Keller SA, Hernandez-Hopkins D, Vider J, Ponomarev V, Hyjek E, Schattner EJ, Cesarman E (2006) NF-κB is essential for progression of KSHV- and EBV-infected lymphomas in vivo. Blood 107:3295–3302

Kieser A, Kilger E, Gires O, Ueffing M, Kolch W, Hammerschmidt W (1997) Epstein-Barr-virus latent membrane protein-1 triggers ap-1 activity via the c-jun n-terminal kinase cascade. EMBO J 16:6478–6485

Kilger E, Kieser A, Baumann M, Hammerschmidt W (1998) Epstein-Barr virus-mediated B-cell proliferation is dependent upon latent membrane protein 1, which simulates an activated CD40 receptor. EMBO J 17:1700–1709

Klein U, Gloghini A, Gaidano G, Chadburn A, Cesarman E, Dalla-Favera R, Carbone A (2003) Gene expression profile analysis of AIDS-related primary effusion lymphoma (PEL) suggests a plasmablastic derivation and identifies PEL-specific transcripts. Blood 101:4115–4121

Kulwichit W, Edwards RH, Davenport EM, Baskar JF, Godfrey V, Raab-Traub N (1998) Expression of the Epstein-Barr virus latent membrane protein 1 induces B cell lymphoma in transgenic mice. Proc Natl Acad Sci USA 95:11963–11968

Lee H, Guo J, Li M, Choi JK, DeMaria M, Rosenzweig M, Jung JU (1998a) Identification of an immunoreceptor tyrosine-based activation motif of K1 transforming protein of Kaposi's sarcoma-associated herpesvirus. Mol Cell Biol 18:5219–5228

Lee H, Veazey R, Williams K, Li M, Guo J, Neipel F, Fleckenstein B, Lackner A, Desrosiers RC, Jung JU (1998b) Deregulation of cell growth by the K1 gene of Kaposi's sarcoma-associated herpesvirus. Nat Med 4:435–440

Lenzmeier BA, Giebler HA, Nyborg JK (1998) Human T-cell leukemia virus type 1 Tax requires direct access to DNA for recruitment of CREB binding protein to the viral promoter. Mol Cell Biol 18:721–731

Leung K, Nabel GJ (1988) HTLV-1 transactivator induces interleukin-2 receptor expression through an NF-κB-like factor. Nature 333:776–778

Li M, Lee H, Guo J, Neipel F, Fleckenstein B, Ozato K, Jung JU (1998) Kaposi's sarcoma-associated herpesvirus viral interferon regulatory factor. J Virol 1998:5433–5440

Li M, Lee H, Yoon D-W, Albrecht J-C, Fleckenstein B, Jung J (1997) Kaposi's sarcoma-associated herpesvirus encodes a functional cyclin. J Virol 71:1984–1991

Liu C, Okruzhnov Y, Li H, Nicholas J (2001) Human herpesvirus 8 (HHV-8)-encoded cytokines induce expression of and autocrine signaling by vascular endothelial growth factor (VEGF) in HHV-8-infected primary-effusion lymphoma cell lines and mediate VEGF-independent antiapoptotic effects. J Virol 75:10933–10940

Low W, Harries M, Ye H, Du MQ, Boshoff C, Collins M (2001) Internal ribosome entry site regulates translation of Kaposi's sarcoma-associated herpesvirus FLICE inhibitory protein. J Virol 75:2938–2945.

Lubyova B, Kellum MJ, Frisancho AJ, Pitha PM (2004) Kaposi's sarcoma-associated herpesvirus-encoded vIRF-3 stimulates the transcriptional activity of cellular IRF-3 and IRF-7. J Biol Chem 279:7643–7654

Mann DJ, Child ES, Swanton C, Laman H, Jones N (1999) Modulation of p27Kip1 levels by the cyclin encoded by Kaposi's sarcoma-associated herpesvirus. EMBO J 18:654–663

Maruyama K, Fukushima T, Kawamura K, Mochizuki S (1990) Chromosome and gene rearrangements in immortalized human lymphocytes infected with human T-lymphotropic virus type I. Cancer Res 50:5697S–5702S

Matolcsy A, Nador RG, Cesarman E, Knowles DM (1998) Immunoglobulin VH gene mutational analysis suggests that primary effusion lymphomas derive from different stages of B cell maturation. Am J Pathol 153:1609-1614

Matta H, Chaudhary PM (2004) Activation of alternative NF-κB pathway by human herpes virus 8-encoded Fas-associated death domain-like IL-1 beta-converting enzyme inhibitory protein (vFLIP). Proc Natl Acad Sci USA 101:9399–9404

Mesri EA, Cesarman E, Arvanitakis L, Rafii S, Moore, M. A. S., Posnett DN, Knowles DM, Asch AS (1996) Human herpesvirus-8/Kaposi's sarcoma-associated herpesvirus is a new transmissible virus that infects B cells. J Exp Med 183:2385–2390

Molden J, Chang Y, You Y, Moore PS, Goldsmith MA (1997) A Kaposi's sarcoma-associated herpesvirus-encoded cytokine homolog (vIL-6) activates signaling through the shared gp130 receptor subunit. J Biol Chem 272:19625–19631

Moore PS, Boschoff C, Weiss RA, Chang Y (1996) Molecular mimicry of human cytokine and cytokine response pathway genes by KSHV. Science 274:1739–1744

Moore PS, Chang Y (1995) Detection of herpesvirus-like DNA sequences in Kaposi's sarcoma lesions from persons with and without HIV infection. N Engl J Med 332:1181–1185

Mori N, Yamada Y, Ikeda S, Yamasaki Y, Tsukasaki K, Tanaka Y, Tomonaga M, Yamamoto N, Fujii M (2002) Bay 11-7082 inhibits transcription factor NF-κB and induces apoptosis of HTLV-I-infected T-cell lines and primary adult T-cell leukemia cells. Blood 100:1828–1834

Moses AV, Fish KN, Ruhl R, Smith PP, Strussenberg JG, Zhu L, Chandran B, Nelson JA (1999) Long-term infection and transformation of dermal microvascular endothelial cells by human herpesvirus 8. J Virol 73:6892–6902.

Nador RG, Cesarman E, Chadburn A, Dawson DB, Ansari MQ, Said J, Knowles DM (1996) Primary effusion lymphoma: a distinct clinicopathologic entity associated with the Kaposi's sarcoma-associated herpesvirus. Blood 88:645–656

Nador RG, Cesarman E, Knowles DM, Said JW (1995) Herpes-like DNA sequences in a body-cavity-based lymphoma in an HIV-negative patient (Letter to the Editor). N Engl J Med 333:943

Nerenberg M, Hinrichs SH, Reynolds RK, Khoury G, Jay G (1987) The tat gene of human T-lymphotropic virus type 1 induces mesenchymal tumors in transgenic mice. Science 237:1324–1329

Nicholas J, Ruvolo VR, Burns WH, Sandford G, Wan XY, Ciufo D, Hendrickson SB, Guo HG, Hayward GS, Reitz MS (1997) Kaposi's sarcoma-associated human herpesvirus-8 encodes homologues of macrophage inflammatory protein-1 and interleukin-6. Nat Med 3:287–292

Nicot C, Tie F, Giam CZ (1998) Cytoplasmic forms of human T-cell leukemia virus type 1 Tax induce NF-κB activation. J Virol 72:6777–6784

Panyutich EA, Said JW, Miles SA (1998) Infection of primary dermal microvascular endothelial cells by Kaposi's sarcoma-associated herpesvirus. AIDS 12:467–472

Petropoulos L, Hiscott J (1998) Association between HTLV-1 Tax and IκBα is dependent on the IκBα phosphorylation state. Virology 252:189–199

Picchio GR, Sabbe RE, Gulizia RJ, McGrath M, Herndier BG, Mosier DE (1997) The KSHV/HHV8-infected BCBL-1 lymphoma line causes tumors in SCID mice but fails to transmit virus to a human peripheral blood mononuclear cell graft. Virology 238:22–29

Pise-Masison CA, Radonovich M, Sakaguchi K, Appella E, Brady JN (1998) Phosphorylation of p53: a novel pathway for p53 inactivation in human T-cell lymphotropic virus type 1-transformed cells. J Virol 72:6348–6355

Poiesz BJ, Ruscetti FW, Gazdar AF, Bunn PA, Minna JD, Gallo RC (1980) Detection and isolation of type C retrovirus particles from fresh and cultured lymphocytes of a patient with cutaneous T-cell lymphoma. Proc Natl Acad Sci USA 77:7415-7419

Prakash O, Swamy OR, Peng X, Tang ZY, Li L, Larson JE, Cohen JC, Gill J, Farr G, Wang S, Samaniego F (2005) Activation of Src kinase Lyn by the Kaposi sarcoma-associated herpesvirus K1 protein: implications for lymphomagenesis. Blood 105:3987–3994

Radkov SA, Kellam P, Boshoff C (2000) The latent nuclear antigen of Kaposi sarcoma-associated herpesvirus targets the retinoblastoma-E2F pathway and with the oncogene hras transforms primary rat cells. Nat Med 6:1121–1127

Reitz MS, Jr., Popovic M, Haynes BF, Clark SC, Gallo RC (1983) Relatedness by nucleic acid hybridization of new isolates of human T-cell leukemia-lymphoma virus (HTLV) and demonstration of provirus in uncultured leukemic blood cells. Virology 126:688–672

Renne R, Zhong W, Herndier B, McGrath M, Abbey N, Kedes D, Ganem D (1996) Lytic growth of Kaposi's sarcoma-associated herpesvirus (human herpesvirus 8) in culture. Nat Med 2:342–346

Rickinson AB, Kieff E (1996). Epstein-Barr virus. In "Virology" (Fields, B.N., Knipe, D.M., and Howley, P.M., eds.), pp. 2397–2446. Lippincott-Raven Publishers, Philadelphia.

Rivas C, Thlick AE, Parravicini C, Moore PS, Chang Y (2001) Kaposi's sarcoma-associated herpesvirus LANA2 is a B-cell-specific latent viral protein that inhibits p53. J Virol 75:429–438.

Rodriguez A, Jung EJ, Flemington EK (2001a) Cell cycle analysis of Epstein-Barr virus-infected cells following treatment with lytic cycle-inducing agents. J Virol 75:4482–4489

Rodriguez A, Jung EJ, Yin Q, Cayrol C, Flemington EK (2001b) Role of c-myc regulation in Zta-mediated induction of the cyclin-dependent kinase inhibitors p21 and p27 and cell growth arrest. Virology 284:159–169

Ruben S, Poteat H, Tan TH, Kawakami K, Roeder R, Haseltine W, Rosen CA (1988) Cellular transcription factors and regulation of IL-2 receptor gene expression by HTLV-I tax gene product. Science 241:89–92

Said JW, Chien K, Tasaka T, Koeffler HP (1997) Ultrastructural characterization of human herpesvirus 8 (Kaposi's sarcoma-associated herpesvirus) in Kaposi's sarcoma lesions: electron microscopy permits distinction from cytomegalovirus (CMV). J Pathol 182:273–281

Said JW, Tasaka T, Takeuchi S, Asou H, de Vos S, Cesarman E, Knowles DM, Koeffler HP (1996) Primary effusion lymphoma in women: report of two cases of Kaposi's sarcoma-herpes virus-associated effusion-based lymphoma in human immunodeficiency virus-negative women. Blood 88:3124–3128

Sanada I, Tanaka R, Kumagai E, Tsuda H, Nishimura H, Yamaguchi K, Kawano F, Fujiwara H, Takatsuki K (1985) Chromosomal aberrations in adult T cell leukemia: relationship to the clinical severity. Blood 65:649–654

Sharp TV, Wang HW, Koumi A, Hollyman D, Endo Y, Ye H, Du MQ, Boshoff C (2002) K15 protein of Kaposi's sarcoma-associated herpesvirus is latently expressed and binds to HAX-1, a protein with antiapoptotic function. J Virol 76:802–816.

Si H, Robertson ES (2006) Kaposi's sarcoma-associated herpesvirus-encoded latency-associated nuclear antigen induces chromosomal instability through inhibition of p53 function. J Virol 80:697–709

Staudt MR, Kanan Y, Jeong JH, Papin JF, Hines-Boykin R, Dittmer DP (2004) The tumor microenvironment controls primary effusion lymphoma growth in vivo. Cancer Res 64:4790–4799

Sun SC, Ballard DW (1999) Persistent activation of NF-κB by the tax transforming protein of HTLV-1: hijacking cellular IκB kinases. Oncogene 18:6948–6958

Szekely L, Chen F, Teramoto N, Ehlin-Henriksson B, Pokrovskaja K, Szeles A, Manneborg-Sandlund A, Lowbeer M, Lennette ET, Klein G (1998) Restricted expression of Epstein-Barr virus (EBV)-encoded, growth transformation-associated antigens in an EBV- and human herpesvirus type 8-carrying body cavity lymphoma line. Gen Virol 79:1445–1452

Talbot SJ, Weiss RA, Kellam P, Boshoff C (1999) Transcriptional analysis of human herpesvirus-8 open reading frames 71, 72, 73, K14, and 74 in a primary effusion lymphoma cell line. Virology 257:84–94

Thome M, Schneider P, Hofmann K, Fickenscher H, Meinl E, Neipel F, Mattmann C, Burns K, Bodmer JL, Schroter M, Scaffidi C, Krammer PH, Peter ME, Tschopp J (1997) Viral FLICE-inhibitory proteins (FLIPs) prevent apoptosis induced by death receptors. Nature 386:517–521

Tomlinson CC, Damania B (2004) The K1 protein of Kaposi's sarcoma-associated herpesvirus activates the Akt signaling pathway. J Virol 78:1918–1927

Uchida T, Kinoshita T, Murate T, Saito H, Hotta T (1998) CDKN2 (MTS1/p16INK4A) gene alterations in adult T-cell leukemia/lymphoma. Leuk Lymphoma 29:27–35

Verschuren EW, Hodgson JG, Gray JW, Kogan S, Jones N, Evan GI (2004) The role of p53 in suppression of KSHV cyclin-induced lymphomagenesis. Cancer Res 64:581–589

Verschuren EW, Klefstrom J, Evan GI, Jones N (2002) The oncogenic potential of Kaposi's sarcoma-associated herpesvirus cyclin is exposed by p53 loss in vitro and in vivo. Cancer Cell 2:229–241

Wang D, Liebowitz D, Kieff E (1985) An EBV membrane protein expressed in immortalized lymphocytes transforms established rodent cells. Cell 43:831–840

Wilson JB, Bell JL, Levine AJ (1996) Expression of Epstein-Barr virus nuclear antigen-1 induces B cell neoplasia in transgenic mice. EMBO J 15:3117–3126

Wu FY, Tang QQ, Chen H, ApRhys C, Farrell C, Chen J, Fujimuro M, Lane MD, Hayward GS (2002) Lytic replication-associated protein (RAP) encoded by Kaposi sarcoma-associated herpesvirus causes p21CIP-1-mediated G1 cell cycle arrest through CCAAT/enhancer-binding protein-α. Proc Natl Acad Sci USA 99:10683–10688

Wu W, Rochford R, Toomey L, Harrington W, Jr., and Feuer G (2005) Inhibition of HHV-8/KSHV infected primary effusion lymphomas in NOD/SCID mice by azidothymidine and interferon-α. Leuk Res 29:545–555

Xiao G, Cvijic ME, Fong A, Harhaj EW, Uhlik MT, Waterfield M, Sun SC (2001) Retroviral oncoprotein Tax induces processing of NF-κB2/p100 in T cells: evidence for the involvement of IKKα. EMBO J 20:6805–6815

Yoshida M (2001) Multiple viral strategies of HTLV-1 for dysregulation of cell growth control. Annu Rev Immunol 19:475–496

Yoshida M, Hattori S, Seiki M (1985) Molecular biology of human T-cell leukemia virus associated with adult T-cell leukemia. Curr Top Microbiol Immunol 115:157–175

Yoshida M, Seiki M, Yamaguchi K, Takatsuki K (1984) Monoclonal integration of human T-cell leukemia provirus in all primary tumors of adult T-cell leukemia suggests causative role of human T-cell leukemia virus in the disease. Proc Natl Acad Sci USA 81:2534–2537

Zimring JC, Goodbourn S, Offermann MK (1998) Human herpesvirus 8 encodes an interferon regulatory factor (IRF) homolog that represses IRF-1-mediated transcription. J Virol 72:701–707

CTMI (2006) 312:289–307

The Use of Antiviral Drugs in the Prevention and Treatment of Kaposi Sarcoma, Multicentric Castleman Disease and Primary Effusion Lymphoma

C. Casper (✉) · A. Wald

University of Washington Virology Research Clinic,
600 Broadway, Suite 400, Seattle, WA 98122, USA
ccasper@u.washington.edu

Abstract Kaposi sarcoma-associated herpesvirus [KSHV, also known as human herpesvirus 8 (HHV-8)] is the most recently identified member of the human herpesvirus family. Kaposi sarcoma (KS), primary effusion lymphoma, and multicentric Castleman disease are all associated with KSHV infection. Although the incidence of KS has declined dramatically in areas with access to highly active antiretroviral therapy, it remains the most common AIDS-associated malignancy in the developed world and is one of the most common cancers in developing nations. Current treatment options for KSHV-associated disease are ineffective, unavailable, or toxic to many affected persons. A growing body of basic science, preclinical, and observational data suggests that antiviral medications may play an important role in the prevention and treatment of KSHV-associated disease.

1 Introduction

Kaposi sarcoma (KS) was one of the initial harbingers of the human immunodeficiency virus (HIV) pandemic in 1981 (Center for Disease Control 1981; Hymes et al. 1981). At that time, one in three patients with HIV in North America and western Europe developed KS. Chemotherapy was the mainstay of treatment, leading to clinical resolution in less than half the cases (Krown 2004). By 1994, when Kaposi sarcoma-associated herpesvirus [KSHV, also known as human herpesvirus 8 (HHV-8)] was first identified as the cause of KS (Chang et al. 1994), the incidence had already begun to decline (Eltom et al. 2002). The widespread introduction of highly active antiretroviral therapy (HAART) in the developed world in 1996 was accompanied by a tenfold reduction in the number of new KS cases (Eltom et al. 2002). The precipitous decline in the prevalence of KS shortly after the identification of KSHV, coupled with the challenges of propagating KSHV in culture and the lack of an animal model, has led to few new advances in the treatment of KSHV-related disease. HAART, either alone or in conjunction with conventional chemotherapy, remains the mainstay of treatment for HIV-associated KS. To date, no large randomized trials support the use of one therapy over another for the treatment or prevention of KS, and the efficacy of these treatments is not known. Furthermore, treatment of HIV-negative KS, or other KSHV-associated diseases such as multicentric Castleman disease (MCD) and primary effusion lymphoma (PEL), is based primarily on anecdotes. Worldwide, KS is the most common cancer in areas where coinfection with HIV and KSHV is highly prevalent, and the frequency with which MCD or PEL occurs in the developing world is not known.

Since the introduction of (minimally toxic) antiviral therapy over 25 years ago, numerous clinical studies have documented the efficacy of DNA synthesis inhibitors against human herpesvirus infections. A substantial body of evidence has accumulated supporting a role for antiviral agents in the treatment and prevention of KSHV-associated disease, providing optimism that novel therapies for these diseases may be found.

2 In Vitro and Animal Model Evidence for the Efficacy of Antiviral Drugs for KSHV

2.1 DNA Synthesis Inhibitors

Antiviral therapy was revolutionized in 1978 with the synthesis of acyclovir, a drug that could selectively inhibit viral DNA replication with minimal toxicity to humans (Schaeffer et al. 1978). Acyclovir is a synthetic guanine analog, which after triphosphorylation is incorporated into the 3′ end of actively replicating DNA, leading to chain termination. The specificity of acyclovir activity is attributed to the high avidity of acyclovir for many human herpesvirus thymidine kinases, which are capable of efficient phosphorylation, and the relative inefficiency of cellular kinases in performing the initial phosphorylation of the drug. Although all wild-type human herpesviruses are thought to possess the thymidine kinase enzyme, not all accept acyclovir as a substrate, leading to significant differences in the ability of acyclovir to inhibit DNA synthesis. The valine ester derivative of acyclovir, valacyclovir, as well as the chemically similar penciclovir and its oral prodrug, famciclovir, have similar mechanisms of action and substrate specificities. Ganciclovir and its valine ester derivative, valganciclovir, compete with cellular dGTP for incorporation into the growing viral DNA chain. These drugs differ in their avidity for the viral kinase, with increased activity against cytomegalovirus (CMV). Foscarnet prevents the binding of pyrophosphate to the viral DNA polymerase through inhibition of the phosphotransferase enzyme, resulting in termination of replication in a thymidine kinase-independent manner. Finally, cidofovir is a cytosine analog that only requires diphosphorylation to cause chain termination after incorporation by the viral DNA polymerase, independent of the action of thymidine kinase.

The nucleotide sequence of KSHV suggests that the virus encodes both a thymidine kinase (TK) (ORF 21) and a phosphotransferase (PT) (ORF 36). These regions share several domains that are conserved among human herpesviruses; the KSHV TK and PT have sequence homologies of nearly 25% when compared with Epstein Barr virus (EBV), and 12%–16% of the nucleotides are shared with HSV and VZV (Cannon et al. 1999). Additionally, domains within the KSHV TK and PT regions show homology to other human herpesvirus sequences that are known to confer susceptibility to ganciclovir. 293 cells transfected with KSHV ORF 21 or ORF 36 are capable of phosphorylating ganciclovir (Cannon et al. 1999). Based on these early observations, a number of in vitro experiments have been performed to determine whether antiherpetic antivirals inhibit KSHV replication.

Although "classic" in vitro antiviral susceptibility testing relies on observing a reduction in viral cytopathic effect (CPE) in cell culture, the inability to observe CPE in KSHV-infected cell lines necessitates novel approaches. The identification of an immortalized cell line that is persistently infected with latent KSHV (BCBL-1), and the recognition that KSHV could be stimulated to lytic replication by the addition of 12-*O*-tetradecanoyl phorbol-13-acetate (TPA) (Renne et al. 1996), has allowed for limited studies of antiviral activity. The amount of KSHV viral DNA produced after stimulating replication with TPA, as quantified by electrophoresis and Southern blotting, was reduced in the presence of ganciclovir (Kedes and Ganem 1997; Medveczky et al. 1997; Neyts and De Clercq 1997), foscarnet (Kedes and Ganem 1997; Medveczky et al. 1997; Neyts and De Clercq 1997), and cidofovir (Kedes and Ganem 1997; Medveczky et al. 1997), with acyclovir having minimal effect (Kedes and Ganem 1997; Medveczky et al. 1997; Neyts and De Clercq 1997). The estimated concentrations at which 50% of viral DNA was inhibited (IC_{50}) ranged from 60 to 80 μM for acyclovir [peak serum concentration 100 μM after administration of 10 mg/kg intravenously (IV)]), from 2 to 5 μM for ganciclovir (peak serum concentration 50 μM after administration of 5 mg/kg IV or 900 mg oral valganciclovir), from 0.05 to 1 μM for cidofovir (peak serum concentration 30–70 μM after administration of 5 mg / kg IV), and from 80to 500 μM for foscarnet (peak serum concentration approximately 1,000–6,000 μM after administration of 90 mg/kg IV). Similar results were described after examining the amount of viral DNA produced with real-time polymerase chain reaction (Friedrichs et al. 2004). Estimates of inhibitory concentrations using flow cytometry to identify cells expressing the lytic envelope protein K8.1 differed significantly, however, with cidofovir (EC_{50} of 3.9 μM) and foscarnet (EC_{50} 74.7 μM) being identified as the only active agents. Other drugs had estimated inhibitory concentrations that would not be sufficiently exceeded with typical administration (acyclovir 38.3 μM and ganciclovir 53.7 μM).

It is not known whether the degree to which antiviral drugs inhibit lytic reactivation from latency in an immortalized cell line mimics antiviral activity in vivo. However, two other systems have been established to examine antiviral activity against KSHV. Human microvascular endothelial cells are infected with KSHV in patients with KS (Dupin et al. 1999) and may be important in the initiation of tumorigenesis (Carroll et al. 2004). In a primary cell line of human microvascular endothelial cells, foscarnet completely suppressed the production of infectious KSHV after primary infection with KSHV (Krug et al. 2004). In another approach, immunocompromised mice implanted with human fetal thymus and liver grafts (SCID-hu Thy/Liv) (McCune et al. 1988) allow productive infection with KSVH after direct inoculation with virus harvested from BCBL-1 cells (Dittmer et al. 1999). The intraperitoneal administration of ganciclovir both prevented the establishment of infection and also decreased the amount of latent and lytic KSHV mRNA detected after infection was established.

2.2 Antiretroviral Therapy

The incidence of KS in the United States began to decline in 1987 (Eltom et al. 2002), shortly after the approval of zidovudine for the treatment of HIV. More dramatic reductions were seen after 1995, when the use of protease inhibitors (PI) became common (Eltom et al. 2002). The mechanism of this effect has been thought to be multifactorial, including the deaths of many persons infected with HIV in the early epidemic, the immune reconstitution seen with antiretroviral therapy, or perhaps a direct effect of the drugs against KSHV. Both direct and indirect effects of antiretroviral therapy on KSHV have been studied in vitro and are summarized below.

2.2.1 Reverse Transcriptase Inhibitors

Two experiments have cloned the KSHV TK and used bacterial expression vectors to assess the ability of this enzyme to phosphorylate HIV nucleoside reverse transcriptase inhibitors (NRTI). The thymidine analogs zidovudine (Gustafson et al. 2000; Lock et al. 2002) and stavudine (Lock et al. 2002) both were phosphorylated by KSHV TK and competitively inhibited the incorporation of radiolabeled thymidine.

2.2.2 Protease Inhibitors

HIV PIs prohibit the assembly of HIV virions by occupying the active site of an aspartyl protease. Herpesviruses use a serine protease during lytic replication to begin assembly of viral particles, which would not be affected by HIV PIs (Shimba et al. 2004). Coculturing BCBL-1 cells with indinavir had no effect on the expression of latent or lytic KHSV genes before or after stimulation with TPA, supporting that the effect of these drugs on KS was not through a direct effect on KSHV replication (Sgadari et al. 2003).

It is plausible that several mechanisms may mediate both direct and indirect effects of PIs on KSHV replication. For example, the use of PIs is associated with a reduction in the amount of HIV-1 Tat, which has been shown to increase KSHV replication (Huang et al. 2001). Ritonavir suppressed the ability of HIV-1 Tat to induce transcriptional activation of nuclear factor κB (NF-kB), which is thought to lead to the development of KS (Pati et al. 2002).

In addition to affecting KSHV replication, PIs may prevent the development of KS from KSHV-infected cells. Coculturing KSHV-infected immortalized cells with ritonavir enhanced their apoptosis (Pati et al. 2002). The use of the SCID-hu mouse model confirmed that ritonavir (Pati et al. 2002), saquinavir (Sgadari et al. 2002), and indinavir (Sgadari et al. 2002) inhibited KS tumor formation after infection with KSHV. Both saquinavir and ritonavir prevented the invasion of human vascular endothelial cells through the basement membrane, an important initial step in angiogenesis and KS tumorigenesis.

Finally, the immune reconstitution seen with HAART may have profound effects in controlling KSHV infection. Cytotoxic lymphocyte (CTL) responses have been examined for seven KSHV proteins, with an increase in CTL to two KSHV proteins in subjects receiving HAART (Wilkinson et al. 2002). Lysis of KSHV latently infected cells by NK cells is deficient in AIDS patients with progressive KS but is restored after HAART (Sirianni et al. 2002). Taken together, these studies support an indirect effect of ART on KSHV through improved cellular immunity.

2.3 Summary of Preclinical Work

Taken together, the body of in vitro and animal model work supports the ability of certain antiviral agents to suppress KSHV replication and prevent establishment of productive infection. Furthermore, the HIV PIs may block some of the early steps that are thought to be essential to initiating tumorigenesis. The translatability of these findings to treating or preventing disease in humans may be tempered by a number of factors.

First, agents that suppress DNA replication would not be expected to have an effect on early KSHV gene products, which are produced by a protein in the KSHV virion and not by viral DNA polymerase, or latent KSHV genes that are transcribed by human DNA polymerase. Many of these gene products may play an important role in the development of human disease, such as the immediate-early gene products viral interleukin 6 (vIL-6) or the viral G protein-coupled receptor (vGPCR), and the latency-associated nuclear antigen (ORF 73). Gene array technology confirms that cidofovir has little effect on early gene products, including vIL-6 and vGPCR, but effectively reduces production of a number of late structural genes (Lu et al. 2004). Additional experiments show that acyclovir, foscarnet, and cidofovir all reduce the ability of BCBL-1 cells induced with TPA to express the late gene product of K8.1, a virion glycoprotein (Zoeteweij et al. 1999). These drugs did not alter the expression of the early gene product ORF 59, a nonstructural protein that is thought to bind DNA polymerase to assist in transcription. Similarly, it has recently been described that in BCBL-1 cells valproic acid induces both lytic replication of KSHV and apoptosis (Klass et al. 2005). Coculturing the cells with foscarnet or ganciclovir inhibited the production of the late structural gene ORF26 but had no effect on vIL-6 or vGPCR. As discussed below, the clinical relevance of these findings is unclear. Antiviral medications have been shown to be associated with improvements in symptoms from MCD, which directly follow reductions in KSHV viral load (Casper et al. 2004)

Second, the role that KSHV replication plays in clinical disease is not clear. In the cases of KS or PEL, it could be argued that KSHV latent and lytic gene products are essential for malignant transformation, but once transformation has occurred their role is minimized. Examination of KS tissue shows that only 5% of cells harbor lytic virus; whether this activity is required for tumor maintenance is not known.

A small number of studies have evaluated the utility of antiviral therapy in the prevention and treatment of KSHV-related disease, as discussed below.

3 Prevention of Kaposi Sarcoma with Antiviral Agents

Additional indirect evidence for antiviral efficacy against KSHV derives from cohort studies conducted early in the HIV epidemic (Table 1). Patients treated with either antiretroviral agents or DNA synthesis inhibitors early in the HIV epidemic were witnessed to have significant reductions in the risk of developing KS.

3.1 Herpesvirus Antiviral Medications

Among 3,688 HIV-positive men in England, the risk of developing KS in persons using intravenous foscarnet or ganciclovir for the treatment of CMV retinitis was reduced by over 60% (Mocroft et al. 1996). Similarly, among 935 HIV-positive American men enrolled in the Multicenter AIDS Cohort Study, the risk of developing KS was reduced by 44% with ganciclovir use and 60% with foscarnet (Glesby et al. 1996). Neither study found acyclovir use to be associated with a lower risk of developing KS. A separate study of high-dose acyclovir for the treatment of HIV showed no reduction in the incidence of KS among persons treated with acyclovir (Ioannidis et al. 1998). A comparison of oral to intravenous ganciclovir administered for the treatment of HIV-associated CMV retinitis found that the risk of developing KS was reduced

Table 1 Clinical studies evaluating the efficacy of antiviral medications in the prevention of Kaposi sarcoma

Study	Antiviral medication(s)	Results
Mocroft (1996)	Foscarnet Ganciclovir Acyclovir	Among 3,688 HIV-positive men in England: • 72% reduction in the chance that persons receiving ganciclovir or foscarnet would develop KS compared with those not receiving either drug (HR 0.38 and 0.39, respectively). • Acyclovir not found to be associated with reduced incidence of KS (HR 1.1).
Glesby (1996)	Foscarnet Ganciclovir Acyclovir	Among 935 HIV-positive men in the US, incidence of KS reduced by: • 60% with foscarnet • 44% with ganciclovir Acyclovir was not found to be associated with reductions in KS.
Ionnidis (1998)	Acyclovir	Meta-analysis of 8 trials including 1,792 patients found no reduction in the odds of developing KS among persons taking acyclovir.
Martin (1999)	Oral vs. intravenous ganciclovir	Risk of developing KS was reduced 75% among HIV-positive men receiving oral ganciclovir for CMV retinitis and 93% in those receiving intravenous ganciclovir.
Joffe (1997)	Zidovudine	Meta-analysis of trials evaluating high-dose zidovudine for the treatment of HIV found 36% reduction in the rate of KS development.
Joffe (1997)	Zidovudine	Analysis of 1,847 men from Multicenter AIDS Cohort Study found no increase in time to KS among persons receiving zidovudine.
Portsmouth (2003)	HAART containing PI vs. NNRTI	Prospective study of 1,204 HIV-positive persons found risk of developing KS was reduced 58% among those receiving NNRTIs and 53% with PIs (difference not statistically significant).
Stebbing (2004)	Ritonavir-containing HAART	No significant difference in the incidence of KS among English HIV-positive men treated with and without ritonavir-containing HAART

KS, Kaposi sarcoma; HR, hazard ratio; CMV, cytomegalovirus; HAART, highly active antiretroviral therapy; PI, protease inhibitor; NNRTI, nonnucleoside reverse transcriptase inhibitor

75% and 93%, respectively, compared with an intravitreal ganciclovir implant alone (Martin et al. 1999). Although these data are encouraging, no randomized trial of antivirals for KS prevention has been reported, and the efficacy in preventing KS in persons without advanced HIV, or preventing MCD or PEL, remains unknown.

3.2 Antiretroviral Therapy

Several randomized, placebo-controlled trials of high-dose zidovudine ($\geq$1000 mg daily) for the treatment of HIV demonstrated a 36% reduction in risk of developing KS compared with persons receiving placebo alone (reviewed in Joffe et al. 1997). Participants in these trials were extremely heterogeneous with respect to their stage of HIV and indication for zidovudine. An analysis of incident KS in the prospective Multicenter AIDS Cohort Study found no reduction in KS associated with zidovudine use, after adjustment for the indication for initiating ART and changes in CD4 count (Joffe et al. 1997).

Subsequent analyses have examined the effect of combination antiretroviral therapy on the incidence of KS. The large Swiss HIV Cohort Study found that the standardized incidence ratio (SIR) for KS among HIV-positive persons was reduced nearly 10-fold for those who used HAART before developing KS (SIR for HAART users 25.3 vs. 23.9 for nonusers) (Clifford et al. 2005). The risk of developing KS was reduced 26% among 1,204 HIV-positive English men who used NRTIs alone, 53% in those on HAART regimens with nonnucleoside reverse transcriptase inhibitors (NNRTI), and 58% among men taking PIs after adjustment for age, ethnicity, and nadir CD4 count before developing KS (Portsmouth et al. 2003). In contrast to what might be expected from in vitro data, there was no statistically significant difference in the effect of NNRTIs vs. PIs in preventing KS, and regimens containing ritonavir were no more effective in KS prevention (Stebbing et al. 2004). These findings suggest that immune reconstitution may be an important component of the improvements in KS that have been well documented after initiation of HAART.

3.3 Conclusions About KS Prevention

The consistent and compelling data from international cancer registries documenting reductions in the incidence of KS with the introduction of antiviral therapy, coupled with supportive basic science research, argue that KS can be prevented in those at highest risk for its development. To date, apart from immunosuppression, the only risk factor for the development of KS has been the detection of KSHV in the peripheral blood (Whitby et al. 1995; Campbell et al. 2000; Broccolo et al. 2002; Gill et al. 2002; Lorenzen et al. 2002; Cannon et al. 2003). As discussed below, the treatment of KSHV-infected patients with either PIs (Leao et al. 2000) or ganciclovir (Casper et al. 2004) is associated with reductions in KSHV viral load. Prospective clinical trials are needed to determine whether antiviral therapy reduces the incidence of epidemic, classic, or endemic KS in KSHV-viremic persons.

4 Treatment of KSHV-Associated Disease with Antivirals

4.1 Kaposi Sarcoma

Shortly after their introduction, case reports indicated that the use of PIs was associated with remission of AIDS-associated KS (Conant et al. 1997; Murphy et al. 1997; Diz Dios et al. 1998; Jung et al. 1998; Krischer et al. 1998; Lebbe et al. 1998; Martinelli et al. 1998; Tavio et al. 1998). Subsequent larger observational series evaluated regimens with both PIs and NNRTIs (Table 2). Among 78 patients with AIDS-associated KS, HAART with either a PI or a NNRTI prolonged the time to receipt of additional topical or systemic therapy, compared with treatment courses in these same patients before initiating HAART (Bower et al. 1999). Twelve of 14 (86%) of patients previously treated with ART or chemotherapy had remissions in their KS documented at a median of 4 months after initiating PI-based ART; of note, the two patients who failed to respond experienced similar increases in CD4 count but had persistent HIV viremia (Cattelan et al. 2001). These observations were corroborated by a separate study of 26 KS patients, of whom 22 (85%) achieved at least a partial response (Pellet et al. 2001). KSHV viremia was also significantly reduced among persons on PIs, which was independently predictive of KS improvement (Leao et al. 2000; Pellet et al. 2001). Finally, two studies documenting the relapse (Bani-Sadr et al. 2003) and new onset (Rey et al. 2001) of KS after switching from PI-based to NNRTI-based HAART suggest that the efficacy of all ART in treating KS may not be the same. To date, no randomized prospective trials have evaluated the frequency and predictors of KS regression among persons treated with HAART (Krown 2004).

The efficacy of DNA synthesis inhibitors in treating KS is less well established. The largest trial to date used intravenous cidofovir to treat seven patients with advanced KS (T_1 disease or >50 lesions), including five HIV-positive patients (4 on HAART), and two patients with classic KS (Little et al. 2003). No patient had either a complete or a partial response to cidofovir, and the median time to progression was 8.1 weeks.

Table 2 Clinical studies evaluating the efficacy of antiviral medications in the treatment of Kaposi sarcoma

Study	Treatment	Findings
Bower (1999)	HAART with PI or NNRT vs. no HAART	• 78 patients with HIV-associated KS had the time from first to second topical or systemic treatment of KS evaluated before and after starting HAART. • Before HAART, median time to second treatment was approximately 9 months, which increased significantly to 1.7 years after initiation of HAART. • No significant difference in time to second treatment among persons on NNRTIs vs. PIs • Persistent HIV viremia while on HAART was the strongest predictor of KS development.
Cattelan (2001)	HAART with PI	• 86% of 14 patients with HIV-associated KS had response to HAART with PI.s • Median time to response was 6 months. • Response predicted by absence of HIV viremia
Pellet (2001)	HAART with PI	• 85% of 26 patients with HIV-associated KS had response to HAART with PIs. • Median time to response was 251 days • Response predicted by lower KSHV viral load and increases in CD4 counts
Badi-Sadr (2003)	HAART	• 5 patients with HIV-associated KS had regression of KS documented with initiation of PI-containing HAART. • All 5 patients relapsed at a median of 11 months when switched to NNRTI-containing HAART, despite little change in CD4 count. • Relapse after NNRTI was not associated with HIV virologic failure.
Rey (2001)	HAART with PI vs. efavirenz	• One of 62 HIV-positive patients developed KS after switching from PI to efavirenz.
Little (2003)	Cidofovir	• None of 5 patients with HIV-associated KS or 2 patients with classic KS responded to cidofovir. • Median time to progression was 8.1 weeks

KS, Kaposi sarcoma; HAART, highly active antiretroviral therapy; PI, protease inhibitor; NNRTI, nonnucleoside reverse transcriptase inhibitor

4.2
Primary Effusion Lymphoma

The relative rarity of PEL makes definitive evaluations of its treatment challenging, and evidence for the efficacy of antiviral therapy comes from case studies alone (Table 3). The first report described a 38-year-old man with HIV-associated PEL, admitted with hemophagocytic syndrome, who was successfully treated with ganciclovir and chemotherapy (Pastore et al. 2000). Subsequently, additional reports described the successful treatment of PEL with intravenous cidofovir and interferon α (Hocqueloux et al. 2001) or intrapleural cidofovir in three elderly Italian HIV-negative patients (Luppi et al. 2005).

4.3
Multicentric Castleman Disease

Of all the KSHV-associated diseases, MCD may be associated with the highest degree of active lytic replication (Katano et al. 2000, 2001). In support of this, clinical flares of MCD are consistently accompanied by KSHV viremia, which resolve with effective therapy (Grandadam et al. 1997; Bottieau et al. 2000; Oksenhendler et al. 2000; Corbellino et al. 2001; Boivin et al. 2002; Berezne et al. 2004; Casper et al. 2004). Our group initially showed that treatment of three patients with KSHV/HIV-associated MCD with ganciclovir or valganciclovir led to rapid resolution of clinical symptoms and viremia (Casper et al. 2004). Two of the three experienced disease-free intervals of over 1 year. In contrast, a subsequent series showed that five patients with HIV-associated MCD failed to respond to the combination of cidofovir and chemotherapy (Berezne et al. 2004). The role of antiretroviral therapy in the treatment of MCD is also poorly defined. One series of seven cases showed no relationship between ART, CD4 count, and the course of MCD (Aaron et al. 2002). A second, however, suggested that the initiation of ART in three patients with MCD could lead to a fulminant and fatal course (Zietz et al. 1999) (Table 3).

5
Conclusions and Future Directions

In the decade since the first description of KSHV, the combination of basic science research, retrospective epidemiologic studies, and clinical observations supports a role for antiviral therapy in the prevention and treatment of KSHV-related disease. Similar to what has been described with Epstein-Barr-associated malignancy, antiviral therapy is likely to have the greatest impact on the prevention of oncogenesis. In areas of the world where KS remains

Table 3 Treatment of primary effusion lymphoma and multicentric Castleman disease with antiviral medications

Study	Disease	Therapy	Findings
Pastore (2000)	PEL	Chemotherapy and ganciclovir	• Report of HIV-positive patient with hemophagocytic syndrome and PEL who responded to intravenous and oral ganciclovir, HAART, cyclophosphamide, vincristine, and prednisone
Hocquelooux (2001)	PEL	Cidofovir, interferon-α, HAART	• Report of HIV-positive patient with PEL who responded to combination therapy including intravenous cidofovir
Luppi (2005)	PEL	Cidofovir	• Series of 3 patients with classic KS received 2–3 doses of intrapleural cidofovir for thoracic PEL. • All 3 had remission, as documented by repeat radiologic studies of the chest.
Casper (2004)	MCD	Ganciclovir or valganciclovir	• 3 patients with HIV-associated MCD experienced clinical and virologic responses to treatment with intravenous ganciclovir or oral valganciclovir.
Berezne (2004)	MCD	Cidofovir and chemotherapy	• 5 patients with HIV-associated MCD experienced recurrent symptoms despite cidofovir and either vinblastine or etoposide.
Aaron (2002)	MCD	HAART with PI	• 7 patients with MCD had recurrent flares despite increases in CD4 counts and HIV virologic suppression in 6/7. • 4/7 had comorbid KS, which resolved in 3/4. • 1 patient failed to respond to chemotherapy, ganciclovir, cidofovir, foscarnet, splenectomy, or corticosteroids.
Zietz (1999)	MCD	HAART	• 3 patients developed symptoms of MCD and died shortly after initiating unspecified HAART regimen.

PEL, primary effusion lymphoma; MCD, multicentric Castleman disease; HAART, highly active antiretroviral therapy; PI, protease inhibitor; NNRTI, nonnucleoside reverse transcriptase inhibitor a prevalent disease, future studies should be directed at the identification of persons at highest risk for developing KS from asymptomatic KSHV infection and determining the optimal antiviral prevention strategies. Although far less common, antiviral therapy may be most promising in the prevention and treatment of the more "active" KSHV diseases, PEL or MCD. Finally, it is clear that even within a class of antiviral drugs, there may be significant heterogeneity in the clinical response to their use for a given disease. As such, comparative trials of antiviral or antiretroviral therapy would be of great benefit to those afflicted with KSHV-associated disease.

Acknowledgements Supported by funding from the National Institutes of Health, K23 AI054162 and U19 AI31448.

References

Aaron L, Lidove O, Yousry C, Roudiere L, Dupont B, Viard JP (2002) Human herpesvirus 8-positive Castleman disease in human immunodeficiency virus-infected patients: the impact of highly active antiretroviral therapy. Clin Infect Dis 35:880–882

Bani-Sadr F, Fournier S, Molina JM (2003) Relapse of Kaposi's sarcoma in HIV-infected patients switching from a protease inhibitor to a non-nucleoside reverse transcriptase inhibitor-based highly active antiretroviral therapy regimen. AIDS 17:1580–1581

Berezne A, Agbalika F, Oksenhendler E, Casper C, Nichols WG, Huang M-L, Corey L, Wald A (2004) Failure of cidofovir in HIV-associated multicentric Castleman disease. Blood 103:4368–4369

Boivin G, Cote S, Cloutier N, Abed Y, Maguigad M, Routy JP (2002) Quantification of human herpesvirus 8 by real-time PCR in blood fractions of AIDS patients with Kaposi's sarcoma and multicentric Castleman's disease. J Med Virol 68:399–403

Bottieau E, Colebunders R, Schroyens W, Van Droogenbroeck J, De Droogh E, Depraetere K, De Raeve H, Van Marck E (2000) Multicentric Castleman's disease in 2 patients with HIV infection, unresponsive to antiviral therapy. Acta Clin Belg 55:97–101

Bower M, Fox P, Fife K, Gill J, Nelson M, Gazzard B (1999) Highly active anti-retroviral therapy (HAART) prolongs time to treatment failure in Kaposi's sarcoma. Aids 13:2105–2111

Broccolo F, Bossolasco S, Careddu AM, Tambussi G, Lazzarin A, Cinque P (2002) Detection of DNA of lymphotropic herpesviruses in plasma of human immunodeficiency virus-infected patients: frequency and clinical significance. Clin Diagn Lab Immunol 9:1222–1228

Campbell TB, Borok M, Gwanzura L, MaWhinney S, White IE, Ndemera B, Gudza I, Fitzpatrick L, Schooley RT (2000) Relationship of human herpesvirus 8 peripheral blood virus load and Kaposi's sarcoma clinical stage. AIDS 14:2109–2116

Cannon JS, Hamzeh F, Moore S, Nicholas J, Ambinder RF (1999) Human herpesvirus 8-encoded thymidine kinase and phosphotransferase homologues confer sensitivity to ganciclovir. J Virol 73:4786–4793

Cannon MJ, Dollard SC, Black JB, Edlin BR, Hannah C, Hogan SE, Patel MM, Jaffe HW, Offermann MK, Spira TJ, Pellett PE, Gunthel CJ (2003) Risk factors for Kaposi's sarcoma in men seropositive for both human herpesvirus 8 and human immunodeficiency virus. AIDS 17:215–222

Carroll PA, Brazeau E, Lagunoff M (2004) Kaposi's sarcoma-associated herpesvirus infection of blood endothelial cells induces lymphatic differentiation. Virology 328:7-18

Casper C, Nichols WG, Huang ML, Corey L, Wald A (2004) Remission of HHV-8 and HIV-associated multicentric Castleman disease with ganciclovir treatment. Blood 103:1632–1634

Cattelan A, Calabro M, Gasperini P, Aversa S, Zanchetta M, Meneghetti F, De Rossi AA, Chieco-Bianchi L (2001) Acquired immunodeficiency syndrome-related Kaposi's sarcoma regression after highly active antiretroviral therapy: biologic correlates of clinical outcome. J Natl Cancer Inst Monogr 28:44–49

Center for Disease Control (1981) Kaposi's sarcoma and *Pneumocystis* pneumonia among homosexual men–New York City and California. MMWR Morb Mortal Wkly Rep 30:305–308

Chang Y, Cesarman E, Pessin MS, Lee F, Culpepper J, Knowles DM, Moore PS (1994) Identification of herpesvirus-like DNA sequences in AIDS-associated Kaposi's sarcoma. Science 266:1865–1869

Clifford GM, Polesel J, Rickenbach M, on behalf of the Swiss HIV Cohort Study, Dal Maso L, Keiser O, Kofler A, Rapiti E, Levi F, Jundt G, Fisch T, Bordoni A, De Weck D, Franceschi S (2005) Cancer Risk in the Swiss HIV Cohort Study: associations with immunodeficiency, smoking, and highly active antiretroviral therapy 10.1093/jnci/dji072. J Natl Cancer Inst 97:425–432

Conant MA, Opp KM, Poretz D, Mills RG (1997) Reduction of Kaposi's sarcoma lesions following treatment of AIDS with ritonovir. AIDS 11:1300–1301

Corbellino M, Bestetti G, Scalamogna C, Calattini S, Galazzi M, Meroni L, Manganaro D, Fasan M, Moroni M, Galli M, Parravicini C (2001) Long-term remission of Kaposi sarcoma-associated herpesvirus-related multicentric Castleman disease with anti-CD20 monoclonal antibody therapy. Blood 98:3473–3475

Dittmer D, Stoddart C, Renne R, Linquist-Stepps V, Moreno ME, Bare C, McCune JM, Ganem D (1999) Experimental transmission of Kaposi's sarcoma-associated herpesvirus (KSHV/HHV-8) to SCID-hu Thy/Liv mice. J Exp Med 190:1857–1868

Diz Dios P, Ocampo Hermida A, Miralles Alvarez C, Vazquez Garcia E, Martinez Vazquez C (1998) Regression of AIDS-related Kaposi's sarcoma following ritonavir therapy. Oral Oncol 34:236–238

Dupin N, Fisher C, Kellam P, Ariad S, Tulliez M, Franck N, van Marck E, Salmon D, Gorin I, Escande JP, Weiss RA, Alitalo K, Boshoff C (1999) Distribution of human herpesvirus-8 latently infected cells in Kaposi's sarcoma, multicentric Castleman's disease, and primary effusion lymphoma. Proc Natl Acad Sci USA 96:4546–4551

Eltom MA, Jemal A, Mbulaiteye SM, Devesa SS, Biggar RJ (2002) Trends in Kaposi's sarcoma and non-Hodgkin's lymphoma incidence in the United States from 1973 through 1998. J Natl Cancer Inst 94:1204–1210

Friedrichs C, Neyts J, Gaspar G, Clercq ED, Wutzler P (2004) Evaluation of antiviral activity against human herpesvirus 8 (HHV-8) and Epstein-Barr virus (EBV) by a quantitative real-time PCR assay. Antiviral Res 62:121–123

Gill J, Bourboulia D, Wilkinson J, Hayes P, Cope A, Marcelin AG, Calvez V, Gotch F, Boshoff C, Gazzard B (2002) Prospective study of the effects of antiretroviral therapy on kaposi sarcoma-associated herpesvirus infection in patients with and without kaposi sarcoma. J Acquir Immune Defic Syndr 31:384–390

Glesby MJ, Hoover DR, Weng S, Graham NM, Phair JP, Detels R, Ho M, Saah AJ (1996) Use of antiherpes drugs and the risk of Kaposi's sarcoma: data from the Multicenter AIDS Cohort Study. J Infect Dis 173:1477–1480

Grandadam M, Dupin N, Calvez V, Gorin I, Blum L, Kernbaum S, Sicard D, Buisson Y, Agut H, Escande JP, Huraux JM (1997) Exacerbations of clinical symptoms in human immunodeficiency virus type 1-infected patients with multicentric Castleman's disease are associated with a high increase in Kaposi's sarcoma herpesvirus DNA load in peripheral blood mononuclear cells. J Infect Dis 175:1198–1201

Gustafson EA, Schinazi RF, Fingeroth JD (2000) Human herpesvirus 8 open reading frame 21 is a thymidine and thymidylate kinase of narrow substrate specificity that efficiently phosphorylates zidovudine but not ganciclovir. J Virol 74:684–692

Hocqueloux L, Agbalika F, Oksenhendler E, Molina JM (2001) Long-term remission of an AIDS-related primary effusion lymphoma with antiviral therapy. AIDS 15:280–282

Huang LM, Chao MF, Chen MY, Shih H, Chiang YP, Chuang CY, Lee CY (2001) Reciprocal regulatory interaction between human herpesvirus 8 and human immunodeficiency virus type 1. J Biol Chem 276:13427–13432

Hymes KB, Cheung T, Greene JB, Prose NS, Marcus A, Ballard H, William DC, Laubenstein LJ (1981) Kaposi's sarcoma in homosexual men-a report of eight cases. Lancet 2:598–600

Ioannidis JP, Collier AC, Cooper DA, Corey L, Fiddian AP, Gazzard BG, Griffiths PD, Contopoulos-Ioannidis DG, Lau J, Pavia AT, Saag MS, Spruance SL, Youle MS (1998) Clinical efficacy of high-dose acyclovir in patients with human immunodeficiency virus infection: a meta-analysis of randomized individual patient data. J Infect Dis 178:349–359

Joffe MM, Hoover DR, Jacobson LP, Kingsley L, Chmiel JS, Visscher BR (1997) Effect of treatment with zidovudine on subsequent incidence of Kaposi's sarcoma. Clin Infect Dis 25:1125–1133

Jung C, Bogner JR, Goebel F (1998) Resolution of severe Kaposi's sarcoma after initiation of antiretroviral triple therapy. Eur J Med Res 3:439–442

Katano H, Sato Y, Itoh H, Sata T (2001) Expression of human herpesvirus 8 (HHV-8)-encoded immediate early protein, open reading frame 50, in HHV-8-associated diseases. J Hum Virol 4:96–102

Katano H, Sato Y, Kurata T, Mori S, Sata T (2000) Expression and localization of human herpesvirus 8-encoded proteins in primary effusion lymphoma, Kaposi's sarcoma, and multicentric Castleman's disease. Virology 269:335–344

Kedes DH, Ganem D (1997) Sensitivity of Kaposi's sarcoma-associated herpesvirus replication to antiviral drugs. Implications for potential therapy. J Clin Invest 99:2082–2086

Klass CM, Krug LT, Pozharskaya VP, Offermann MK (2005) The targeting of primary effusion lymphoma cells for apoptosis by inducing lytic replication of human herpesvirus 8 while blocking virus production. Blood: 2004-2009-3569

Krischer J, Rutschmann O, Hirschel B, Vollenweider-Roten S, Saurat JH, Pechere M (1998) Regression of Kaposi's sarcoma during therapy with HIV-1 protease inhibitors: a prospective pilot study. J Am Acad Dermatol 38:594–598

Krown SE (2004) Highly active antiretroviral therapy in AIDS-associated Kaposi's sarcoma: implications for the design of therapeutic trials in patients with advanced, symptomatic Kaposi's sarcoma. J Clin Oncol 22:399–402

Krug LT, Pozharskaya VP, Yu Y, Inoue N, Offermann MK (2004) Inhibition of infection and replication of human herpesvirus 8 in microvascular endothelial cells by α interferon and phosphonoformic acid. J Virol 78:8359–8371

Leao JC, Kumar N, McLean KA, Porter SR, Scully CM, Swan AV, Teo CG (2000) Effect of human immunodeficiency virus-1 protease inhibitors on the clearance of human herpesvirus 8 from blood of human immunodeficiency virus-1-infected patients. J Med Virol 62:416–420

Lebbe C, Blum L, Pellet C, Blanchard G, Verola O, Morel P, Danne O, Calvo F (1998) Clinical and biological impact of antiretroviral therapy with protease inhibitors on HIV-related Kaposi's sarcoma. AIDS 12: F45–F49

Little RF, Merced-Galindez F, Staskus K, Whitby D, Aoki Y, Humphrey R, Pluda JM, Marshall V, Walters M, Welles L, Rodriguez-Chavez IR, Pittaluga S, Tosato G, Yarchoan R (2003) A pilot study of cidofovir in patients with Kaposi sarcoma. J Infect Dis 187:149–153

Lock MJ, Thorley N, Teo J, Emery VC (2002) Azidodeoxythymidine and didehydrodeoxythymidine as inhibitors and substrates of the human herpesvirus 8 thymidine kinase. J Antimicrob Chemother 49:359–366

Lorenzen T, Albrecht D, Paech V, Meyer T, Hoffmann C, Stoehr A, Degen O, Stellbrink HJ, Meigel WN, Arndt R, Plettenberg A (2002) HHV-8 DNA in blood and the development of HIV-associated Kaposi's sarcoma in the era of HAART—a prospective evaluation. Eur J Med Res 7:283–286

Lu M, Suen J, Frias C, Pfeiffer R, Tsai M-H, Chuang E, Zeichner SL (2004) Dissection of the Kaposi's sarcoma-associated herpesvirus gene expression program by using the viral DNA replication inhibitor cidofovir. J Virol 78:13637–13652

Luppi M, Trovato R, Barozzi P, Vallisa D, Rossi G, Re A, Ravazzini L, Potenza L, Riva G, Morselli M, Longo G, Cavanna L, Roncaglia R, Torelli G (2005) Treatment of herpesvirus associated primary effusion lymphoma with intracavity cidofovir. Leukemia 19:473–476

Martin DF, Kuppermann BD, Wolitz RA, Palestine AG, Li H, Robinson CA (1999) Oral ganciclovir for patients with cytomegalovirus retinitis treated with a ganciclovir implant. Roche Ganciclovir Study Group. N Engl J Med 340:1063–1070

Martinelli C, Zazzi M, Ambu S, Bartolozzi D, Corsi P, Leoncini F (1998) Complete regression of AIDS-related Kaposi's sarcoma-associated human herpesvirus-8 during therapy with indinavir. AIDS 12:1717–1719

McCune JM, Namikawa R, Kaneshima H, Shultz LD, Lieberman M, Weissman IL (1988) The SCID-hu mouse: murine model for the analysis of human hematolymphoid differentiation and function. Science 241:1632–1639

Medveczky MM, Horvath E, Lund T, Medveczky PG (1997) In vitro antiviral drug sensitivity of the Kaposi's sarcoma-associated herpesvirus. AIDS 11:1327–1332
Mocroft A, Youle M, Gazzard B, Morcinek J, Halai R, Phillips AN (1996) Anti-herpesvirus treatment and risk of Kaposi's sarcoma in HIV infection. Royal Free/Chelsea and Westminster Hospitals Collaborative Group. AIDS 10:1101–1105
Murphy M, Armstrong D, Sepkowitz KA, Ahkami RN, Myskowski PL (1997) Regression of AIDS-related Kaposi's sarcoma following treatment with an HIV-1 protease inhibitor. AIDS 11:261–262
Neyts J, De Clercq E (1997) Antiviral drug susceptibility of human herpesvirus 8. Antimicrob Agents Chemother 41:2754–2756
Oksenhendler E, Carcelain G, Aoki Y, Boulanger E, Maillard A, Clauvel JP, Agbalika F (2000) High levels of human herpesvirus 8 viral load, human interleukin-6, interleukin-10, and C reactive protein correlate with exacerbation of multicentric castleman disease in HIV-infected patients. Blood 96:2069–2073
Pastore RD, Chadburn A, Kripas C, Schattner EJ (2000) Novel association of haemophagocytic syndrome with Kaposi's sarcoma-associated herpesvirus-related primary effusion lymphoma. Br J Haematol 111:1112–1115
Pati S, Pelser CB, Dufraine J, Bryant JL, Reitz MS, Jr., Weichold FF (2002) Antitumorigenic effects of HIV protease inhibitor ritonavir: inhibition of Kaposi sarcoma. Blood 99:3771–3779
Pellet C, Chevret S, Blum L, Gauville C, Hurault M, Blanchard G, Agbalika F, Lascoux C, Ponscarme D, Morel P, Calvo F, Lebbe C (2001) Virologic and immunologic parameters that predict clinical response of AIDS-associated Kaposi's sarcoma to highly active antiretroviral therapy. J Invest Dermatol 117:858–863
Portsmouth S, Stebbing J, Gill J, Mandalia S, Bower M, Nelson M, Gazzard B (2003) A comparison of regimens based on non-nucleoside reverse transcriptase inhibitors or protease inhibitors in preventing Kaposi's sarcoma. AIDS 17: F17–F22
Renne R, Lagunoff M, Zhong W, Ganem D (1996) The size and conformation of Kaposi's sarcoma-associated herpesvirus (human herpesvirus 8) DNA in infected cells and virions. J Virol 70:8151–8154
Rey D, Schmitt MP, Partisani M, Hess-Kempf G, Krantz V, de Mautort E, Bernard-Henry C, Priester M, Cheneau C, Lang JM (2001) Efavirenz as a substitute for protease inhibitors in HIV-1-infected patients with undetectable plasma viral load on HAART: a median follow- up of 64 weeks. J Acquir Immune Defic Syndr 27:459–462
Schaeffer HJ, Beauchamp L, de Miranda P, Elion GB, Bauer DJ, Collins P (1978) 9-(2-Hydroxyethoxymethyl) guanine activity against viruses of the herpes group. Nature 272:583–585
Sgadari C, Barillari G, Toschi E, Carlei D, Bacigalupo I, Baccarini S, Palladino C, Leone P, Bugarini R, Malavasi L, Cafaro A, Falchi M, Valdembri D, Rezza G, Bussolino F, Monini P, Ensoli B (2002) HIV protease inhibitors are potent anti-angiogenic molecules and promote regression of Kaposi sarcoma. Nat Med 8:225–232
Sgadari C, Monini P, Barillari G, Ensoli B (2003) Use of HIV protease inhibitors to block Kaposi's sarcoma and tumour growth. Lancet Oncol 4:537–547
Shimba N, Nomura AM, Marnett AB, Craik CS (2004) Herpesvirus protease inhibition by dimer disruption. J Virol 78:6657–6665

Sirianni MC, Vincenzi L, Topino S, Giovannetti A, Mazzetta F, Libi F, Scaramuzzi D, Andreoni M, Pinter E, Baccarini S, Rezza G, Monini P, Ensoli B (2002) NK cell activity controls human herpesvirus 8 latent infection and is restored upon highly active antiretroviral therapy in AIDS patients with regressing Kaposi's sarcoma. Eur J Immunol 32:2711–2720

Stebbing J, Portsmouth S, Nelson M, Mandalia S, Kandil H, Alexander N, Davies L, Brock C, Bower M, Gazzard B (2004) The efficacy of ritonavir in the prevention of AIDS-related Kaposi's sarcoma. Int J Cancer 108:631–633

Tavio M, Nasti G, Spina M, Errante D, Vaccher E, Tirelli U (1998) Highly active antiretroviral therapy in HIV-related Kaposi's sarcoma. Ann Oncol 9:923

Whitby D, Howard MR, Tenant-Flowers M, Brink NS, Copas A, Boshoff C, Hatzioannou T, Suggett FE, Aldam DM, Denton AS, et al. (1995) Detection of Kaposi sarcoma associated herpesvirus in peripheral blood of HIV-infected individuals and progression to Kaposi's sarcoma Lancet 346:799–802

Wilkinson J, Cope A, Gill J, Bourboulia D, Hayes P, Imami N, Kubo T, Marcelin A, Calvez V, Weiss R, Gazzard B, Boshoff C, Gotch F (2002) Identification of Kaposi's sarcoma-associated herpesvirus (KSHV)-specific cytotoxic T-lymphocyte epitopes and evaluation of reconstitution of KSHV-specific responses in human immunodeficiency virus type 1-infected patients receiving highly active antiretroviral therapy. J Virol 76:2634–2640

Zietz C, Bogner JR, Goebel FD, Lohrs U (1999) An unusual cluster of cases of Castleman's disease during highly active antiretroviral therapy for AIDS. N Engl J Med 340:1923–1924

Zoeteweij JP, Eyes ST, Orenstein JM, Kawamura T, Wu L, Chandran B, Forghani B, Blauvelt A (1999) Identification and rapid quantification of early- and late-lytic human herpesvirus 8 infection in single cells by flow cytometric analysis: characterization of antiherpesvirus agents. J Virol 73:5894–5902

CTMI (2006) 312:309–326

Interactions Between HIV-1 Tat and KSHV

Y. Aoki (✉) · G. Tosato

Development, Astellas Pharma Inc., 17-1 Hasune 3-Chome,
Itabashi-ku, 174-8612 Japan
yoshiyasu.aoki@jp.astellas.com

Abstract Since the advent of the HIV-1 pandemic, a close association between HIV-1 infection and the development of selected types of cancers has been brought to light. The discovery of Kaposi sarcoma-associated herpesvirus (KSHV) has led to significant advances in uncovering the virological and molecular mechanisms involved in the pathogenesis of AIDS-related malignancies. Extensive evidence indicates that HIV-1 *trans*-activating protein Tat plays an oncogenic role in the development of KSHV-associated neoplasms. Comprehensive knowledge of the functions of Tat-1 together with the KSHV genes will contribute to a better understanding of the pathogenesis of virus-associated cancers and the interaction of viruses with their hosts.

1 Introduction

HIV-1 infection predisposes to the development of specific types of cancer. Most neoplasms that arise during states of immunodeficiency are associated with oncogenic virus infections, such as Epstein-Barr virus (EBV), human papillomavirus (HPV) and Kaposi sarcoma (KS)-associated herpesvirus (KSHV). For this reason, a diminished immune surveillance against viruses and virus-infected tumor cells is believed to contribute to the development of these malignancies.

The introduction of highly active antiretroviral therapy (HAART) has dramatically changed the landscape of HIV-1 disease. HAART brings about a substantial and sustained decrease in peripheral blood HIV-1 load, as well as an increase in $CD4^+$ T cells and a significant decrease in the incidence of AIDS-defining KS (Mocroft et al. 2000; Nasti et al. 2003; Palella et al. 1998). Patients with advanced KS treated with chemotherapy concurrently with HAART have achieved long-term survival (Holkova et al. 2001; Mocroft et al. 2000; Nasti et al. 2003; Palella et al. 1998), and some KS lesions regress after initiation of HAART (Aoki et al. 2004a; Krown 2004).

Importantly, however, the onset of AIDS-KS is not always associated with reduced peripheral blood $CD4^+$ T cell counts. In addition, KS is more aggressive in AIDS than in other immunodeficiency states, including solid organ transplant recipients, implying that factors other than increased KSHV replication due to immunodeficiency are involved. Also, HPV-associated cervical cancer develops in HIV-infected patients irrespective of host immune status. However, disease progression is enhanced by HIV, suggesting that HIV-1 acts as a cofactor promoting cancer growth (Clarke et al. 2002). Experimental evidence has shown that the *trans*-activating protein Tat from HIV-1 has unusual properties for a transcriptional factor in that it can up-egulate expression of a number of viral and cellular genes, act as a proangiogenic factor, enhance KSHV transmission to target cells, and cooperate with KSHV-derived proteins in promoting tumorigenesis (Albini et al. 1996b; Ambrosino et al. 1997; Aoki et al. 2004a; Badou et al. 2000; Bennasser et al. 2002; Scala et al. 1994; Watson et al. 1999). This review discusses pathogenetic mechanisms underlying KSHV-associated neoplasms, taking into account direct and indirect effects of HIV-1 Tat on tumorigenesis.

2
HIV-1 Tat

HIV-1 encodes three structural proteins (matrix, capsid, and nucleocapsid), two envelope proteins (gp120 and gp41), three enzymes (protease, reverse transcriptase, and integrase), two regulatory proteins (Tat and Rev), and four accessory proteins (Nef, Vpr, Vpu, and Vif) (Frankel et al. 1998). After two decades of extensive investigation, remarkable progress has been achieved in the understanding of the molecular mechanisms mediating the various activities of HIV-1-derived proteins, particularly in characterizing the Tat protein in relation to cancer development (Watson et al. 1999).

The HIV-1 Tat protein, the *trans*-activating factor of HIV-1, is a polypeptide comprising 86–101 amino acids (aa), which is encoded by 2 exons (Kuppuswamy et al. 1989; Watson et al. 1999). The first 72 amino acids, encoded by the first exon, possess the full *trans*-activating activity of Tat (Garcia et al. 1988). Structure-function studies have distinguished the first 72 residues of Tat into five major domains (Fig. 1). The cysteine-rich region mediates the formation of metal-linked dimers in vitro and is essential for Tat function (Frankel et al. 1988; Garcia et al. 1988; Ruben et al. 1989). The basic region is important for the nuclear localization of Tat and contributes to Tat *trans*-activating function (Calnan et al. 1991; Hauber et al. 1989; Rubartelli et al. 1998). Within the basic region, which comprises residues 49–57, a nuclear localization signal, GRKKR, has been identified (Ruben et al. 1989). As a result, Tat accumulation in the nucleus is rapid, occurring in a few minutes (Watson et al. 1999). The carboxy-terminal amino acids encoded by the second exon contain an RGD motif, which is important for Tat binding to $\alpha_V\beta_3$ and $\alpha_5\beta$ integrins (Barillari et al. 1993). It was proposed that through this RGD-mediated integrin engagement, the carboxy-terminal region is indispensable for Tat function as a proangiogenic factor in vitro and in vivo (Mitola et al. 2000).

Tat is a potent transactivator of HIV-1 gene expression and is essential for viral replication. Biochemical studies have demonstrated that Tat binds to the HIV-1 promoter located immediately downstream of the transcriptional start-site at the 5′ end of the nascent viral RNA transcript at the *trans*-activation responsive (TAR) element (Delling et al. 1991; Roy et al. 1990). Despite its lacking a secretory signal Tat is released from HIV-1 infected cells, and extracellular Tat has the ability to enter the uninfected cell and transactivate endogenous genes, such as tumor necrosis factor, interleukin (IL)-2, and IL-6 (Ambrosino et al. 1997; Chen et al. 1997; Ehret et al. 2001; Rubartelli et al.

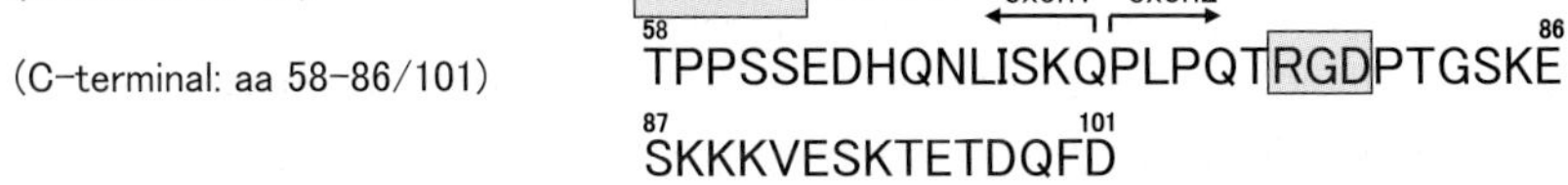

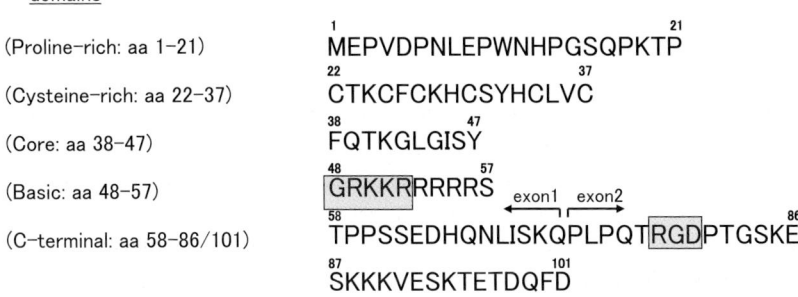

Fig. 1 Domain structure of HIV-1 Tat. The complete amino acid sequence of the Tat polypeptide is shown, and the five major domains are indicated. The first 72 aa are encoded by exon 1 and the rest are encoded by exon 2. The nuclear localization signal and the RGD motif are *shaded*

1998; Westendorp et al. 1994). The addition of hydrophobic groups to Tat was shown to increase the protein uptake by cells (Chen et al. 1995). Through this and perhaps other ill-defined mechanisms, Tat has been reported to *trans*-activate other cellular genes (Frankel et al. 1998; Rubartelli et al. 1998) and to regulate collagen expression and cell survival-related proteins including p53 and Bcl-2 (Li et al. 1995; Zauli et al. 1995). Moreover, Tat can carry biologically active proteins into cells in a receptor- and transporter-independent fashion and can elicit different biological responses in distinct target cells (Rubartelli et al. 1998; Schwarze et al. 2000; Vives et al. 2003). This shuttle function is mediated principally by Tat's basic domain.

Tat has a net positive charge and, owing to its strong electrostatic interaction, binds to negatively charged molecules, including cell-associated heparan sulfate proteoglycans, immunoglobulin-like domains of vascular endothelial growth factor (VEGF) receptor (VEGFR)-2, and chemokine receptors (Albini et al. 1996b; Rusnati et al. 2000; Tao et al. 1993). Binding of Tat to VEGFR-2 promotes angiogenesis in vivo, which is strongly potentiated by heparin (Albini et al. 1994). Many proangiogenic factors are heparin-binding proteins, including fibroblast growth factors and VEGF, and their activity is dependent on the presence of heparin or heparan sulfate (Albini et al. 1996a, 1996b ; Klagsbrun et al. 1991; Tessler et al. 1994; Yayon et al. 1991). As is the case with fibroblast growth factor (Sommer et al. 1989), the tight binding to heparin may protect Tat from proteolytic degradation and disclose the proangiogenic activity of Tat at concentrations at which Tat alone is not effective (Albini et al. 1996a).

There is evidence supporting a role for Tat as a cofactor in the pathogenesis of AIDS-associated KS. In vitro, Tat is a growth factor for KS spindle cells, and transgenic overexpression of Tat in mice promotes formation of KS-like lesions (Ensoli et al. 1990; Huang et al. 1993). When human KS cell lines are injected into mice, tumor cells grow faster in the presence of circulating HIV-1 Tat (Prakash et al. 1997). However, this effect may not be restricted to KS as Tat could indirectly promote tumor cell growth by promoting tumor angiogenesis. Tat also appears to affect cell adherence and anchorage-dependent cell growth through its RGD integrin-binding motif. Close contact of AIDS-KS with normal vascular cells is facilitated by Tat via a specific interaction of the Tat RGD sequence with integrins $\alpha_v\beta_3$ and $\alpha_5\beta_1$, a function that is augmented by Tat basic domain (Barillari et al. 1993).

In addition to displaying proangiogenic activity in several systems, Tat was reported to induce apoptosis of primary human brain microvascular endothelial cells, which resulted in increased endothelial cell permeability (Kim et al. 2003). The mechanism by which Tat induces apoptosis in selected endothelial cells is not presently known.

3
KSHV Infection

The discovery of KSHV within KS lesions from HIV-seropositive and -seronegative individuals (Moore et al. 1995) and a number of subsequent studies have provided strong evidence for an essential role of KSHV in the pathogenesis of KS (Boshoff et al. 2002). KSHV is also believed to play a role in the pathogenesis of primary effusion lymphoma (PEL) and a proportion of cases of multicentric Castleman disease (MCD) (Aoki et al. 2000a, 2004a, 2000c, 2001c). Histologically, KS lesions consist of a proliferation of spindle cells mixed with inflammatory cells, and are characterized by the formation of hypervascular structures and enhanced vascular permeability (Antman et al. 2000). In KS lesions, KSHV is present in the vast majority (>90%) of spindle cells and in the surrounding lymphatic or neoangiogenic vessels (Dupin et al. 1999; Katano et al. 2000, 1999). However, the virus is not detectable in well-formed established vascular endothelium or in neighboring normal dermis (Dupin et al. 1999). This has been attributed to the fact that KSHV is mostly latent in KS tissues, and in part to the difficulty at transmitting KSHV infection to normal noninfected cells.

This limited transmission rate of KSHV is supported by in vitro observations in various assay systems. Endothelial cell infection with KSHV in vitro has been explored in detail because most studies concur that KS cells are of endothelial lineage, as inferred by their expression of endothelium-specific molecules (Karp et al. 1996). In general, there are two avenues for viral transmission: direct contact between target and virus-producing cells (cell-mediated transmission) and cell-free transmission by viral particles (Dimitrov 2004; Spear et al. 2003). Most groups using cell-free KSHV transmission system have prepared highly concentrated KSHV particles harvested from phorbol acetate-induced PEL cell lines, mixed viral preparations with adhesive agents such as Polybrene, and added the mixture onto the endothelial cells (Akula et al. 2002; Flore et al. 1998; Moses et al. 1999; Poole et al. 2002; Tomescu et al. 2003; Vieira et al. 2001). Transmission efficiencies have varied significantly among cell types, and long-term infection has often been difficult (Grundhoff et al. 2004). Cell-mediated KSHV transmission was successfully demonstrated by direct coculture of KSHV-infected PEL cells with primary human vascular endothelial cells (HUVECs) (Sakurada et al. 2001). With this system, KSHV was detected in HUVECs up to 30 days after the removal of PEL cells. In this model, however, 10-fold more PEL cells than HUVECs were required for successful virus transmission (Sakurada et al. 2001), suggesting that the virus may not infect endothelial cells efficiently. When cells from KS lesions are placed into culture, KSHV is lost in a few passages while outgrowth of virus-negative spindle cells is quite rapid. As a consequence, all known KS-derived cell lines are virus negative.

Although impaired immune surveillance may result in increased KSHV replication in infected patients, there is a significantly higher incidence of KS in AIDS than in other immunodeficiency states. Compared to the general population, immunosuppressed patients after solid organ transplantation display an increase up to 500-fold in the incidence of KS, whereas male homosexual AIDS patients display an increase of KS incidence up to 20,000-fold. It is thus possible that HIV-1-related factors promote KSHV transmission. We have found that Tat can accelerate KSHV transmission into endothelial cells (Aoki et al. 2004a). The essential fragment mediating this effect is the 13-amino acid basic region of Tat (Aoki et al. 2004a). In vitro experiments indicate that Tat can increase KSHV transmission into cells at protein concentrations 10-fold higher than those reported present in sera from HIV-1-infected individuals (Aoki et al. 2004a). However, Tat concentrations in lymph nodes of HIV-infected individuals are expected to be substantially higher than those found in sera from HIV-1-positive individuals because HIV-1 load in lymphoid tissues exceeds by orders of magnitude that found in the bloodstream (Cavert et al. 1997; Hockett et al. 1999; Pantaleo et al. 1998, 1993). The occurrence of most intense HIV-1 replication in lymphoid tissues and the association of primary KSHV infection with the development of lymph nodal KS in an HIV-1-infected patient (Oksenhendler et al. 1998) suggest that lymphoid tissues are a likely preferential site of KSHV spread. This hypothesis that KS precursor cells or KSHV-infected cells originate at a distant site and reach the site of KS development, most commonly the skin, is supported by recent studies in transplant recipients (Barozzi et al. 2003) and in the experimental setting (Yao et al. 2003).

4 HIV-1 Versus HIV-2

Although the incidence of various malignancies is clearly increased in HIV-1-infected individuals, there are no definitive cancer statistics for HIV-2-infected individuals (Bock et al. 2001). In contrast to HIV-1 infection, HIV-2 infection is associated with a longer asymptomatic period, lower plasma viral loads, and lower rates of heterosexual and perinatal transmission than those of HIV-1 (Kanki et al. 1994; Marlink et al. 1994; Popper et al. 1999; Schim van der Loeff et al. 1999). HIV-2 infection is not commonly associated with KS development, even when accompanied by KSHV infection (Ariyoshi et al. 1998; Weiss et al. 2000). The incidence of KS is higher in African countries where HIV-1 is prevalent than in countries where HIV-2 is more common, despite similar seroprevalence of KSHV. HIV-1 is pandemic, but HIV-2 is more endemic, with stable prevalence rates in most countries. HIV-2 transmission appears to differ from that of HIV-1. Heterosexual transmission is responsible for the majority of HIV-2 infections worldwide, although a few reports of homosexual HIV-2 transmission have been documented (Cilla et al. 2001; Schim van der Loeff et al. 1999). In addition, HIV-2 transmission is estimated to occur at a much lower rate than that of HIV-1: Sexual and vertical forms of HIV-2 transmission are about five- to ninefold and 10- to 20-fold reduced relative to HIV-1, respectively (Reeves et al. 2002). Thus the difference between routes of HIV-1 and HIV-2 viral transmission may be related, in part, to differences in KS incidence between HIV-1- versus HIV-2-infected individuals (Ariyoshi et al. 1998).

Another potential explanation for the epidemiological findings is that Tat proteins encoded by HIV-1 and HIV-2 (Tat-1 and Tat-2, respectively) differ in their ability to serve as cofactors in KSHV-induced KS development. Tat-2 is a polypeptide consisting of 130 residues encoded by 2 exons. Like Tat-1, a basic region of Tat-2 encoded by exon 1 binds to a TAR-RNA-binding domain at the 5′ end of viral transcripts and stimulates viral transcription (Chang et al. 1992). However, exon 2 of HIV-2 Tat increases the binding affinity of Tat-2 to HIV-2 TAR RNA and contributes to *trans* -activation of HIV-2 LTR (Chang et al. 1992; Pagtakhan et al. 1995; Rhim et al. 1994). Exon 2 of Tat-2 encodes distinct domains, including the amino-terminal, cysteine-rich, and core regions (Pagtakhan et al. 1995; Rhim et al. 1994). Although the latter two regions have greater than 65% amino acid sequence homology with the corresponding regions of Tat-1, the amino-terminal region of Tat-2 has only 10% homology to Tat-1 (Pagtakhan et al. 1997). In addition, Tat-2 lacks an RGD motif, which is important for Tat-1 integrin binding (Bock et al. 2001). Although Tat-1 and Tat-2 show considerable similarity, Tat-2 is unable to effectively transactivate the HIV-1 promoter (Berkhout et al. 1990). The Tat-1 protein, however, can effectively transactivate the wild-type HIV-2 promoter (Rhim et al. 1995). Tat-1 is not a stronger transactivator than Tat-2, as fusion protein studies suggest that Tat-2 is the more potent transactivator if it can be adequately recruited to the promoter (Rhim et al. 1993). Detailed analysis of structural and functional differences between Tat-1 and Tat-2 may help understand the mechanisms by which HIV-1 acts as a potent cofactor in the development of certain cancers.

5
Interactions Between Tat-1 and KSHV

The critical role of KSHV infection in KS pathogenesis is well established, but many aspects of this process are incompletely understood. KSHV encodes numerous open reading frames with striking homology to cell regulatory genes (Choi et al. 2001). However, KSHV gene expression in KS lesions is highly restricted, and the pattern of viral mRNA in most KS tumor cells is consistent with latent infection (Katano et al. 2000).

Viral FLICE-inhibitory protein (vFLIP: ORF K13/71) is one of the few viral proteins to be expressed in latently infected KS spindle cells and PEL cell lines (Godfrey et al. 2005; Guasparri et al. 2004; Katano et al. 2000). Cells expressing vFLIP are protected against apoptosis induced by Fas or tumor necrosis factor receptor-1 (Djerbi et al. 1999). A recent study showed that vFLIP is largely responsible for constitutive nuclear factor (NF)-κB activation in PEL cells (Guasparri et al. 2004). As in other virus-associated lymphomas, constitutive activation of both the NF-κB and signal transducer and activator of transcription-3 appears to be essential for PEL cell survival (Aoki et al. 2003a; Keller et al. 2000). In addition to vFLIP, viral G protein-coupled receptor (vGPCR: ORF74) and ORF K1 can activate NF-κB. However, these viral transcripts are predominantly expressed during the lytic phase (Cannon et al. 2003; Cesarman et al. 2000; Samaniego et al. 2001). Tat-1 is a potent NF-κB activator (Demarchi et al. 1996), which can bind directly to NF-κB enhancer sequence (Dandekar et al. 2004) and may thus contribute to PEL cell survival in HIV-1-infected individuals.

Although most HIV-1-infected cells are not co-infected with KSHV, experimental results suggest that selected KSHV-encoded proteins may influence the life cycle of HIV-1. KSHV Rta (ORF50) was shown to increase HIV-1 infectivity (Varthakavi et al. 2002). Latency-associated nuclear antigen (LANA: ORF73) was reported to activate HIV-1 long terminal repeat (Hyun et al. 2001). In an HIV-1-infected patient with MCD, circulating levels of viral IL-6 (vIL-6: ORF K2) correlated directly with circulating levels of HIV-1 RNA (Aoki et al. 2001b). In addition, it is possible that HIV-1-derived proteins can activate KSHV lytic replication, although the role of Tat-1 alone is controversial (Inoue et al. 2003; Varthakavi et al. 2002).

KSHV and Tat-1 may also cooperate to modify cellular gene expression. One important target of exogenous Tat-1 is the vascular or lymphatic endothelium. KSHV infection of endothelial cells promotes the expression of VEGF-A and VEGF-C as well as their receptors VEGFR-2 and VEGFR-3 (Hong et al. 2004; Masood et al. 2002; Wang et al. 2004). Tat-1 also activates VEGFR-2 and VEGFR-3, and by this mechanism, can induce angiogenesis and stimulate KS spindle cells (Albini et al. 1996b). In addition, Tat-1 induces the expression of adhesion molecules and matrix metalloproteinases, which play a critical role in KS progression (Aoki et al. 2003c; Lafrenie et al. 1997). vIL-6, which is not expressed in most KS cells but is elevated in sera from HIV-1-positive patients with KS (Aoki et al. 2000a, 2001c), directly activates KS cells (Klouche et al. 2002) and stimulates VEGF-A production in vitro and in vivo (Aoki et al. 1999a). VEGF-A is an essential growth factor for PEL cell survival in vivo (Aoki et al. 1999b, 2001a, 2000b). Tat-1 has been shown to induce the expression of IL-6 and IL-10, which serve as autocrine growth factors for MCD and PEL, respectively (Beck et al. 1994; Jones et al. 1999). Transgenic vGPCR mice develop KS-like tumors (Guo et al. 2003). A cell line derived from one such tumor expresses vGPCR and forms tumors in nude mice (Yang et al. 2000). Transfection of the tat-1 gene into these tumor cells increases NF-kB and NF-AT activation levels, and accelerates tumor formation (Guo et al. 2004).

In certain hematopoietic cells, Tat-1 can upregulate the expression of the chemokine receptor CXCR4 (Gibellini et al. 2003). Its ligand, stromal-derived factor-1 (SDF-1), which is constitutively expressed by skin capillary endothelium, can trigger specific arrest of KSHV-infected cells under physiological shear flow conditions (Yao et al. 2003). By triggering specific adhesion of circulating KSHV-infected cells and favoring their entry into the extravascular cutaneous space, endothelial cell-associated SDF-1 in cutaneous capillaries may contribute to the occurrence of KS in the skin.

Taken together, these results suggest that Tat-1 can contribute to the high incidence and aggressive nature of KS in AIDS, presumably by promoting KSHV infection. The impact of HAART on the occurrence and severity of KS may be due, at least in part, to suppression of Tat-1 protein expression.

6 Conclusions

Based on the advances in our understanding of HIV-1-related disorders, HIV-1 Tat emerges as a very important protein for its role not only in HIV-1 infection but also in cancer development. Understanding the direct and indirect roles of Tat-1 in the pathogenesis of KSHV-infected neoplasms could provide means to help us design novel strategies for prevention and treatment of virus-induced pathology. Tat-1 targeting may represent yet another avenue for future clinical application in certain KSHV-related disorders.

A role for HIV-1 in the pathogenesis of KS is supported by the observation that the introduction of HAART has dramatically decreased the incidence of KS. In contrast, KSHV-positive MCD in AIDS often progresses to fatal disease in spite of HAART (Aoki et al. 2001b, 2001c; Dupin et al. 2000; Zietz et al. 1999). Furthermore, the impact of HAART on AIDS-associated EBV-related lymphomas and HPV-related cervical cancer is variable (Clarke et al. 2002; Kirk et al. 2001; Levine et al. 2000), suggesting that additional factors are involved in the pathogenesis of these neoplasms. Further advances in identifying these additional factors may lead us to uncover novel facets of HIV-1 infection and its role in the initiation and progression of malignant disorders.

References

Akula SM, Pamod NP, Wang FZ, Chandran B (2002) Integrin a3b1 (CD 49c/29) is a cellular receptor for Kaposi's sarcoma-associated herpesvirus (KSHV/HHV-8) entry into the target cells. Cell 108:407–419

Albini A, Benelli R, Presta M, Rusnati M, Ziche M, Rubartelli A, Paglialunga G, Bussolino F, Noonan D (1996a) HIV-tat protein is a heparin-binding angiogenic growth factor. Oncogene 12:289–297

Albini A, Fontanini G, Masiello L, Tacchetti C, Bigini D, Luzzi P, Noonan DM, Stetler-Stevenson WG (1994) Angiogenic potential in vivo by Kaposi's sarcoma cell-free supernatants and HIV-1 tat product: inhibition of KS-like lesions by tissue inhibitor of metalloproteinase-2. AIDS 8:1237–1244

Albini A, Soldi R, Giunciuglio D, Giraudo E, Benelli R, Primo L, Noonan D, Salio M, Camussi G, Rockl W, Bussolino F (1996b) The angiogenesis induced by HIV-1 tat protein is mediated by the Flk-1/KDR receptor on vascular endothelial cells. Nat Med 2:1371–1375

Ambrosino C, Ruocco MR, Chen X, Mallardo M, Baudi F, Trematerra S, Quinto I, Venuta S, Scala G (1997) HIV-1 Tat induces the expression of the interleukin-6 (IL6) gene by binding to the IL6 leader RNA and by interacting with CAAT enhancer-binding protein beta (NF-IL6) transcription factors. J Biol Chem 272:14883–14892

Antman K, Chang Y (2000) Kaposi's sarcoma. N Engl J Med 342: 1027–1038

Aoki Y, Feldman GM, Tosato G (2003a) Inhibition of STAT3 signaling induces apoptosis and decreases survivin expression in primary effusion lymphoma. Blood 101:1535–1542

Aoki Y, Jaffe ES, Chang Y, Jones K, Teruya-Feldstein J, Moore PS, Tosato G (1999a) Angiogenesis and hematopoiesis induced by Kaposi's sarcoma-associated herpesvirus-encoded interleukin-6. Blood 93:4034–4043

Aoki Y, Jones KD, Tosato G (2000a) Kaposi's sarcoma-associated herpesvirus-encoded interleukin-6. J Hematother Stem Cell Res 9:137–145

Aoki Y, Tosato G (1999b) Role of vascular endothelial growth factor/vascular permeability factor in the pathogenesis of Kaposi's sarcoma-associated herpesvirus-infected primary effusion lymphomas. Blood 94:4247–4254

Aoki Y, Tosato G (2001a) Vascular endothelial growth factor/vascular permeability factor in the pathogenesis of primary effusion lymphomas. Leuk Lymphoma 41:229–237

Aoki Y, Tosato G (2004a) HIV-1 Tat enhances Kaposi sarcoma-associated herpesvirus (KSHV) infectivity. Blood 104:810–814

Aoki Y, Tosato G, Fonville TW, Pittaluga S (2001b) Serum viral interleukin-6 in AIDS-related multicentric Castleman's disease. Blood 97:2526–2527

Aoki Y, Tosato G, Nambu Y, Iwamoto A, Yarchoan R (2000b) Detection of vascular endothelial growth factor in AIDS-related primary effusion lymphomas. Blood 95:1109–1110

Aoki Y, Yarchoan R, Braun J, Iwamoto A, Tosato G (2000c) Viral and cellular cytokines in AIDS-related malignant lymphomatous effusions. Blood 96:1599–1601

Aoki Y, Yarchoan R, Wyvill K, Okamoto S, Little RF, Tosato G (2001c) Detection of viral interleukin-6 in Kaposi sarcoma-associated herpesvirus-linked disorders. Blood 97:2173–2176

Ariyoshi K, Schim van der Loeff M, Cook P, Whitby D, Corrah T, Jaffar S, Cham F, Sabally S, O'Donovan D, Weiss RA, Schulz TF, Whittle H (1998) Kaposi's sarcoma in the Gambia, West Africa is less frequent in human immunodeficiency virus type 2 than in human immunodeficiency virus type 1 infection despite a high prevalence of human herpesvirus 8. J Hum Virol 1:193–199

Badou A, Bennasser Y, Moreau M, Leclerc C, Benkirane M, Bahraoui E (2000) Tat protein of human immunodeficiency virus type 1 induces interleukin-10 in human peripheral blood monocytes: implication of protein kinase C-dependent pathway. J Virol 74:10551–10562

Barillari G, Gendelman R, Gallo RC, Ensoli B (1993) The Tat protein of human immunodeficiency virus type 1, a growth factor for AIDS Kaposi sarcoma and cytokine-activated vascular cells, induces adhesion of the same cell types by using integrin receptors recognizing the RGD amino acid sequence. Proc Natl Acad Sci USA 90:7941–7945

Barozzi P, Luppi M, Facchetti F, Mecucci C, Alu M, Sarid R, Rasini V, Ravazzini L, Rossi E, Festa S, Crescenzi B, Wolf DG, Schulz TF, Torelli G (2003) Post-transplant Kaposi sarcoma originates from the seeding of donor-derived progenitors. Nat Med 9:554–561

Beck JT, Hsu SM, Wijdenes J, Bataille R, Klein B, Vesole D, Hayden K, Jagannath S, Barlogie B (1994) Brief report: alleviation of systemic manifestations of Castleman's disease by monoclonal anti-interleukin-6 antibody. N Engl J Med 330:602–605

Bennasser Y, Bahraoui E (2002) HIV-1 Tat protein induces interleukin-10 in human peripheral blood monocytes: involvement of protein kinase C-βII and -δ. FASEB J 16:546–554

Berkhout B, Gatignol A, Silver J, Jeang KT (1990) Efficient trans-activation by the HIV-2 Tat protein requires a duplicated TAR RNA structure. Nucleic Acids Res 18:1839–1846

Bock PJ, Markovitz DM (2001) Infection with HIV-2. AIDS 15 Suppl 5: S35–45

Boshoff C, Weiss R (2002) AIDS-related malignancies. Nat Rev Cancer 2:373–382

Calnan BJ, Biancalana S, Hudson D, Frankel AD (1991) Analysis of arginine-rich peptides from the HIV Tat protein reveals unusual features of RNA-protein recognition. Genes Dev 5:201–210

Cannon M, Philpott NJ, Cesarman E (2003) The Kaposi's sarcoma-associated herpesvirus G protein-coupled receptor has broad signaling effects in primary effusion lymphoma cells. J Virol 77:57–67

Cavert W, Notermans DW, Staskus K, Wietgrefe SW, Zupancic M, Gebhard K, Henry K, Zhang ZQ, Mills R, McDade H, Schuwirth CM, Goudsmit J, Danner SA, Haase AT (1997) Kinetics of response in lymphoid tissues to antiretroviral therapy of HIV-1 infection. Science 276:960–964

Cesarman E, Mesri EA, Gershengorn MC (2000) Viral G protein-coupled receptor and Kaposi's sarcoma: a model of paracrine neoplasia? J Exp Med 191:417–422

Chang YN, Jeang KT (1992) The basic RNA-binding domain of HIV-2 Tat contributes to preferential trans-activation of a TAR2-containing LTR. Nucleic Acids Res 20:5465–5472

Chen LL, Frankel AD, Harder JL, Fawell S, Barsoum J, Pepinsky B (1995) Increased cellular uptake of the human immunodeficiency virus-1 Tat protein after modification with biotin. Anal Biochem 227:168–175

Chen P, Mayne M, Power C, Nath A (1997) The Tat protein of HIV-1 induces tumor necrosis factor-α production. Implications for HIV-1-associated neurological diseases. J Biol Chem 272:22385–22388

Choi J, Means RE, Damania B, Jung JU (2001) Molecular piracy of Kaposi's sarcoma associated herpesvirus. Cytokine Growth Factor Rev 12:245–257

Cilla G, Rodes B, Perez-Trallero E, Arrizabalaga J, Soriano V (2001) Molecular evidence of homosexual transmission of HIV type 2 in Spain. AIDS Res Hum Retroviruses 17:417–422

Clarke B, Chetty R (2002) Postmodern cancer: the role of human immunodeficiency virus in uterine cervical cancer. Mol Pathol 55:19–24

Dandekar DH, Ganesh KN, Mitra D (2004) HIV-1 Tat directly binds to NFκB enhancer sequence: role in viral and cellular gene expression. Nucleic Acids Res 32:1270–1278

Delling U, Roy S, Sumner-Smith M, Barnett R, Reid L, Rosen CA, Sonenberg N (1991) The number of positively charged amino acids in the basic domain of Tat is critical for trans-activation and complex formation with TAR RNA. Proc Natl Acad Sci USA 88:6234–6238

Demarchi F, d'Adda di Fagagna F, Falaschi A, Giacca M (1996) Activation of transcription factor NF-κB by the Tat protein of human immunodeficiency virus type 1. J Virol 70:4427–4437

Dimitrov DS (2004) Virus entry: molecular mechanisms and biomedical applications. Nat Rev Microbiol 2:109–122

Djerbi M, Screpanti V, Catrina AI, Bogen B, Biberfeld P, Grandien A (1999) The inhibitor of death receptor signaling, FLICE-inhibitory protein defines a new class of tumor progression factors. J Exp Med 190:1025–1032

Dupin N, Diss TL, Kellam P, Tulliez M, Du MQ, Sicard D, Weiss RA, Isaacson PG, Boshoff C (2000) HHV-8 is associated with a plasmablastic variant of Castleman disease that is linked to HHV-8-positive plasmablastic lymphoma. Blood 95:1406–1412

Dupin N, Fisher C, Kellam P, Ariad S, Tulliez M, Franck N, van Marck E, Salmon D, Gorin I, Escande JP, Weiss RA, Alitalo K, Boshoff C (1999) Distribution of human herpesvirus-8 latently infected cells in Kaposi's sarcoma, multicentric Castleman's disease, and primary effusion lymphoma. Proc Natl Acad Sci USA 96:4546–4551

Ehret A, Li-Weber M, Frank R, Krammer PH (2001) The effect of HIV-1 regulatory proteins on cellular genes: derepression of the IL-2 promoter by Tat. Eur J Immunol 31:1790–1799

Ensoli B, Barillari G, Salahuddin SZ, Gallo RC, Wong-Staal F (1990) Tat protein of HIV-1 stimulates growth of cells derived from Kaposi's sarcoma lesions of AIDS patients. Nature 345:84–86

Flore O, Rafii S, Ely S, O'Leary JJ, Hyjek EM, Cesarman E (1998) Transformation of primary human endothelial cells by Kaposi's sarcoma-associated herpesvirus. Nature 394:588–592

Frankel AD, Chen L, Cotter RJ, Pabo CO (1988) Dimerization of the tat protein from human immunodeficiency virus: a cysteine-rich peptide mimics the normal metal-linked dimer interface. Proc Natl Acad Sci USA 85:6297–6300

Frankel AD, Young JA (1998) HIV-1: fifteen proteins and an RNA. Annu Rev Biochem 67:1-25

Garcia JA, Harrich D, Pearson L, Mitsuyasu R, Gaynor RB (1988) Functional domains required for tat-induced transcriptional activation of the HIV-1 long terminal repeat. EMBO J 7:3143–3147

Gibellini D, Re MC, Vitone F, Rizzo N, Maldini C, La Placa M, Zauli G (2003) Selective up-regulation of functional CXCR4 expression in erythroid cells by HIV-1 Tat protein. Clin Exp Immunol 131:428–435

Godfrey A, Anderson J, Papanastasiou A, Takeuchi Y, Boshoff C (2005) Inhibiting primary effusion lymphoma by lentiviral vectors encoding short hairpin RNA. Blood 105:2510–2518

Grundhoff A, Ganem D (2004) Inefficient establishment of KSHV latency suggests an additional role for continued lytic replication in Kaposi sarcoma pathogenesis. J Clin Invest 113:124–136

Guasparri I, Keller SA, Cesarman E (2004) KSHV vFLIP Is essential for the survival of infected lymphoma cells. J Exp Med 199:993–1003

Guo HG, Pati S, Sadowska M, Charurat M, Reitz M (2004) Tumorigenesis by human herpesvirus 8 vGPCR is accelerated by human immunodeficiency virus type 1 Tat. J Virol 78:9336–9342

Guo HG, Sadowska M, Reid W, Tschachler E, Hayward G, Reitz M (2003) Kaposi's sarcoma-like tumors in a human herpesvirus 8 ORF74 transgenic mouse. J Virol 77:2631–2639

Hauber J, Malim MH, Cullen BR (1989) Mutational analysis of the conserved basic domain of human immunodeficiency virus tat protein. J Virol 63:1181–1187

Hockett RD, Kilby JM, Derdeyn CA, Saag MS, Sillers M, Squires K, Chiz S, Nowak MA, Shaw GM, Bucy RP (1999) Constant mean viral copy number per infected cell in tissues regardless of high, low, or undetectable plasma HIV RNA. J Exp Med 189:1545–1554

Holkova B, Takeshita K, Cheng DM, Volm M, Wasserheit C, Demopoulos R, Chanan-Khan A (2001) Effect of highly active antiretroviral therapy on survival in patients with AIDS-associated pulmonary Kaposi's sarcoma treated with chemotherapy. J Clin Oncol 19:3848–3851

Hong YK, Foreman K, Shin JW, Hirakawa S, Curry CL, Sage DR, Libermann T, Dezube BJ, Fingeroth JD, Detmar M (2004) Lymphatic reprogramming of blood vascular endothelium by Kaposi sarcoma-associated herpesvirus. Nat Genet 36:683–685

Huang SK, Martin FJ, Jay G, Vogel J, Papahadjopoulos D, Friend DS (1993) Extravasation and transcytosis of liposomes in Kaposi's sarcoma-like dermal lesions of transgenic mice bearing the HIV tat gene. Am J Pathol 143:10–14

Hyun TS, Subramanian C, Cotter MA, 2nd, Thomas RA, Robertson ES (2001) Latency-associated nuclear antigen encoded by Kaposi's sarcoma-associated herpesvirus interacts with Tat and activates the long terminal repeat of human immunodeficiency virus type 1 in human cells. J Virol 75:8761–8771

Inoue N, Winter J, Lal RB, Offermann MK, Koyano S (2003) Characterization of entry mechanisms of human herpesvirus 8 by using an Rta-dependent reporter cell line. J Virol 77:8147–8152

Jones KD, Aoki Y, Chang Y, Moore PS, Yarchoan R, Tosato G (1999) Involvement of interleukin-10 (IL-10) and viral IL-6 in the spontaneous growth of Kaposi's sarcoma herpesvirus-associated infected primary effusion lymphoma cells. Blood 94:2871–2879

Kanki PJ, Travers KU, S MB, Hsieh CC, Marlink RG, Gueye NA, Siby T, Thior I, Hernandez-Avila M, Sankale JL, et al. (1994) Slower heterosexual spread of HIV-2 than HIV-1. Lancet 343:943–946

Karp JE, Pluda JM, Yarchoan R (1996) AIDS-related Kaposi's sarcoma. A template for the translation of molecular pathogenesis into targeted therapeutic approaches. Hematol Oncol Clin North Am 10:1031–1049

Katano H, Sato Y, Kurata T, Mori S, Sata T (1999) High expression of HHV-8-encoded ORF73 protein in spindle-shaped cells of Kaposi's sarcoma. Am J Pathol 155:47–52

Katano H, Sato Y, Kurata T, Mori S, Sata T (2000) Expression and localization of human herpesvirus 8-encoded proteins in primary effusion lymphoma, Kaposi's sarcoma, and multicentric Castleman's disease. Virology 269:335–344

Keller SA, Schattner EJ, Cesarman E (2000) Inhibition of NF-κB induces apoptosis of KSHV-infected primary effusion lymphoma cells. Blood 96:2537–2542

Kim TA, Avraham HK, Koh YH, Jiang S, Park IW, Avraham S (2003) HIV-1 Tat-mediated apoptosis in human brain microvascular endothelial cells. J Immunol 170:2629–2637

Kirk O, Pedersen C, Cozzi-Lepri A, Antunes F, Miller V, Gatell JM, Katlama C, Lazzarin A, Skinhoj P, Barton SE (2001) Non-Hodgkin lymphoma in HIV-infected patients in the era of highly active antiretroviral therapy. Blood 98:3406–3412

Klagsbrun M, Baird A (1991) A dual receptor system is required for basic fibroblast growth factor activity. Cell 67:229–231

Klouche M, Brockmeyer N, Knabbe C, Rose-John S (2002) Human herpesvirus 8-derived viral IL-6 induces PTX3 expression in Kaposi's sarcoma cells. AIDS 16: F9–18

Krown SE (2004) Highly active antiretroviral therapy in AIDS-associated Kaposi's sarcoma: implications for the design of therapeutic trials in patients with advanced, symptomatic Kaposi's sarcoma. J Clin Oncol 22:399–402

Kuppuswamy M, Subramanian T, Srinivasan A, Chinnadurai G (1989) Multiple functional domains of Tat, the trans-activator of HIV-1, defined by mutational analysis. Nucleic Acids Res 17:3551–3561

Lafrenie RM, Wahl LM, Epstein JS, Yamada KM, Dhawan S (1997) Activation of monocytes by HIV-Tat treatment is mediated by cytokine expression. J Immunol 159:4077–4083

Levine AM, Seneviratne L, Espina BM, Wohl AR, Tulpule A, Nathwani BN, Gill PS (2000) Evolving characteristics of AIDS-related lymphoma. Blood 96:4084–4090

Li CJ, Wang C, Friedman DJ, Pardee AB (1995) Reciprocal modulations between p53 and Tat of human immunodeficiency virus type 1. Proc Natl Acad Sci USA 92:5461–5464

Marlink R, Kanki P, Thior I, Travers K, Eisen G, Siby T, Traore I, Hsieh CC, Dia MC, Gueye EH, et al. (1994) Reduced rate of disease development after HIV-2 infection as compared to HIV-1. Science 265:1587–1590

Masood R, Cesarman E, Smith DL, Gill PS, Flore O (2002) Human herpesvirus-8-transformed endothelial cells have functionally activated vascular endothelial growth factor/vascular endothelial growth factor receptor. Am J Pathol 160:23–29

Mitola S, Soldi R, Zanon I, Barra L, Gutierrez MI, Berkhout B, Giacca M, Bussolino F (2000) Identification of specific molecular structures of human immunodeficiency virus type 1 Tat relevant for its biological effects on vascular endothelial cells. J Virol 74:344–353

Mocroft A, Katlama C, Johnson AM, Pradier C, Antunes F, Mulcahy F, Chiesi A, Phillips AN, Kirk O, Lundgren JD (2000) AIDS across Europe, 1994–98: the EuroSIDA study. Lancet 356:291–296

Moore PS, Chang Y (1995) Detection of herpesvirus-like DNA sequences in Kaposi's sarcoma in patients with and without HIV infection. N Engl J Med 332:1181–1185

Moses AV, Fish KN, Ruhl R, Smith PP, Strussenberg JG, Zhu L, Chandran B, Nelson JA (1999) Long-term infection and transformation of dermal microvascular endothelial cells by human herpesvirus 8. J Virol 73:6892–6902

Nasti G, Talamini R, Antinori A, Martellotta F, Jacchetti G, Chiodo F, Ballardini G, Stoppini L, Di Perri G, Mena M, Tavio M, Vaccher E, D'Arminio Monforte A, Tirelli U (2003) AIDS-related Kaposi's Sarcoma: evaluation of potential new prognostic factors and assessment of the AIDS Clinical Trial Group Staging System in the Haart Era–the Italian Cooperative Group on AIDS and Tumors and the Italian Cohort of Patients Naive From Antiretrovirals. J Clin Oncol 21:2876–2882

Oksenhendler E, Cazals-Hatem D, Schulz TF, Barateau V, Grollet L, Sheldon J, Clauvel JP, Sigaux F, Agbalika F (1998) Transient angiolymphoid hyperplasia and Kaposi's sarcoma after primary infection with human herpesvirus 8 in a patient with human immunodeficiency virus infection. N Engl J Med 338:1585–1590

Pagtakhan AS, Tong-Starksen SE (1995) Function of exon 2 in optimal trans-activation by Tat of HIV type 2. AIDS Res Hum Retroviruses 11:1367–1372

Pagtakhan AS, Tong-Starksen SE (1997) Interactions between Tat of HIV-2 and transcription factor Sp1. Virology 238:221–230

Palella FJ, Jr., Delaney KM, Moorman AC, Loveless MO, Fuhrer J, Satten GA, Aschman DJ, Holmberg SD (1998) Declining morbidity and mortality among patients with advanced human immunodeficiency virus infection. HIV Outpatient Study Investigators. N Engl J Med 338:853–860

Pantaleo G, Cohen OJ, Schacker T, Vaccarezza M, Graziosi C, Rizzardi GP, Kahn J, Fox CH, Schnittman SM, Schwartz DH, Corey L, Fauci AS (1998) Evolutionary pattern of human immunodeficiency virus (HIV) replication and distribution in lymph nodes following primary infection: implications for antiviral therapy. Nat Med 4:341–345

Pantaleo G, Graziosi C, Demarest JF, Butini L, Montroni M, Fox CH, Orenstein JM, Kotler DP, Fauci AS (1993) HIV infection is active and progressive in lymphoid tissue during the clinically latent stage of disease. Nature 362:355–358

Poole LJ, Yu Y, Kim PS, Zheng QZ, Pevsner J, Hayward GS (2002) Altered patterns of cellular gene expression in dermal microvascular endothelial cells infected with Kaposi's sarcoma-associated herpesvirus. J Virol 76:3395–3420

Popper SJ, Sarr AD, Travers KU, Gueye-Ndiaye A, Mboup S, Essex ME, Kanki PJ (1999) Lower human immunodeficiency virus (HIV) type 2 viral load reflects the difference in pathogenicity of HIV-1 and HIV-2. J Infect Dis 180:1116–1121

Prakash O, Teng S, Ali M, Zhu X, Coleman R, Dabdoub RA, Chambers R, Aw TY, Flores SC, Joshi BH (1997) The human immunodeficiency virus type 1 Tat protein potentiates zidovudine-induced cellular toxicity in transgenic mice. Arch Biochem Biophys 343:173–180

Reeves JD, Doms RW (2002) Human immunodeficiency virus type 2. J Gen Virol 83:1253–1265

Rhim H, Rice AP (1993) TAR RNA binding properties and relative transactivation activities of human immunodeficiency virus type 1 and 2 Tat proteins. J Virol 67:1110–1121

Rhim H, Rice AP (1994) Exon2 of HIV-2 Tat contributes to transactivation of the HIV-2 LTR by increasing binding affinity to HIV-2 TAR RNA. Nucleic Acids Res 22:4405–4413

Rhim H, Rice AP (1995) HIV-1 Tat protein is able to efficiently transactivate the HIV-2 LTR through a TAR RNA element lacking both dinucleotide bulge binding sites. Virology 206:673–678

Roy S, Delling U, Chen CH, Rosen CA, Sonenberg N (1990) A bulge structure in HIV-1 TAR RNA is required for Tat binding and Tat-mediated trans-activation. Genes Dev 4:1365–1373

Rubartelli A, Poggi A, Sitia R, Zocchi MR (1998) HIV-I Tat: a polypeptide for all seasons. Immunol Today 19:543–545

Ruben S, Perkins A, Purcell R, Joung K, Sia R, Burghoff R, Haseltine WA, Rosen CA (1989) Structural and functional characterization of human immunodeficiency virus tat protein. J Virol 63:1-8

Rusnati M, Taraboletti G, Urbinati C, Tulipano G, Giuliani R, Molinari-Tosatti MP, Sennino B, Giacca M, Tyagi M, Albini A, Noonan D, Giavazzi R, Presta M (2000) Thrombospondin-1/HIV-1 tat protein interaction: modulation of the biological activity of extracellular Tat. FASEB J 14:1917–1930

Sakurada S, Katano H, Sata T, Ohkuni H, Watanabe T, Mori S (2001) Effective human herpesvirus 8 infection of human umbilical vein endothelial cells by cell-mediated transmission. J Virol 75:7717–7722

Samaniego F, Pati S, Karp JE, Prakash O, Bose D (2001) Human herpesvirus 8 K1-associated nuclear factor-κB-dependent promoter activity: role in Kaposi's sarcoma inflammation? J Natl Cancer Inst Monogr: 15–23

Scala G, Ruocco MR, Ambrosino C, Mallardo M, Giordano V, Baldassarre F, Dragonetti E, Quinto I, Venuta S (1994) The expression of the interleukin 6 gene is induced by the human immunodeficiency virus 1 TAT protein. J Exp Med 179:961–971

Schim van der Loeff MF, Aaby P (1999) Towards a better understanding of the epidemiology of HIV-2. AIDS 13 Suppl A: S69–84

Schwarze SR, Hruska KA, Dowdy SF (2000) Protein transduction: unrestricted delivery into all cells? Trends Cell Biol 10:290–295

Sommer A, Rifkin DB (1989) Interaction of heparin with human basic fibroblast growth factor: protection of the angiogenic protein from proteolytic degradation by a glycosaminoglycan. J Cell Physiol 138:215–220

Spear PG, Longnecker R (2003) Herpesvirus entry: an update. J Virol 77:10179–10185

Tao J, Frankel AD (1993) Electrostatic interactions modulate the RNA-binding and transactivation specificities of the human immunodeficiency virus and simian immunodeficiency virus Tat proteins. Proc Natl Acad Sci USA 90:1571–1575

Tessler S, Rockwell P, Hicklin D, Cohen T, Levi BZ, Witte L, Lemischka IR, Neufeld G (1994) Heparin modulates the interaction of VEGF165 with soluble and cell associated flk-1 receptors. J Biol Chem 269:12456–12461

Tomescu C, Law WK, Kedes DH (2003) Surface downregulation of major histocompatibility complex class I, PE-CAM, and ICAM-1 following de novo infection of endothelial cells with Kaposi's sarcoma-associated herpesvirus. J Virol 77:9669–9684

Varthakavi V, Smith RM, Deng H, Sun R, Spearman P (2002) Human immunodeficiency virus type-1 activates lytic cycle replication of Kaposi's sarcoma-associated herpesvirus through induction of KSHV Rta. Virology 297:270–280

Vieira J, O'Hearn P, Kimball L, Chandran B, Corey L (2001) Activation of Kaposi's sarcoma-associated herpesvirus (human herpesvirus 8) lytic replication by human cytomegalovirus. J Virol 75:1378–1386

Vives E, Richard JP, Rispal C, Lebleu B (2003) TAT peptide internalization: seeking the mechanism of entry. Curr Protein Pept Sci 4:125–132

Wang HW, Trotter MW, Lagos D, Bourboulia D, Henderson S, Makinen T, Elliman S, Flanagan AM, Alitalo K, Boshoff C (2004) Kaposi sarcoma herpesvirus-induced cellular reprogramming contributes to the lymphatic endothelial gene expression in Kaposi sarcoma. Nat Genet 36:687–693

Watson K, Edwards RJ (1999) HIV-1-trans-activating (Tat) protein: both a target and a tool in therapeutic approaches. Biochem Pharmacol 58:1521–1528

Weiss R, Boshoff C (2000) Addressing controversies over Kaposi's sarcoma. J Natl Cancer Inst 92:677–679

Westendorp MO, Li-Weber M, Frank RW, Krammer PH (1994) Human immunodeficiency virus type 1 Tat upregulates interleukin-2 secretion in activated T cells. J Virol 68:4177–4185

Yang TY, Chen SC, Leach MW, Manfra D, Homey B, Wiekowski M, Sullivan L, Jenh CH, Narula SK, Chensue SW, Lira SA (2000) Transgenic expression of the chemokine receptor encoded by human herpesvirus 8 induces an angioproliferative disease resembling Kaposi's sarcoma. J Exp Med 191:445–454

Yao L, Salvucci O, Cardones AR, Hwang ST, Aoki Y, De La Luz Sierra M, Sajewicz A, Pittaluga S, Yarchoan R, Tosato G (2003) Selective expression of stromal-derived factor-1 in the capillary vascular endothelium plays a role in Kaposi sarcoma pathogenesis. Blood 102:3900–3905

Yayon A, Klagsbrun M, Esko JD, Leder P, Ornitz DM (1991) Cell surface, heparin-like molecules are required for binding of basic fibroblast growth factor to its high affinity receptor. Cell 64:841–848

Zauli G, Gibellini D, Caputo A, Bassini A, Negrini M, Monne M, Mazzoni M, Capitani S (1995) The human immunodeficiency virus type-1 Tat protein upregulates Bcl-2 gene expression in Jurkat T-cell lines and primary peripheral blood mononuclear cells. Blood 86:3823–3834
Zietz C, Bogner JR, Goebel FD, Lohrs U (1999) An unusual cluster of cases of Castleman's disease during highly active antiretroviral therapy for AIDS. N Engl J Med 340:1923–1924

Subject Index

Current Topics in Microbiology and Immunology

Volumes published since 1989 (and still available)

Vol. 268: **Zwickl, Peter; Baumeister, Wolfgang (Eds.):** The Proteasome-Ubiquitin Protein Degradation Pathway. 2002. 17 figs. X, 213 pp. ISBN 3-540-43096-2

Vol. 269: **Koszinowski, Ulrich H.; Hengel, Hartmut (Eds.):** Viral Proteins Counteracting Host Defenses. 2002. 47 figs. XII, 325 pp. ISBN 3-540-43261-2

Vol. 270: **Beutler, Bruce; Wagner, Hermann (Eds.):** Toll-Like Receptor Family Members and Their Ligands. 2002. 31 figs. X, 192 pp. ISBN 3-540-43560-3

Vol. 271: **Koehler, Theresa M. (Ed.):** Anthrax. 2002. 14 figs. X, 169 pp. ISBN 3-540-43497-6

Vol. 272: **Doerfler, Walter; Böhm, Petra (Eds.):** Adenoviruses: Model and Vectors in Virus-Host Interactions. Virion and Structure, Viral Replication, Host Cell Interactions. 2003. 63 figs., approx. 280 pp. ISBN 3-540-00154-9

Vol. 273: **Doerfler, Walter; Böhm, Petra (Eds.):** Adenoviruses: Model and Vectors in VirusHost Interactions. Immune System, Oncogenesis, Gene Therapy. 2004. 35 figs., approx. 280 pp. ISBN 3-540-06851-1

Vol. 274: **Workman, Jerry L. (Ed.):** Protein Complexes that Modify Chromatin. 2003. 38 figs., XII, 296 pp. ISBN 3-540-44208-1

Vol. 275: **Fan, Hung (Ed.):** Jaagsiekte Sheep Retrovirus and Lung Cancer. 2003. 63 figs., XII, 252 pp. ISBN 3-540-44096-3

Vol. 276: **Steinkasserer, Alexander (Ed.):** Dendritic Cells and Virus Infection. 2003. 24 figs., X, 296 pp. ISBN 3-540-44290-1

Vol. 277: **Rethwilm, Axel (Ed.):** Foamy Viruses. 2003. 40 figs., X, 214 pp. ISBN 3-540-44388-6

Vol. 278: **Salomon, Daniel R.; Wilson, Carolyn (Eds.):** Xenotransplantation. 2003. 22 figs., IX, 254 pp. ISBN 3-540-00210-3

Vol. 279: **Thomas, George; Sabatini, David; Hall, Michael N. (Eds.):** TOR. 2004. 49 figs., X, 364 pp. ISBN 3-540-00534X

Vol. 280: **Heber-Katz, Ellen (Ed.):** Regeneration: Stem Cells and Beyond. 2004. 42 figs., XII, 194 pp. ISBN 3-540-02238-4

Vol. 281: **Young, John A. T. (Ed.):** Cellular Factors Involved in Early Steps of Retroviral Replication. 2003. 21 figs., IX, 240 pp. ISBN 3-540-00844-6

Vol. 282: **Stenmark, Harald (Ed.):** Phosphoinositides in Subcellular Targeting and Enzyme Activation. 2003. 20 figs., X, 210 pp. ISBN 3-540-00950-7

Vol. 283: **Kawaoka, Yoshihiro (Ed.):** Biology of Negative Strand RNA Viruses: The Power of Reverse Genetics. 2004. 24 figs., IX, 350 pp. ISBN 3-540-40661-1

Vol. 284: **Harris, David (Ed.):** Mad Cow Disease and Related Spongiform Encephalopathies. 2004. 34 figs., IX, 219 pp. ISBN 3-540-20107-6

Vol. 285: **Marsh, Mark (Ed.):** Membrane Trafficking in Viral Replication. 2004. 19 figs., IX, 259 pp. ISBN 3-540-21430-5

Vol. 286: **Madshus, Inger H. (Ed.):** Signalling from Internalized Growth Factor Receptors. 2004. 19 figs., IX, 187 pp. ISBN 3-540-21038-5

Vol. 287: **Enjuanes, Luis (Ed.):** Coronavirus Replication and Reverse Genetics. 2005. 49 figs., XI, 257 pp. ISBN 3-540-21494-1

Vol. 288: **Mahy, Brain W. J. (Ed.):** Foot-and-Mouth-Disease Virus. 2005. 16 figs., IX, 178 pp. ISBN 3-540-22419X

Vol. 289: **Griffin, Diane E. (Ed.):** Role of Apoptosis in Infection. 2005. 40 figs., IX, 294 pp. ISBN 3-540-23006-8

Vol. 290: **Singh, Harinder; Grosschedl, Rudolf (Eds.):** Molecular Analysis of B Lymphocyte Development and Activation. 2005. 28 figs., XI, 255 pp. ISBN 3-540-23090-4

Vol. 291: **Boquet, Patrice; Lemichez Emmanuel (Eds.)** Bacterial Virulence Factors and Rho GTPases. 2005. 28 figs., IX, 196 pp. ISBN 3-540-23865-4

Vol. 292: **Fu, Zhen F (Ed.):** The World of Rhabdoviruses. 2005. 27 figs., X, 210 pp. ISBN 3-540-24011-X

Vol. 293: **Kyewski, Bruno; Suri-Payer, Elisabeth (Eds.):** CD4+CD25+ Regulatory T Cells: Origin, Function and Therapeutic Potential. 2005. 22 figs., XII, 332 pp. ISBN 3-540-24444-1

Vol. 294: **Caligaris-Cappio, Federico, Dalla Favera, Ricardo (Eds.):** Chronic Lymphocytic Leukemia. 2005. 25 figs., VIII, 187 pp. ISBN 3-540-25279-7

Vol. 295: **Sullivan, David J.; Krishna Sanjeew (Eds.):** Malaria: Drugs, Disease and Post-genomic Biology. 2005. 40 figs., XI, 446 pp. ISBN 3-540-25363-7

Vol. 296: **Oldstone, Michael B. A. (Ed.):** Molecular Mimicry: Infection Induced
Autoimmune Disease. 2005. 28 figs., VIII, 167 pp. ISBN 3-540-25597-4

Vol. 297: **Langhorne, Jean (Ed.):** Immunology and Immunopathogenesis of Malaria. 2005. 8 figs., XII, 236 pp. ISBN 3-540-25718-7

Vol. 298: **Vivier, Eric; Colonna, Marco (Eds.):** Immunobiology of Natural Killer Cell Receptors. 2005. 27 figs., VIII, 286 pp. ISBN 3-540-26083-8

Vol. 299: **Domingo, Esteban (Ed.):** Quasispecies: Concept and Implications. 2006. 44 figs., XII, 401 pp. ISBN 3-540-26395-0

Vol. 300: **Wiertz, Emmanuel J.H.J.; Kikkert, Marjolein (Eds.):** Dislocation and Degradation of Proteins from the Endoplasmic Reticulum. 2006. 19 figs., VIII, 168 pp. ISBN 3-540-28006-5

Vol. 301: **Doerfler, Walter; Böhm, Petra (Eds.):** DNA Methylation: Basic Mechanisms. 2006. 24 figs., VIII, 324 pp. ISBN 3-540-29114-8

Vol. 302: **Robert N. Eisenman (Ed.):** The Myc/Max/Mad Transcription Factor Network. 2006. 28 figs. XII, 278 pp. ISBN 3-540-23968-5

Vol. 303: **Thomas E. Lane (Ed.):** Chemokines and Viral Infection. 2006. 14 figs. XII, 154 pp. ISBN 3-540-29207-1

Vol. 304: **Stanley A. Plotkin (Ed.):** Mass Vaccination: Global Aspects -- Progress and Obstacles. 2006. 40 figs. X, 270 pp. ISBN 3-540-29382-5

Vol. 305: **Radbruch, Andreas; Lipsky, Peter E. (Eds.):** Current Concepts in Autoimmunity. 2006. 29 figs. IIX, 276 pp. ISBN 3-540-29713-8

Vol. 306: **William M. Shafer (Ed.):** Antimicrobial Peptides and Human Disease. 2006. 12 figs. XII, 262 pp. ISBN 3-540-29915-7

Vol. 307: **John L. Casey (Ed.):** Hepatitis Delta Virus. 2006. 22 figs. XII, 228 pp. ISBN 3-540-29801-0

Vol. 308: **Honjo, Tasuku; Melchers, Fritz (Eds.):** Gut-Associated Lymphoid Tissues. 2006. 24 figs. XII, 204 pp. ISBN 3-540-30656-0

Vol. 309: **Polly Roy (Ed.):** Reoviruses: Entry, Assembly and Morphogenesis. 2006. 43 figs. XX, 261 pp. ISBN 3-540-30772-9

Vol. 310: **Doerfler, Walter; Böhm, Petra (Eds.):** DNA Methylation: Development, Genetic Disease and Cancer. 2006. 25 figs. X, 284 pp. ISBN 3-540-31180-7

Vol. 311: **Pulendran, Bali; Ahmed, Rafi (Eds.):** From Innate Immunity to Immunological Memory. 2006. 13 figs. X, 177 pp. ISBN 3-540-32635-9

Printing: Krips bv, Meppel
Binding: Stürtz, Würzburg